SI metric conversion factors

Quantity	To convert from	To	Multiply by
Area	square feet (ft²)	square meter (m²)	0.0929
Bending moment or torque	(pound-force) (feet)	newton-meters	1.356
Degree (heat)	Fahrenheit (°F)	Celsius (°C)	$T = (°F - 32)/1.8$
Force	pound-force (lbf) kilogram-force (kgf) kyne	newton (N)	4.448
Force per unit length	pound-force per feet	newtons per meter (N/m)	14.594
Length	foot (ft) inch (in)	meters (m) centimeter (cm)	0.3048 0.0254
Mass	pound (lbm)	kilogram (kg)	0.4536
Mass density	pound-mass per cubic foot (lbm/ft³)	kilogram per cubic meter (kg/m³)	16.021
Stress (pressure)	pound-force per square inch (lbf/in²)	pascal (Pa)	6.894×10^3
Velocity	foot per second (ft/s)	meter per second (m/s)	0.3048
Viscosity	pound-force per inches centipoise	newtons per meter (N/m)	10^{-3}
	pound force second per square foot (lbf · s/ft²)	(Pa · s)	47.88

AIRCRAFT STRUCTURES

AIRCRAFT STRUCTURES

Second Edition

David J. Peery, Ph.D.

Late Professor of Aeronautical Engineering
The Pennsylvania State University

J. J. Azar, Ph.D.

Professor of Petroleum and Mechanical Engineering
University of Tulsa

McGraw-Hill Book Company

New York   St. Louis   San Francisco   Auckland   Bogotá   Hamburg
Johannesburg   London   Madrid   Mexico   Montreal   New Delhi
Panama   Paris   São Paulo   Singapore   Sydney   Tokyo   Toronto

To the memory of
Dr. David Peery

This book was set in Times Roman by Santype-Byrd.
The editors were Diane D. Heiberg and J. W. Maisel;
the production supervisor was Charles Hess.
New drawings were done by J & R Services, Inc.
The cover was designed by Kao & Kao Associates.
Halliday Lithograph Corporation was printer and binder.

-0. FEB 1985

AIRCRAFT STRUCTURES

4567890 HDHD 8987654

ISBN 0-07-049196-8

Library of Congress Cataloging in Publication Data

Peery, David J.
    Aircraft structures.

    Bibliography: p.
    Includes index.
    1. Airframes.   I. Azar, Jamal J., date
II. Title
TL671.6.P4   1982   629.134'31   81-17196
ISBN 0-07-049196-8   AACR2

CONTENTS

PREFACE

The purpose of this book is twofold: (1) to provide the reader with fundamental concepts in the analysis and design of flight structures, and (2) to develop unified analytical tools for the prediction and assessment of structural behavior regardless of field application.

The text is written in a manner which allows the reader to develop a working knowledge in both the classical and the modern computer techniques of structural analysis. In addition, it helps the reader develop a thorough understanding of the important factors which must be considered in the design of structural components.

The scope covers areas that the authors feel are essential fundamentals from which the reader may progress to the analysis and design of more complex and larger structural systems. The definitions of structural systems and its constituents, loads, supports and reactions, the concepts in statics, and the fundamental principles of mechanics are covered in the first portion of the book. The basic elasticity relations (stress-strain, strain displacements, etc.), and material behavior and selection are discussed. Load analysis of flight vehicles and the analysis and design of specific flight-vehicle structural components are presented. Fatigue analysis, thermal stress analysis, and instability analysis of structures are also included. Energy methods, finite-difference methods, and the stiffness matrix methods in the deflection analysis of structures are provided. Numerous examples are solved in every section of the book to close the gap between the theoretical developments and its application in solving practical problems.

The book is written for a first course in structural analysis and design. It is intended for senior-level students. The essential prerequisites are strength of materials and a basic course in calculus. The book also serves as an excellent reference for the practicing engineer.

J. J. Azar

ONE

STATIC ANALYSIS OF STRUCTURES

1.1 INTRODUCTION

The full understanding of both the terminology in statics and the fundamental principles of mechanics is an essential prerequisite to the analysis and design of structures. Therefore, this chapter is devoted to the presentation and the application of these fundamentals.

1.2 STRUCTURAL SYSTEM

Any deformable solid body which is capable of carrying loads and transmitting these loads to other parts of the body is referred to as a *structural system*. The constituents of such systems are beams, plates, shells, or a combination of the three.

Bar elements, such as shown in Fig. 1.1, are one-dimensional structural members which are capable of carrying and transmitting bending, shearing, torsional, and axial loads or a combination of all four.

Bars which are capable of carrying only axial loads are referred to as *axial rods* or *two-force* members. Structural systems constructed entirely out of axial rods are called *trusses* and frequently are used in many atmospheric, sea, and land-based structures, since simple tension or compression members are usually the lightest for transmitting forces.

Plate elements, such as shown in Fig. 1.2, are two-dimensional extensions of bar elements. Plates made to carry only in-plane axial loads are called *mem-*

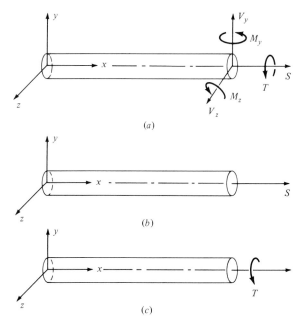

Figure 1.1 Bar elements. (a) General bar; (b) axial rod; (c) torsional rod.

branes. Those which are capable of carrying only in-plane shearing loads are referred to as *shear panels*; frequently these are found in missile fins, aircraft wings, and tail surfaces.

Shells are curved plate elements which occupy a space. Fuselages, building domes, pressure vessels, etc., are typical examples of shells.

1.3 LOAD CLASSIFICATION

Loads which act on a structural system may be generally classified in accordance with their causes. Those which are produced by surface contact are called *surface loads.* Dynamic and/or static pressures are examples of surface loads. If the area of contact is very small, then the load is said to be *concentrated;* otherwise, it is called a *distributed load.* (See Fig. 1.3.)

Loads which depend on body volume are called *body loads.* Inertial, magnetic, and gravitational forces are typical examples. Generally, these loads are assumed to be distributed over the entire volume of the body.

Loads also may be categorized as dynamic, static, or thermal. *Dynamic loads* are time-dependent, whereas *static loads* are independent of time. *Thermal loads* are created on a restrained structure by a uniform and/or nonuniform temperature change.

Regardless of the classification of the externally imposed loads, a structural member, in general, resists these loads internally in the form of bending, axial, shear, and torsional actions or a combination of the four.

In order to present definitions for internal loads, pass a hypothetical plane so

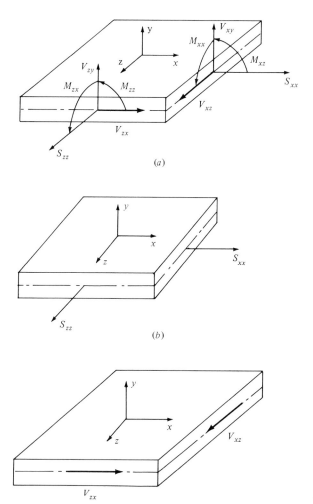

Figure 1.2 Plate elements. (a) General plate element; (b) membrane element; (c) shear panel.

that it cuts the face of a structural member perpendicular to the member axis, as shown in Fig. 1.4. Thus, a *bending moment* may be defined as a force whose vector representation lies in and parallel to the plane of the cut, while a *torque* is a force whose vector representation is normal to that cut. On the other hand, *shear load* is a force which lies in and is parallel to the plane of the cut, while *axial load* is a force which acts normal to the plane of the cut.

1.4 SUPPORTS AND REACTIONS

The primary function of *supports* is to provide, at some points of a structural system, physical restraints that limit the freedom of movement to only that

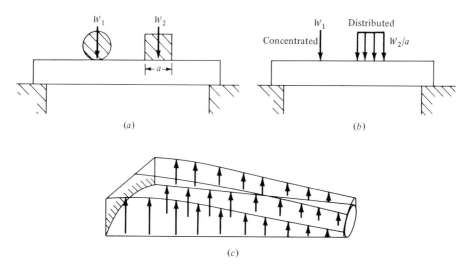

(a) (b)

(c)

Figure 1.3 Concentrated and distributed loads. (a) Actual loads; (b) idealized loads; (c) wing pressure load.

intended in the design. The types of supports that occur in ordinary practice are shown in Fig. 1.5.

The forces induced at points of support are called *reactions*. For example, a *hinge support* is designed to allow only rotation at the point of connection, and

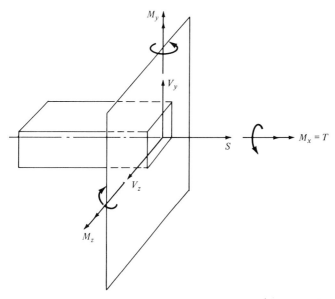

Figure 1.4 S = axial load, lb; $M_x = T$ = torque, lb·in; M_y, M_z = bending moment(s), lb·in; V_x, V_y = shear loads, lb.

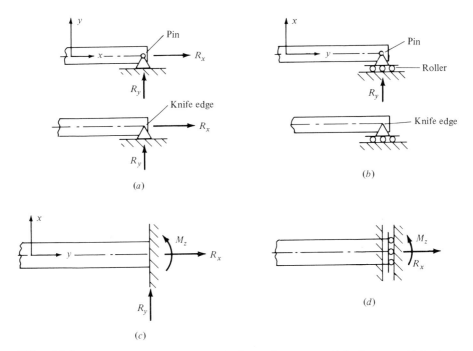

Figure 1.5 Support types. (a) Hinge support; (b) hinge-roller support; (c) fixed support; (d) fixed-roller support.

thus reactive forces (reactions) are developed in the other directions where movements are not allowed.

Likewise, a *hinge-roller support* allows rotation and a translation in only the x direction, and hence there exists one reactive force (reaction) in the y direction. A *fixed* support normally is designed to provide restraints against rotation and all translations; therefore, reactive forces and moments (reactions) are developed along the directions where movements are not permitted.

1.5 EQUATIONS OF STATIC EQUILIBRIUM

One of the first steps in the design of a structural system is the determination of internal loads acting on each system member. Any solid body in space or any part cut out of the body is said to be in a state of *stable static equilibrium* if it simultaneously satisfies:

$$\Sigma \, F_i = 0$$
$$\Sigma \, M_i = 0 \qquad (i = x, y, z) \qquad (1.1)$$

where $\Sigma \, F_i = 0$ implies that the vector sum of all forces acting on the system or on part of it must add to zero in any one direction along a chosen set of system axes x, y, and z; likewise, $\Sigma \, M_i = 0$ means that the vector sum of all moments at any preselected point must add to zero around any one of the chosen set of axes x, y, and z.

While Eq. (1.1) applies for general space structures, for the case of planar-type structures it reduces to

$$\Sigma F_i = 0$$
$$\Sigma M_z = 0 \qquad (i = x, y) \qquad (1.2)$$

Note that only six independent equations exist for any free body in space and three independent equations exist for a coplanar free body. If, for example, an attempt is made to find four unknown forces in a coplanar free body by using the two force equations and moment equations at two selected points, the four equations cannot be solved because they are not independent (i.e., one of the equations can be derived from the other three). The following equations cannot be solved for the numerical values of the three unknowns because they are not independent:

$$F_1 + F_2 + F_3 = 3$$
$$F_1 + F_2 + 2F_3 = 4$$
$$2F_1 + 2F_2 + 3F_3 = 7$$

In matrix form these are as follows.

$$\begin{bmatrix} 1 & 1 & 1 \\ 1 & 1 & 2 \\ 2 & 2 & 3 \end{bmatrix} \begin{bmatrix} F_1 \\ F_2 \\ F_3 \end{bmatrix} = \begin{bmatrix} 3 \\ 4 \\ 7 \end{bmatrix}$$

The third equation may be obtained by adding the first two equations, and consequently it does not represent an independent condition. The dependence of these equations is more readily established if an attempt is made to find the inverse of the 3×3 matrix of coefficients. It can be shown easily that the determinant of the matrix is zero; hence equation dependence does, in fact, exist. A matrix whose determinant is zero is said to be *singular* and therefore cannot be inverted.

1.6 STATICALLY DETERMINATE AND INDETERMINATE STRUCTURES

A structure is said to be *determinate* if all its external reactions and the internal loads on its members can be obtained by utilizing only the static equations of equilibrium. Otherwise the structure is said to be statically *indeterminate*. In the latter, or what is commonly referred to as a *redundant* structure, there are more unknown forces than the number of independent equations of statics which can be utilized. The additional equations required for the analysis of redundant structures can be obtained by considering the deformations (displacements) in the

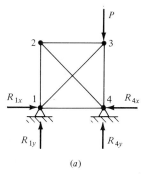 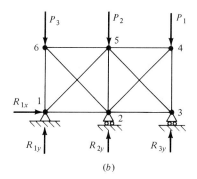

Figure 1.6 Indeterminate structures.

structure. This is studied in detail in later chapters. External reactions, internal loads, or a combination of both may cause a structural system to be statically redundant. The number of redundancies in a structure is governed by the number of external reactions and/or the number of members that may be taken out without the stability of the structure being affected. For example, in Fig. 1.6*a* if member 1–3 and/or reaction R_{4x} is removed, then the structure becomes statically determinate and maintains its stability. Likewise, in Fig. 1.6*b* if members 2–4 and 2–6 and either reaction R_{2y} or R_{3y} are removed, then the structure becomes determinate and stable. If, on the other hand, additional members such as 3–5 and 1–5 are removed, then the structure becomes a mechanism or unstable. Mechanisms cannot resist loads and therefore are not used as structural systems.

1.7 APPLICATIONS

While the set of [Eqs. (1.1) and (1.2)] is simple and well known, it is very important for a student to acquire proficiency in the application of these equations to various types of structural systems. Several typical structural systems are analyzed as illustrative examples.

Example 1.1 Find the internal loads acting on each member of the structure shown in Fig. 1.7.

SOLUTION First, disassemble the structure as shown in Fig. 1.8 and make a free-body diagram for each member. Since members 1–2 and 4–6 are two-force members (axial rods), the forces acting on them are along the line joining the pin joints of these members. All directions of forces are chosen arbitrarily and must be reversed if a negative value is obtained; that is, F_{3x} is assumed to act to the right on the horizontal member and therefore must act to the left since its magnitude came to be negative.

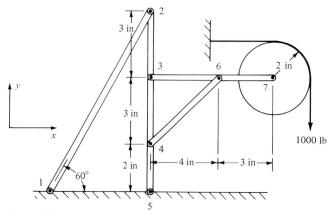

Figure 1.7

For the pulley

$$\Sigma M_7 = 0 \ (+\curvearrowright)$$

$$1000 \times 2 - 2T = 0$$

$$T = 1000 \text{ lb}$$

$$\Sigma F_y = 0 \ (\xrightarrow{+})$$

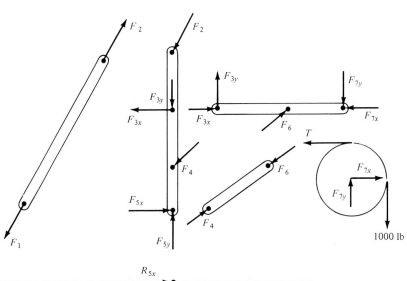

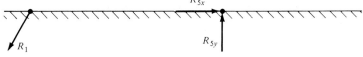

Figure 1.8 Disassembled structure.

$$F_{7x} - 1000 = 0$$

$$F_{7x} = 1000 \text{ lb}$$

$$\Sigma F_y = 0 \ (+ \uparrow)$$

$$F_{7y} - 1000 = 0$$

$$F_{7y} = 1000 \text{ lb}$$

For member 3-6-7,

$$\Sigma M_3 = 0 \ (+ \curvearrowright)$$

$$1000 \times 7 - 2.4F_4 = 0$$

$$F_4 = 2917 \text{ lb}$$

$$\Sigma F_x = 0 \ (\xrightarrow{+})$$

$$F_{3x} + 2917 \cos 36.9° - 1000 = 0$$

$$F_{3x} = -1333 \text{ lb}$$

$$\Sigma F_y = 0 \ (+ \uparrow)$$

$$F_{3y} + 2917 \sin 36.9° - 1000 = 0$$

$$F_{3y} = -751 \text{ lb}$$

Since the magnitudes of F_{3x} and F_{3y} came to be negative, the assumed direction must be reversed. A common practice is to cross out (rather than erase) the original arrows.

For member 2-3-4-5,

$$\Sigma M_5 = 0 \ (+ \curvearrowright)$$

$$1333 \times 5 - 2917 \times 2 \cos 36.9° - 4F_1 = 0$$

$$F_1 = 500 \text{ lb}$$

$$\Sigma F_x = 0 \ (\xrightarrow{+})$$

$$F_{5x} - 2917 \cos 36.9° + 1333 - 500 \cos 60° = 0$$

$$F_{5x} = 1250 \text{ lb}$$

$$\Sigma F_y = 0 \ (+ \uparrow)$$

$$F_{5y} - 2917 \sin 36.9° + 751 - 500 \sin 60° = 0$$

$$F_{5y} = 1434 \text{ lb}$$

Now all internal loads have been obtained without the use of the entire structure as a free body. The solution is checked by applying the three equations of equilibrium to the entire structure:

$$\Sigma F_x = 0 \ (\xrightarrow{+})$$
$$1250 - 1000 - 500 \cos 60° = 0$$

$$\Sigma F_y = 0 \ (+\uparrow)$$
$$1433 - 1000 - 500 \sin 60° = 0$$

$$\Sigma M_5 = 0 \ (+\curvearrowright)$$
$$1000 \times 9 - 1000 \times 7 - 500 \times 4 = 0$$

This equilibrium check should be made wherever possible to detect any errors that might have occurred during the analysis.

Example 1.2 Find the internal load in member 5 of the coplanar truss structure shown in Fig. 1.9.

SOLUTION Several methods are available for analyzing truss structures; two are discussed and applied in solving this example.

(a) *Method of joints.* In the analysis of a truss by the method of joints, the two equations of static equilibrium, $\Sigma F_x = 0$ and $\Sigma F_y = 0$, are applied for each joint as a free body. Two unknown forces may be obtained for each joint. Since each member is an axial rod (two-force member), it exerts equal and opposite forces on the joints at its ends. The joints of a truss must be analyzed in sequence by starting at a joint which has only two members meeting with unknown forces. Then the joints are analyzed in the proper sequence until all joints have been considered, if necessary.

To find the unknown reactions, consider the entire structure as a free body.

$$\Sigma M_4 = 0 \ (+\curvearrowright)$$
$$2000 \times 10 + 4000 \times 10 + 1000 \times 30 - 20R_{6y} = 0$$
$$R_{6y} = 4500 \text{ lb}$$

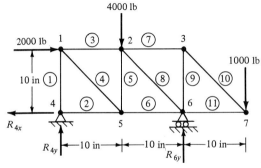

Figure 1.9 Truss structure.

$$\Sigma F_x = 0 \ (\overset{+}{\rightarrow})$$

$$2000 - R_{4x}$$

$$R_{4x} = 2000 \text{ lb}$$

$$\Sigma F_y = 0 \ (+\uparrow)$$

$$R_{4y} - 4000 - 1000 + 4500 = 0$$

$$R_{4y} = 500 \text{ lb}$$

The directions of unknown forces in each member are assumed, as in the previous example, and vectors are changed on the sketch when they are found to be negative. (See Fig. 1.10.)

Isolating joint 4, we have

$$\Sigma F_x = 0 \ (\overset{+}{\rightarrow})$$

$$F_2 - 2000 = 0$$

$$F_2 = 2000 \text{ lb}$$

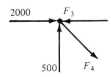

$$\Sigma F_y = 0 \ (+\uparrow)$$

$$500 - F_1 = 0$$

$$F_1 = 500 \text{ lb}$$

Isolating joint 1 gives

$$\Sigma F_y = 0 \ (+\uparrow)$$

$$500 - F_4 \sin 45° = 0$$

$$F_4 = 707 \text{ lb}$$

$$\Sigma F_x = 0 \ (\overset{+}{\rightarrow})$$

$$2000 + 707 \cos 45° - F_3 = 0$$

$$F_3 = 2500 \text{ lb}$$

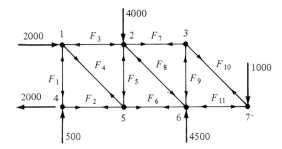

Figure 1.10

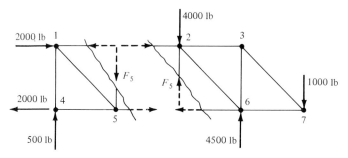

Figure 1.11

Finally, isolating joint 5 to obtain the load in member 5, we get

$$\Sigma F_y = 0 \ (+\uparrow)$$

$$707 \sin 45° - F_5 = 0$$

$$F_5 = 500 \text{ lb}$$

                                             707

                          2000 F_5

                                          F_6

    Arrows acting toward a joint indicate that a member is in compression, and arrows acting away from a joint indicate tension.

(*b*) *Method of sections.* As in this example, often it is desirable to find the internal loads in certain selected members of a truss without analyzing the entire truss. Usually the method of joints is cumbersome in this case, since the loads in all the members to the left or right of any member must be obtained before the force is found in that particular member. An analysis by the method of sections will yield the internal load in any preselected member by a single operation, without the necessity of finding loads in the other members. Instead of considering the joints as free bodies, a cross section is taken through the truss and the part of the truss on either side is considered as a free body. The cross section is chosen so that it cuts the members for which the forces are desired and preferably only three members.

    In our example, the internal load in member 5 is desired; the free body is as shown in Fig. 1.11. The load in member 5 may be found by summing forces in the *y* direction on either part of the cut truss. Considering the left part as a free body, we get

$$\Sigma F_y = 0 \ (+\uparrow)$$

$$500 - F_5 = 0$$

$$F_5 = 500 \text{ lb}$$

Example 1.3 Analyze the structural system shown in Fig. 1.12.

SOLUTION Quite often, structural systems are made up largely of axial rods but contain some members which are loaded laterally, as shown in Fig. 1.12.

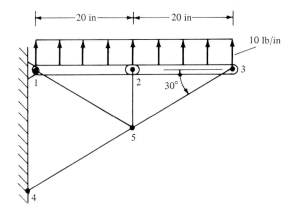

Figure 1.12

These structures usually are classified as trusses, since the analysis is similar to that used for trusses. As shown in Fig. 1.12, members 1-2 and 2-3 are not axial rods, and separate free-body diagrams for these members, as shown in Fig. 1.13a and b, are required. Since each of these members has four unknown reactions, the equations of static equilibrium are not sufficient to find all four unknowns. It is possible, though, to find the vertical reactions $R_{1y} = R_{2y} = R_{2'y} = R_{3y} = 100$ lb and to obtain the relations $R_{1x} = R_{2x}$ and $R_{2'x} = R_{3x}$ from the equations of equilibrium.

When the unknown reactions obtained from members 1-2 and 2-3 are applied to the remaining part of the structure as a free body (Fig. 1.13b), it is apparent that the rest of the analysis is similar to that of the previous examples. All members except 1-2 and 2-3 may now be designed as simple tension or compression members. The horizontal members (1-2 and 2-3) must be designed for bending combined with axial and shear loads.

Example 1–4 Find the reactions at supports A, B, and C, for the landing gear of Fig. 1.14. Members OB and OC are two-force members. Member OA resists bending and torsion, but point A is hinged by a universal joint so that the member can carry torsion but no bending in any direction at this point.

SOLUTION First consider the components of the torsional reaction at point A. The resultant torsional vector T_A, shown in Fig. 1.15b, must be along the member, and it has components T_{AX} and T_{AY}.

Any vector force Q acting in space may be resolved into components along the chosen set of axes by the method of direction cosines.

$$Q_x = Q \cos \alpha$$

$$Q_y = Q \cos \beta$$

$$Q_z = Q \cos \gamma \tag{1.3}$$

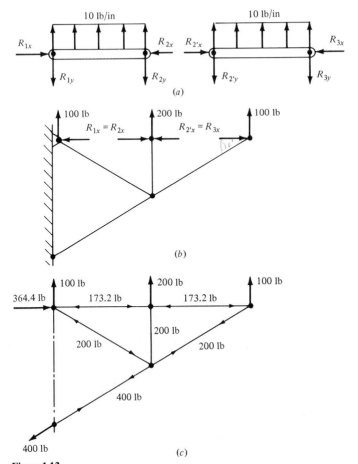

Figure 1.13

Here α, β, and γ are the angles between the x, y, and z axes and the line of action of the force vector Q, respectively.

Thus, from Fig. 1.15b, utilizing Eq. (1.3), we have

$$T_{AY} = T_A \cos \beta = T_A \left(\tfrac{40}{50}\right) = 0.8 T_A$$

$$T_{AX} = T_A \cos \alpha = T_A \left(\tfrac{30}{50}\right) = 0.6 T_A$$

$$T_{AZ} = T_A \cos \gamma = T_A \left(\tfrac{0}{50}\right) = 0$$

In the free-body diagram for the entire structure, shown in Fig. 1.15a, there are six unknown reaction forces. The six equations of static equilibrium are just sufficient to determine these unknown forces. Taking moments about an axis through points A and B gives

$$\Sigma M_{AB} = 4000 \times 36 - 0.8 F_C \times 30 = 0$$

$$F_C = 6000 \text{ lb}$$

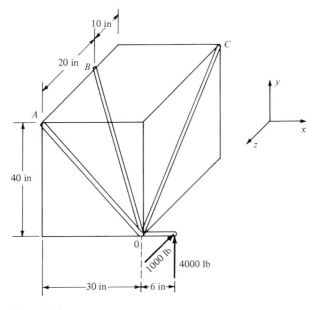

Figure 1.14

The torsional moment T_A may be found by taking moments about line
$0A$. Note that all unknown forces act through this line.

$$\Sigma M_{A0} = 1000 \times 4.8 - T_A = 0$$

$$T_A = 4800 \text{ in} \cdot \text{lb}$$

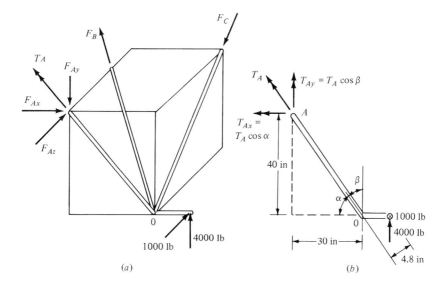

The other forces are obtained from the following equations, which are chosen so that only one unknown appears in each equation.

$$\Sigma M_{OX} = 2880 - 40 F_{AZ} = 0$$

$$F_{AZ} = 72 \text{ lb}$$

The subscript "OX" designates an axis through point 0 in the x direction.

$$\Sigma M_z = 1000 + 72 - 6000 \times 0.6 + 0.371 F_B = 0$$

$$F_B = 6820 \text{ lb}$$

$$\Sigma F_X = F_{AX} - 6820 \times 0.557 = 0$$

$$F_{AX} = 3800 \text{ lb}$$

$$\Sigma F_Y = 4000 + 6820 \times 0.743 - 6000 \times 0.8 - F_{AY} = 0$$

$$F_{AY} = 4270 \text{ lb}$$

Check: $\Sigma M_{AY} = -1000 \times 36 + 6000 \times 0.6 \times 30 - 6820 \times 0.557$

$$\times\, 20 + 3840 = 0$$

Example 1.5 Find the internal loads on all members of the landing gear shown in Fig. 1.16.

SOLUTION For convenience, the reference axes V, D, and S are taken as shown in Fig. 1.16 with the V axis parallel to the oleo strut. Free-body diagrams for the oleo strut and the the horizontal member are shown in Fig. 1.17. Forces perpendicular to the plan of the paper are shown by a circled dot for forces toward the observer and a circled cross for forces away from

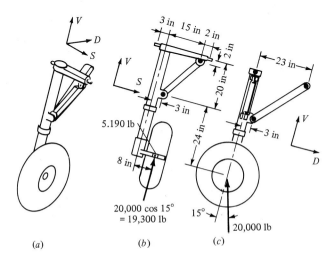

(a)   (b)   (c)

Figure 1.16

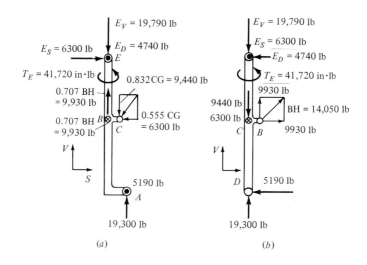

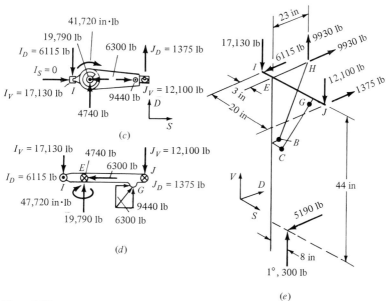

Figure 1.17

the observer. The V component of the 20,000-lb force is

$$20,000 \cos 15° = 19,300 \text{ lb}$$

The D component is

$$20,000 \sin 15° = 5190 \text{ lb}$$

The angle of the side-brace member CG with the V axis is

$$\tan^{-1} \tfrac{12}{18} = 33.7°$$

The V and S components of the force in member CG are

$$CG \cos 33.7° = 0.832CG \qquad CG \sin 33.7° = 0.555CG$$

The drag-brace member BH is at an angle of 45° with the V axis, and the components of the force in this member along the V and D axes are

$$BH \cos 45° = 0.707BH \qquad BH \sin 45° = 0.707BH$$

The six unknown forces acting on the oleo strut a· ᴗ now obtained from the following equations:

$$\Sigma M_{EV} = 5190 \times 8 - T_e = 0$$
$$T_e = 41{,}720 \text{ in} \cdot \text{lb}$$

$$\Sigma M_{ES} = 5190 \times 44 - 0.707BH \times 20 - 0.707BH \times 3 = 0$$
$$BH = 14{,}050 \text{ lb}$$
$$0.707BH = 9930 \text{ lb}$$

$$\Sigma M_{ED} = 0.555CG \times 20 + 0.832CG \times 3 - 19{,}300 \times 8 = 0$$
$$CG = 11{,}350 \text{ lb}$$
$$0.555CG = 6300 \text{ lb}$$
$$0.832CG = 9440 \text{ lb}$$

$$\Sigma F_v = 19{,}300 + 9930 - 9440 - E_v = 0$$
$$E_v = 19{,}700 \text{ lb}$$
$$\Sigma F_s = E_s - 6300 = 0$$
$$E_s = 6300 \text{ lb}$$

$$\Sigma F_d = -5190 + 9930 - E_d = 0$$
$$E_d = 4740 \text{ lb}$$

The horizontal member IJ is now considered as a free body. The forces obtained above are applied to this member, as shown in Fig. 1.17c and d, and the five unknown reactions are obtained as follows:

$$\Sigma F_s = I_s = 0$$
$$\Sigma M_{ID} = 19{,}790 \times 3 + 9440 \times 18 + 6300 \times 2 - 20J_v = 0$$
$$J_v = 12{,}100 \text{ lb}$$

$$\Sigma F_v = 19{,}790 + 9440 - 12{,}100 - I_v = 0$$
$$I_v = 17{,}130 \text{ lb}$$

$$\Sigma M_{IV} = 41{,}720 - 4740 \times 3 + 20J_d = 0$$

$$J_d = -1375 \text{ lb}$$

$$\Sigma F_d = 4740 + 1375 - I_d = 0$$

$$I_d = 6115 \text{ lb}$$

The reactions are now checked by considering the entire structure as a free body, as shown in Fig. 1.17e.

$$\Sigma F_v = 19{,}300 - 17{,}130 - 12{,}100 + 9930 = 0$$

$$\Sigma F_d = -5190 + 1375 - 6115 + 9930 = 0$$

$$\Sigma F_s = 0$$

$$\Sigma M_{IV} = 5190 \times 11 - 1375 \times 20 - 9930 \times 3 = 0$$

$$\Sigma M_{ID} = 19{,}300 \times 11 - 12{,}100 \times 20 + 9930 \times 3 = 0$$

$$\Sigma M_{IJ} = 5190 \times 44 - 9930 \times 23 = 0$$

Example 1.6 Find the loads on the lift and drag-truss members of the externally braced monoplane wing shown in Fig. 1.18. The air load is assumed to be uniformly distributed along the span of the wing. The diagonal drag-truss members are wires, with the tension diagonal effective and the other diagonal carrying no load.

SOLUTION The vertical load of 20 lb/in is distributed to the spars in inverse proportion to the distance between the center of pressure and the spars. The

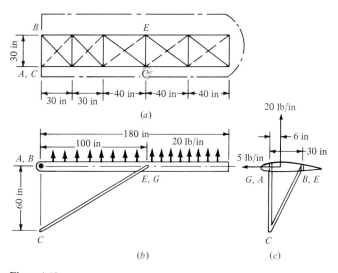

Figure 1.18

load on the front spar is therefore 16 lb/in, and that on the rear spar is 4 lb/in. If the front spar is considered as a free body, as shown in Fig. 1.19a, the vertical forces at A and G may be obtained.

$$\Sigma M_A = -16 \times 180 \times 90 + 100G_z = 0$$

$$G_z = 2590 \text{ lb}$$

$$\frac{G_y}{100} = \frac{2590}{60}$$

$$G_y = 4320 \text{ lb}$$

$$\Sigma F_z = 16 \times 180 - 2590 - A_z = 0$$

$$A_z = 290 \text{ lb}$$

Force A_y cannot be found at this point in the analysis, since the drag-truss members exert forces on the front spar which are not shown in Fig. 1.19a.

If the rear spar is considered as a free body, as shown in Fig. 1.19b, the

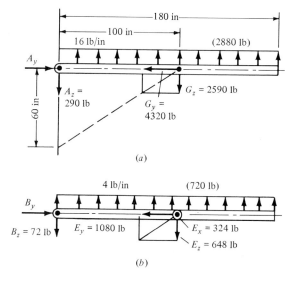

(a)

(b)

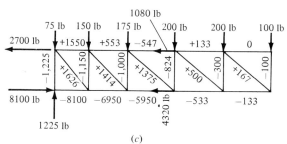

(c)

Figure 1.19

vertical forces at B and E may be obtained:

$$\Sigma M_B = -4 \times 180 \times 90 + 100E_z = 0$$

$$E_z = 648 \text{ lb}$$

$$\frac{E_x}{30} = \frac{E_y}{100} = \frac{648}{60}$$

$$E_x = 324 \text{ lb} \qquad E_y = 1080 \text{ lb}$$

$$\Sigma F_z = 4 \times 180 - 648 - B_z = 0$$

$$B_z = 72 \text{ lb}$$

The loads in the plane of the drag truss can be obtained now. The forward load of 5 lb/in is applied as concentrated loads at the panel points, as shown in Fig. 1.19c. The components of the forces at G and E which lie in the plane of the truss also must be considered. The remaining reactions at A and B and the forces in all drag-truss members now can be obtained by the methods of analysis for coplanar trusses, shown in Fig. 1.19c.

PROBLEMS

1.1 A 5000-lb airplane is in a steady glide with the flight path at an angle θ below the horizontal (see Fig. P1.1). The drag force in the direction of the flight path is 750 lb. Find the lift force L normal to the flight path and the angle θ.

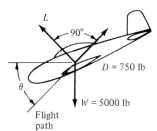

Figure P1.1

1.2 A jet-propelled airplane in steady flight has forces acting as shown in Fig. P1.2. Find the jet thrust T, lift L, and the tail load P.

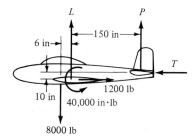

Figure P1.2

1.3 A wind-tunnel model of an airplane wing is suspended as shown in Fig. P1.3 and P1.4. Find the loads in members B, C, and E if the forces at A are $L = 43.8$ lb , $D = 3.42$ lb, and $M = -20.6$ in · lb.

1.4 For the load of Prob. 1.3, find the forces L, D, and M at a point A if the measured forces are $B = 40.2$, $C = 4.16$, and $E = 3.74$ lb.

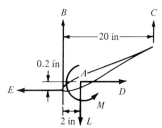

Figure P1.3 and 1.4

1.5 Find the forces at points A and B of the landing gear shown in Fig. P1.5.

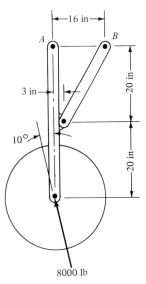

8000 lb   **Figure P1.5**

1.6 Find the forces at points A, B, and C of the structure of the braced-wing monoplane shown in Fig. P1.6.

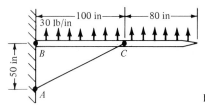

Figure P1.6

1.7 Find the forces V and M at the cut cross section of the beam shown in Fig. P1.7.

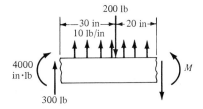

Figure P1.7

1.8 Find the internal loads in all members of the truss structure shown in Fig. P1.8.

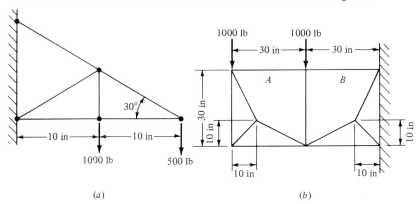

(*a*)

(*b*)

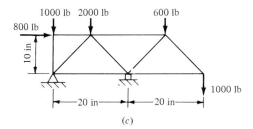

(*c*)

Figure P1.8

1.9 Find the internal loads on all members of the fuselage truss structure shown in Fig. P1.9.

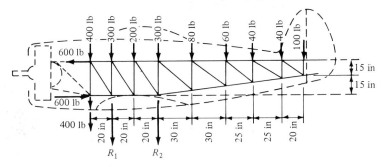

Figure P1.9

1.10 All members of the structure shown in Fig. P1.10 are two-force members, except member *ABC*. Find the reactions on member *ABC* and the loads in other members of the structure.

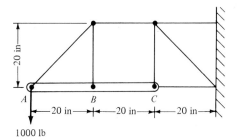

1000 lb

Figure P1.10

1.11 The bending moments about the x and z axes in a plane perpendicular to the spanwise axis of a wing are 400,000 and 100,000 in·lb, as shown in Fig. P1.11. Find the bending moments about the x_1 and z_1 axes which are in the same plane but rotated 10° counterclockwise.

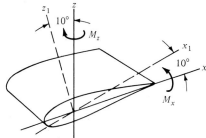

Figure P1.11

1.12 The main beam of the wing shown in Fig. P1.12 has a sweepback angle of 30°. First the moments of 300,000 and 180,000 in·lb are computed about the x and y axes which are parallel and perpendicular to the centerline of the airplane. Find the moments about the x' and y' axes.

1.13 Find the forces acting on all members of the nose-wheel structure shown in Fig. P1.13. Assume the V axis is parallel to the oleo strut.

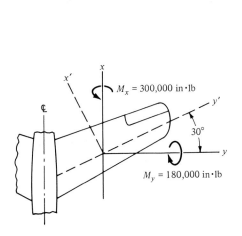

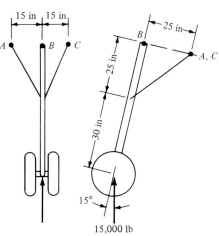

Figure P1.12   **Figure P1.13**

1.14 Analyze the landing gear structure of Example 1.5 for a 15,000-lb load up parallel to the V axis and a 5000-lb load aft parallel to the D axis. The loads are applied at the same point of the axle as the load in Example 1.5.

1.15 Write a computer program to calculate the reactions of the beam structure shown in Fig. P1.15.

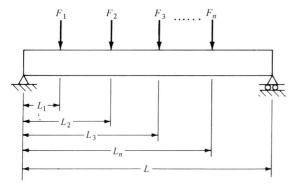

Figure P1.15

1.16 Write a computer program to calculate the internal shear and bending moments at every station of the cantilevered wing shown in Fig. P1.16. Assume the center of pressure is at 25 percent of the chord length measured from the x axis.

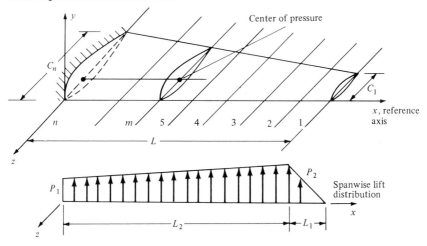

Figure P1.16

TWO

FLIGHT-VEHICLE IMPOSED LOADS

2.1 INTRODUCTION

Before the final selection of member sizes on flight vehicles can be made, all load conditions imposed on the structure must be known. The load conditions are those which are encountered both in flight and on the ground. Since it is impossible to investigate every loading condition which a flight vehicle might encounter in its service lifetime, it is normal practice to select only those conditions that will be critical for every structural member of the vehicle. These conditions usually are determined from past investigation and experience and are definitely specified by the licensing or procuring agencies.

Although the calculations of loads imposed on flight-vehicle structures are the prime responsibility of a special group in an engineering organization called the loads group, a basic general overall knowledge of the loads on vehicles is essential to stress analysts. Therefore, in this chapter we present the fundamentals and terminology pertaining to flight-vehicle loads.

2.2 GENERAL CONSIDERATIONS

Every flight vehicle is designed to safely carry out specific missions. This results in a wide variety of vehicles relative to size, configuration, and performance. Commercial transport aircraft are specifically designed to transport passengers from one airport to another. These types of aircraft are never subjected to violent intentional maneuvers. Military aircraft, however, used in fighter or dive-bomber operations, are designed to resist violent maneuvers. The design conditions usually are determined from the maximum acceleration which the human body can withstand, and the pilot will lose consciousness before reaching the load factor (load factor is related to acceleration) which would cause structural failure of the aircraft.

To ensure safety, structural integrity, and reliability of flight vehicles along with the optimality of design, government agencies, both civil and military, have established definite specifications and requirements in regard to the magnitude of loads to be used in structural design of the various flight vehicles. Terms are defined below which are generally used in the specification of loads on flight vehicles.

The *limit loads* used by civil agencies or *applied loads* used by military agencies are the maximum anticipated loads in the entire service life-span of the vehicle. The *ultimate loads*, commonly referred to as *design loads*, are the limit loads multiplied by a factor of safety (FS):

$$FS = \frac{\text{ultimate load}}{\text{limit load}}$$

Generally, a factor of safety which varies from 1.25 for missile structures to 1.5 for aircraft structures is used in practically every design because of the uncertainties involving

1. The simplifying assumption used in the theoretical analyses
2. The variations in material properties and in the standards of quality control
3. The emergency actions which might have to be taken by the pilot, resulting in loads on the vehicle larger than the specified limit loads.

The limit loads and ultimate loads quite often are prescribed by specifying certain load factors. The *limit-load factor* is a factor by which basic loads on a vehicle are multiplied to obtain the limit loads. Likewise, the *ultimate load factor* is a factor by which basic vehicle loads are multiplied to obtain the ultimate loads; in other words, it is the product of the limit load factor and the factor of safety.

2.3 BASIC FLIGHT LOADING CONDITIONS

One of four basic conditions will probably produce the highest load in any part of the airplane for any flight condition. Usually these conditions are called positive high angle of attack, positive low angle of attack, negative high angle of attack, and negative low angle of attack. All these conditions represent symmetrical flight maneuvers; i.e., there is no motion normal to the plane of symmetry of the airplane.

The positive high angle of attack (PHAA) condition is obtained in a pullout at the highest possible angle of attack on the wing. The lift and drag forces are perpendicular and parallel respectively, to the relative wind, which is shown as horizontal in Fig. 2.1a. The resultant R of these forces always has an aft component with respect to the relative wind, but will usually have a forward component C with respect to the wing chord line, because of the high angle of attack α. The maximum forward component C will be obtained when α has a maximum value. In order to account for uncertainties in obtaining the stalling angle of attack under unsteady flow conditions, most specifications arbitrarily require that a value of α be used which is higher than the wing stalling angle under steady flow conditions. An angle of attack corresponding to a coefficient of lift of 1.25 times the maximum coefficient of lift for steady flow conditions is often used, and aerodynamic data are extrapolated from data measured for steady flow conditions. Experiments show that these high angles of attack and high lift coefficients may be obtained momentarily in a sudden pull-up before the airflow reaches a steady condition, but it is difficult to obtain accurate lift measurements during the unsteady conditions.

In the PHAA condition, the bending moments from the normal forces N, shown in Fig. 2.1a, produce compressive stresses on the upper side of the wing, and the moments from the chordwise forces C produce compressive stresses on the leading edge of the wing. These compressive stresses will be additive in the upper flange of the front spar and the stringers adjacent to it. The PHAA condition, therefore, will be critical for compressive stresses in the upper forward

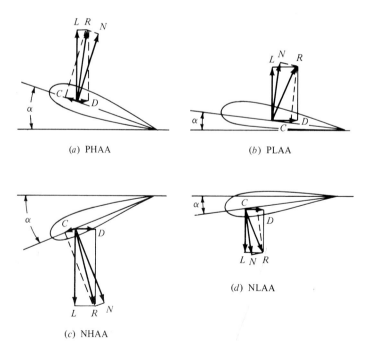

(a) PHAA   (b) PLAA

(c) NHAA   (d) NLAA

Figure 2.1

region of the wing cross section and for tensile stresses in the lower aft region of the wing cross section. For normal wings, in which the aerodynamic pitching-moment coefficient is negative, the line of action of the resultant force R is farther forward on the wing in the PHAA condition than in any other possible flight attitude producing an upload on the wing. The upload on the horizontal tail in this condition usually will be larger than for any other positive flight attitude, since pitching accelerations are normally neglected and the load on the horizontal tail must balance the moments of other aerodynamic forces about the center of gravity of the airplane.

In the positive low angle of attack (PLAA) condition, the wing has the smallest possible angle of attack at which the lift corresponding to the limit-load factor may be developed. For a given lift on the wing, the angle of attack decreases as the indicated airspeed increases, and consequently the PLAA condition corresponds to the maximum indicated airspeed at which the airplane will dive. This limit on the permissible diving speed depends on the type of aircraft, but usually is specified as 1.2 to 1.5 times the maximum indicated speed in level flight, according to the function of the aircraft. Some specifications require that the terminal velocity of the aircraft—the velocity obtained in a vertical dive sustained until the drag equals the airplane weight—be calculated and the limit on the diving speed be determined as a function of the terminal velocity. Even fighter aircraft are seldom designed for a diving speed equal to the terminal velocity, since the terminal velocity of such airplanes is so great that difficult aerodynamic and structural problems are encountered. Aircraft are placarded so that the pilot will not exceed the diving speed limit.

In the PLAA condition, shown in Fig. 2.1b, the chordwise force C is the largest force acting aft on the wing for any positive flight attitude. The wing bending moments in this condition produce the maximum compressive stresses on the upper rear spar flange and adjacent stringers and maximum tensile stresses on the lower front spar flange and adjacent stringers. In this condition, the line of action of the resultant wing force R is farther aft than for any other positive flight condition. The moment of this force about the center of gravity of the airplane has the maximum negative (pitching) value; consequently, the download on the horizontal tail required to balance the moments of other aerodynamic forces will be larger than for any positive flight condition.

The negative high angle of attack (NHAA) condition, shown in Fig. 2.1c, occurs in intentional flight maneuvers in which the air loads on the wing are down or when the airplane strikes sudden downdrafts while in level flight. The load factors for intentional negative flight attitudes are considerably smaller than for positive flight attitudes, because conventional aircraft engines cannot be operated under a negative load factor for very long and because the pilot is in the uncomfortable position of being suspended from the safety belt or harness. Gust load factors are also smaller for negative flight attitudes, since in level flight the weight of the airplane adds to the inertia forces for positive gusts but subtracts from the inertia forces for negative gusts.

In the NHAA condition, usually the wing is assumed to be at the negative stalling angle of attack for steady flow conditions. The assumption used in the PHAA condition—the maximum lift coefficient momentarily exceeds that for steady flow—is seldom used because it is improbable that negative maneuvers will be entered suddenly. The wing bending moments in the negative high angle of attack condition produce the highest compressive stresses in the lower forward region of the wing cross section and the highest tensile stresses in the upper aft region of the wing cross section. The line of action of the resultant force R is farther aft than for any other negative flight attitude, and it will probably produce the greatest balancing upload on the horizontal tail for any negative flight attitude.

The negative low angle of attack (NLAA) condition, shown in Fig. 2.1d, occurs at the diving-speed limit of the airplane. This condition may occur in an intentional maneuver producing a negative load factor or in a negative gust condition. The aft load C is a maximum for any negative flight attitude, the compressive bending stresses have a maximum value in the lower aft region of the wing cross section, and the tensile bending stresses have a maximum value in the upper forward region of the wing cross section. The resultant force R is farther forward than in any other flight attitude, and the download on the horizontal tail will probably be larger than in any other negative flight attitude.

In summary, one of the four basic symmetrical flight conditions is critical for the design of almost every part of the airplane structure. In the stress analysis of a conventional wing, it is necessary to investigate each cross section for each of the four conditions. Then each stringer or spar flange is designed for the maximum tension and the maximum compression obtained in any of the conditions. The probable critical conditions for each region of the cross section are shown in Fig. 2.2.

Some specifications require the investigation of additional conditions of medium-high angle of attack and medium-low angle of attack which may be critical for stringers midway between the spars, but usually these conditions are not considered of sufficient importance to justify the additional work required for the analysis. The wing, of course, must be strong enough to resist loads at medium angles of attack, but normally the wing will have adequate strength if it meets the requirements for the four limiting conditions.

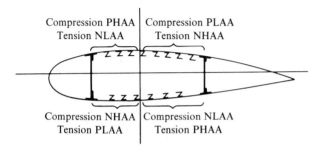

Figure 2.2

For aircrafts such as transport or cargo aircrafts, in which the load may be placed in various positions in the gross-weight condition, it is necessary to determine the balancing tail loads for the most forward and most rearward center-of-gravity positions at which the airplane may be flown at the gross weight. Each of the four flight conditions must be investigated for each extreme position of the center of gravity. For smaller aircraft, in which the useful load cannot be shifted appreciably, there may be only one position of the center of gravity at the gross-weight condition. To account for greater balancing tail loads which may occur for another location of the center of gravity, it may be possible to make some conservative assumption and still compute balancing tail loads for only one location.

The gust load factors on an aircraft are greater when it is flying at the minimum flying weight than they are at the gross-weight condition. While this is seldom critical for the wings, since they have less weight to carry, it is critical for a structure such as the engine mount which carries the same weight at a higher load factor. It is therefore necessary to calculate gust load factors at the minimum weight at which the aircraft will be flown.

For aircraft equipped with wing flaps, other high lift devices, or dive brakes, additional flight loading conditions must be investigated for the flaps extended. These conditions usually are not critical for wing bending stresses, since the specified load factors are not large, but may be critical for wing torsion, shear in the rear spar, or down tail loads, since the negative pitching moments may be quite high. The aft portion of the wing, which forms the flap supporting structure, will be critical for the condition with flaps extended.

Unsymmetrical loading conditions and pitching-acceleration conditions for commercial aircraft are seldom of sufficient importance to justify extensive analysis. Conservative simplifying assumptions usually are specified by the licensing agency for use in the structural design of members which will be critical for these conditions. The additional structural weight required to meet conservative design assumptions is not sufficient to justify a more accurate analysis. Some military aircraft must perform violent evasive maneuvers such as snap rolls, abrupt rolling pullouts, and abrupt pitching motions. The purchasing agency for such airplanes specifies the conditions which should be investigated. Such investigations require the calculation of the mass moment of inertia of the airplane about the pitching, rolling, and yawing axes. The aerodynamic forces on the airplane are calculated and set in equilibrium by inertial forces on the airplane.

2.4 FLIGHT-VEHICLE AERODYNAMIC LOADS

Extensive aerodynamic information is required to investigate the performance, control, and stability of a proposed aircraft. Only the information which is required for the structural analysis is considered here, although normally this would be obtained as part of a much more extensive program. The first aerodynamic data required for the structural analysis are the lift, drag, and pitching-moment force distributions for the complete aircraft with the horizontal tail removed, through the range of angles of attack from the negative stalling angle to the positive stalling angle. While these data can be calculated accurately for a wing with a conventional airfoil section, similar data for the combination of the wing and fuselage or the wing, fuselage, and nacelles are more difficult to calculate accurately from published information because of the uncertain effects of the aerodynamic interference of various components. It is therefore desirable to obtain wind tunnel data on a model of the complete airplane less horizontal tail. It is often necessary, of course, to calculate these data from published information in order to obtain approximate air loads for preliminary design purposes.

Wind tunnel tests of a model of the complete airplane with the horizontal tail removed provide values of the lift, drag, and pitching moment for all angles of attack. Then components of the lift and drag forces with respect to airplane reference axes are determined. The aircraft reference axes may be chosen as shown in Fig. 2.3. The force components are $C_z qS$ and $C_x qS$ along these axes, where $q = \rho V^2/2$ is the dynamic pressure and S is the surface wing area. The nondimensional force coefficients C_z and C_x are obtained by projecting the lift and drag coefficients, respectively, for the airplane less horizontal tail along the reference axes by the following equations:

$$C_z = C_L \cos \theta + C_D \sin \theta \tag{2.1}$$

$$C_x = C_D \cos \theta - C_L \sin \theta \tag{2.2}$$

The angle θ is measured from the flight path to the x axis, as shown in Fig. 2.3, and is equal to the difference between the angle of attack α and the angle of wing incidence i.

The pitching moment about the airplane's center of gravity is obtained from wind tunnel data and is $C_{m_{a-t}} \bar{c} qS$, where $C_{m_{a-t}}$ is the dimensionless pitching-moment coefficient of the airplane less tail and $\bar{c}$ is the mean aerodynamic chord of the wing. The mean aerodynamic chord (MAC) is a wing reference chord which usually is calculated from the wing planform. If every airfoil section along

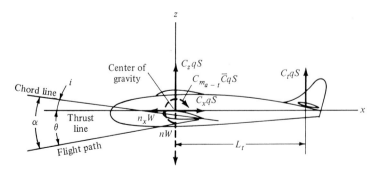

Figure 2.3

the wing span has the same pitching-moment coefficient c_m, the MAC is determined so that the total wing pitching moment is $c_m \bar{c} q S$. For a rectangular wing planform the value of $\bar{c}$ (the MAC) is equal to the wing chord; for a trapezoidal planform of the semiwing, the value of $\bar{c}$ is equal to the chord at the centroid of the trapezoid. The MAC is actually an arbitrary length, and any reference length would be satisfactory if it were used consistently in all wind tunnel tests and calculations. For irregular shapes of planforms, some procuring or licensing agencies require that the mean chord (wing area divided by wing span) be used as the reference chord.

The balancing air load on the horizontal tail, $C_t q S$, is obtained from the assumption that there is no angular acceleration of the airplane. The moments of the forces shown in Fig. 2.3 about the center of gravity are therefore in equilibrium:

$$C_t q S L_t = C_{m_{a-t}} \bar{c} q S$$

or

$$C_t = \frac{\bar{c}}{L_t} C_{m_{a-t}} \tag{2.3}$$

where C_t is a dimensionless tail force coefficient expressed in terms of the wing area and L_t is the distance from the airplane's center of gravity to the resultant air load on the horizontal tail, as shown in Fig. 2.3. Since the pressure distribution on the horizontal tail varies according to the attitude of the airplane, L_t theoretically varies for different loading conditions. This variation is not great, and it is customary to assume L_t contant, by using a conservative forward position of the center of pressure on the horizontal tail. The total aerodynamic force on the airplane in the z direction, $C_{z_a} q S$, is equal to the sum of the force $C_z q S$ on the airplane less tail and the balancing tail load $C_t q S$:

$$C_{z_a} q S = C_z q S + C_t q S$$

or

$$C_{z_a} = C_z + C_t \tag{2.4}$$

For power-on flight conditions, the moment of the propeller or jet thrust about the center of gravity of the airplane should also be considered. This adds another term to the Eq. (2.3).

Now the aerodynamic coefficients can be plotted against the angle of attack α, as shown in Fig. 2.4. If more than one position of the center of gravity is considered in the analysis, it is necessary to calculate the curves for $C_{m_{a-t}}$, C_t, and C_{z_a} for each center-of-gravity position. The right-hand portions of the solid curves shown in Fig. 2.4 represent the aerodynamic characteristics after stalling of the wing. Since stalling reduces the air loads on the wing, these portions of the curves are not used. Instead, the curves are extrapolated, as shown by the dotted lines, in order to approximate the conditions of a sudden pull-up, in which high lift coefficients may exist for a short time. For the PHAA condition, the angle of attack corresponding to the force coefficient of 1.25 times the maximum value of C_{z_a} is used, and the curves are extrapolated to this value, as shown in Fig. 2.4.

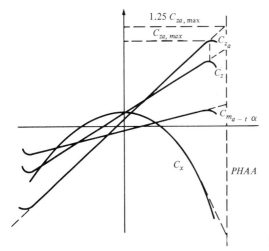

Figure 2.4

2.5 FLIGHT-VEHICLE INERTIA LOADS

The maximum load on any part of a flight-vehicle structure occurs when the vehicle is being accelerated. The loads produced by landing impact, maneuvering, gusts, boost and staging operations, launches, and dockings are always greater than the loads occurring when all the forces on the vehicle are in equilibrium. Before any structural component can be designed, it is necessary to determine the inertia loads acting on the vehicle.

In many of the loading conditions, a flight vehicle may be considered as being in pure translation or pure rotation. The inertia force on any element of mass is equal to the product of the mass and the acceleration and acts in the direction opposite to the acceleration. If the applied loads and inertia forces act on an element as a free body, these forces are in equilibrium. For example, a body of mass m under the action of a force vector F moves so as to satisfy the equation

$$F = ma \tag{2.5}$$

where m is the mass and a is the acceleration relative to a newtonian frame of reference. If a cartesian system of x, y, and z axes is chosen in this frame, then Eq. (2.5) gives, upon resolving into components,

$$F_x = m\ddot{x}, \qquad F_y = m\ddot{y}, \qquad F_z = m\ddot{z} \tag{2.6}$$

where F_x, F_y, and F_z are the components of F along the x, y, and z axes, respectively, and $\ddot{x}$, $\ddot{y}$, and $\ddot{z}$ are the components of acceleration along the x, y, and z axes. In the preceding discussion, all parts of the rigid body were moving in straight, parallel lines and had equal velocities and accelerations. In many engineering problems, it is necessary to consider the inertia forces acting on a rigid body which has other types of motion. In many cases where the elements of a rigid body are moving in curved paths, they are moving in such a way that each element moves in only one plane and all elements move in parallel planes. This type of motion is called *plane motion*, and it occurs, for example, when a vehicle is pitching and yet has no rolling or yawing motion. All elements of the vehicle move in planes parallel to the plane of symmetry. Any type of plane motion can be considered as a rotation about some instantaneous axis perpendicular to the planes of motion, and the following equations for inertia forces are derived on the assumption that the rigid body is rotating about an instantaneous axis perpendicular to a plane of symmetry of the body. The inertia forces obtained may be used for the pitching motion of a vehicle, but when they are utilized for rolling or yawing motions, it is necessary to first obtain the principal axes and moments of inertia of the vehicle.

The rigid mass shown in Fig. 2.5 is rotating about point O with a constant angular velocity ω. The acceleration of any point a distance r from the center of rotation is $\omega^2 r$ and is directed toward the center of rotation. The inertia force acting on an element of mass dM is the product of the mass and the acceleration, or $\omega^2 r\, dM$, and is directed away from the axis of rotation. This inertia force has components $\omega^2 x\, dM$ parallel to the x axis and $\omega^2 y\, dM$ parallel to the y axis. If the x axis is chosen through the center of gravity C, the forces are simplified. The resultant inertia force in the y direction for the entire body is found as follows:

$$F_y = \int \omega^2 y\, dM = \omega^2 \int y\, dM = 0$$

The angular velocity ω is constant for all elements of the body, and the integral is zero because the x axis was chosen through the center of gravity. The inertia force in the x direction is found in the same manner:

$$F_x = \int \omega^2 x\, dM = \omega^2 \int x\, dM = \omega^2 \bar{x} M \qquad (2.7)$$

The term $\bar{x}$ is the distance from the axis of rotation O to the center of gravity C, as shown in Fig. 2.5.

If the body has an angular acceleration α, the element of mass dM has an

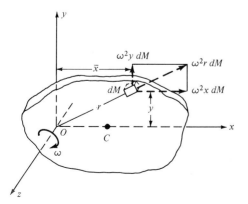

Figure 2.5

additional inertia force $\alpha r \, dM$ acting perpendicular to r and opposite to the direction of acceleration. This force has components $\alpha x \, dM$ in the y direction and $\alpha y \, dM$ in the x direction, as shown in Fig. 2.6. The resultant inertia force on the entire body in the x direction is

$$F_x = \int \alpha y \, dM = \alpha \int y \, dM = 0$$

The resultant inertia force in the y direction is

$$F_y = \int \alpha x \, dM = \alpha \int x \, dM = \alpha \bar{x} M \tag{2.8}$$

The resultant inertia torque about the axis of rotation is found by integrating the terms representing the product of the tangential force on each element $\alpha r \, dM$ and its moment arm r:

$$T_0 = \int \alpha r^2 \, dM = \alpha \int r^2 \, dM = \alpha I_0 \tag{2.9}$$

The term I_0 represents the moment of inertia of the mass about the axis of rotation. It can be shown that this moment of inertia can be transferred to a parallel axis through the center of gravity by use of the following relationship:

$$I_0 = M\bar{x}^2 + I_c \tag{2.10}$$

where I_c is the moment of inertia of the mass about an axis through the center of gravity, obtained as the sum of the products of mass elements dM and the square of their distances r_c from the center of gravity:

$$I_c = \int r_c^2 \, dM$$

By substituting the value of I_0 from Eq. (2.10) in Eq. (2.9), the following expression for the inertia torque is obtained:

$$T_0 = M\bar{x}^2\alpha + I_c\alpha \tag{2.11}$$

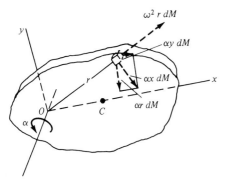

Figure 2.6

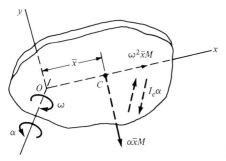

Figure 2.7

The inertia forces obtained in Eqs. (2.7), (2.8), and (2.11) may be represented as forces acting at the center of gravity and the couple $I_c \alpha$, as shown in Fig. 2.7. The force $\alpha \bar{x} M$ and the couple $I_c \alpha$ must both produce moments about point O which are opposite to the direction of α. The force $\omega^2 \bar{x} M$ must act away from point O.

It is seen from Fig. 2.7 that the forces at the centroid represent the product of the mass of the body and the components of acceleration of the center of gravity. In many cases, the axis of rotation is not known, but the components of acceleration of the center of gravity can be obtained. In other cases, the acceleration of one point of the body and the angular velocity and angular acceleration are known. If the point O in Fig. 2.8 has an acceleration a_0, an inertia force at the center of gravity of Ma_0, opposite to the direction of a_0, must be considered in addition to those previously taken into account.

2.6 LOAD FACTORS FOR TRANSLATIONAL ACCELERATION

For flight or landing conditions in which the vehicle has only translational acceleration, every part of the vehicle is acted on by parallel inertia forces which are proportional to the weight of the part. For purposes of analysis, it is convenient to combine these inertia forces with the forces of gravity, by multiplying the weight of each part by a load factor n, and thus to consider the combined weight and inertia forces. When the vehicle is being accelerated upward, the weight and inertia forces add directly. The weight of w of any part and the inertia force wa/g have a sum nw:

$$nw = w + w \frac{a}{g}$$

or
$$n = 1 + \frac{a}{g} \tag{2.12}$$

The combined inertia and gravity forces are considered in the analysis in the same manner as weights which are multiplied by the load factor n.

In the case of an airplane in flight with no horizontal acceleration, as shown in Fig. 2.9, the engine thrust is equal to the airplane drag, and the horizontal

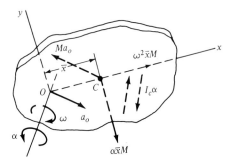

Figure 2.8

components of the inertia and gravity forces are zero. The weight and the inertia force on the airplane act down and will be equal to the lift. The airplane lift L is the resultant of the wing and tail lift forces. The load factor is defined as follows:

$$\text{Load factor} = \frac{\text{lift}}{\text{weight}}$$

or

$$n = \frac{L}{W} \tag{2.13}$$

This value for the load factor can be shown to be the same as that given by Eq. (2.12) by equating the lift nW to the sum of the weight and inertia forces:

$$L = nW = W + W\frac{a}{g}$$

or

$$n = 1 + \frac{a}{g}$$

which corresponds to Eq. (2.12).

Flight vehicles frequently have horizontal acceleration as well as vertical acceleration. The airplane shown in Fig. 2.10 is being accelerated forward, since the engine thrust T is greater than the airplane drag D. Every element of mass in the airplane is thus under the action of a horizontal inertia force equal to the product of its mass and the horizontal acceleration. It is also convenient to consider the horizontal inertia loads as equal to the product of a load factor n_x

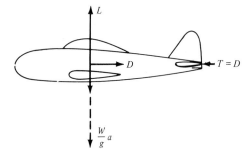

Figure 2.9

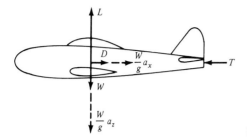

Figure 2.10

and the weights. This horizontal load factor, often called the *thrust load factor*, is obtained from the equilibrium of the horizontal forces shown in Fig. 2.10:

$$n_x W = \frac{a_x}{g} W = T - D$$

or
$$n_x = \frac{T - D}{W} \tag{2.14}$$

A more general case of translational acceleration is shown in Fig. 2.11, in which the airplane thrust line is not horizontal. It is usually convenient to obtain components of forces along x and z axes which are parallel and perpendicular to the airplane thrust line. The combined weight and inertia load on any element has a component along the z axis of the following magnitude:

$$nW = W \cos \theta + W \frac{a_z}{g}$$

or
$$n = \cos \theta + \frac{a_z}{g} \tag{2.15}$$

From summation of all forces along the z axis,

$$L = W \left(\cos \theta + \frac{a_z}{g} \right) \tag{2.16}$$

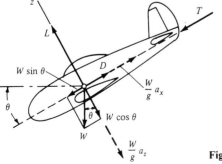

Figure 2.11

By combining Eqs. (2.15) and (2.16),

$$L = Wn$$

or

$$n = \frac{L}{W}$$

which corresponds with the value used in Eq. (2.13) for a level attitude of the airplane.

The thrust load factor for the condition shown in Fig. 2.11 is also similar to that obtained for the airplane in level attitude. Since the thrust and drag forces must be in equilibrium with the components of weight and inertia forces along the x axes, the thrust load factor is obtained as follows:

$$n_x W = \frac{W}{g} a_x - W \sin \theta = T - D$$

or

$$n_x = \frac{T - D}{W}$$

This value is the same as that obtained in Eq. (2.14) for a level attitude of the airplane.

In the case of the airplane landing as shown in Fig. 2.12, the landing load factor is defined as the vertical ground reaction divided by the airplane weight. The load factor in the horizontal direction is similarly defined as the horizontal ground reaction divided by the airplane weight:

$$n_z = \frac{R_z}{W} \tag{2.17}$$

and

$$n_x = \frac{R_x}{W} \tag{2.18}$$

In the airplane analysis, it is necessary to obtain the components of the load factor along axes parallel and perpendicular to the propeller thrust line. However, aerodynamic forces are usually obtained first as lift and drag forces perpendicular and parallel to the direction of flight. If load factors are obtained first along lift and drag axes, they may be resolved into components along other axes, in the same manner as forces are resolved into components. The force acting on any

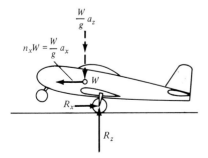

Figure 2.12

weight w is wn, and the component of this force along any axis at an angle θ to the force is $wn \cos \theta$. The component of the load factor is then $n \cos \theta$.

As a general definition, the load factor n along any axis i is such that the product of the load factor and the weight of an element is equal to the sum of the components of the weight and inertia forces along that axis. The weight and inertia forces are always in equilibrium with the external forces acting on the airplane, and the sum of the components of the weight and inertia forces along any axis must be equal and opposite to the sum of the components of the external forces along the axis ΣF_i. The load factor is therefore defined as

$$n_i = \frac{\Sigma F_i}{W} \tag{2.19}$$

where ΣF_i includes all forces except weight and inertia forces.

2.7 VELOCITY–LOAD-FACTOR DIAGRAM

The various loading conditions for an airplane usually are represented on a graph of the limit-load factor n plotted against the indicated airspeed V. This diagram is often called a *V-n diagram*, since the load factor n is related to the acceleration of gravity g. In all such diagrams, the indicated airspeed is used, since all air loads are proportional to q or $\rho V^2/2$. The value of q is the same for the air density ρ and the actual airspeed at altitude as it is for the standard sea-level density ρ_0 and the indicated airspeed, since the indicated airspeed is defined by this relationship. The *V-n* diagram is therefore the same for all altitudes if indicated airspeeds are used. Where compressibility effects are considered, they depend on actual airspeed rather than indicated airspeed and consequently are more pronounced at altitude. Compressibility effects are not considered here.

The aerodynamic forces on an airplane are in equilibrium with the forces of gravity and inertia. If the airplane has no angular acceleration, both the inertia and gravity forces will be distributed in the same manner as the weight of various items of the aircraft and will have resultants acting through the center of gravity of the aircraft. It is convenient to combine the inertia and gravity forces as the product of a load factor n and the weight W, as described previously. The z component of the resultant gravity and inertia force is the force nW acting at the center of gravity of the airplane, as shown in Fig. 2.13. The load factor n is obtained from a summation of forces along the z axis:

$$C_{z_a} qS = nW$$

or

$$n = \frac{C_{z_a} \rho S V^2}{2W} \tag{2.20}$$

The maximum value of the normal force coefficient C_{z_a} may be obtained at various airplane speeds. For level flight at a unit-load factor, the value of V corresponding to $C_{z_a, \max}$ would be the stalling speed of the airplane. In acceler-

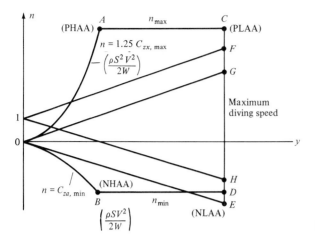

Figure 2.13

ated flight, the maximum coefficient might be obtained at higher speeds. For $C_{z_a,\,max}$ to be obtained at twice the stalling speed, a load factor $n = 4$ would be developed as shown by Eq. (2.20). For a force coefficient of $1.25 C_{z_a,\,max}$ representing the highest angle of attack for which the wing is analyzed, the value of the load factor n is obtained from Eq. (2.20) and may be plotted against the airplane velocity V, as shown by line OA in Fig. 2.13. This line OA represents a limiting condition, since it is possible to maneuver the airplane at speeds and load factors corresponding to points below or to the right of line OA, but it is impossible to maneuver at speeds and load factors corresponding to points above or to the left of line OA because this would represent angles of attack much higher than the stalling angle.

The line AC in Fig. 2.13 represents the limit on the maximum maneuvering load factor for which the airplane is designed. This load factor is determined from the specifications for which the airplane is designed, and the pilot must restrict maneuvers so as not to exceed this load factor. At speeds below that corresponding to point A, it is impossible for the pilot to exceed the limit load factor in any symmetrical maneuver, because the wing will stall at a lower load factor. For airspeeds between those corresponding to points A and C, it is not practical to design the airplane structure so that it could not be overstressed by violent maneuvers. Some types of airplanes may be designed so that the pilot would have to exert large forces on the controls in order to exceed the limit-load factor.

Line CD in Fig. 2.13 represents the limit on the permissible diving speed for the airplane. This value is usually specified as 1.2 to 1.5 times the maximum indicated airspeed in level flight. Line OB corresponds to line OA, except that the wing is at the negative stalling angle of attack, and the air load is down on the wing. The equation for line OB is obtained by substituting the maximum negative value of C_{z_a} into Eq. (2.20). Similarly, line BD corresponds to line AC, except that the limit-load factor specified for negative maneuvers is considerably less than for positive maneuvers.

The aircraft may therefore be maneuvered in such a manner that velocities and load factors corresponding to the coordinates of points within the area *OACDB* may be obtained. The most severe structural load *ng* conditions will be represented by the corners of the diagram, points *A*, *B*, *C*, and *D*. Points *A* and *B* represent PHAA and NHAA conditions. Point *C* represents the PLAA condition in most cases, although the positive gust load condition, represented by point *F*, may occasionally be more severe. The NLAA condition is represented by point *D* or by the negative gust condition, point *E*, depending on which condition produces the greatest negative load factor. The method of obtaining the gust load factors, represented by points *E* and *F*, is explained in the following section.

2.8 GUST LOAD FACTORS

When an airplane is in level flight in calm air, the angle of attack α is measured from the wing chord line to the horizontal. If the airplane suddenly strikes an ascending air current which has a vertical velocity KU, the angle of attack is increased by the angle $\Delta\alpha$, as shown in Fig. 2.14. The angle $\Delta\alpha$ is small, and the angle in radians may be considered as equal to its tangent:

$$\Delta\alpha = \frac{KU}{V} \tag{2.21}$$

The change in the airplane normal force coefficient C_{z_a}, resulting from a change in angle of attack $\Delta\alpha$, may be obtained from the curve of C_{z_a} versus α of Fig. 2.4. This curve is approximately a straight line, and it has a slope β which may be considered constant:

$$\beta = \frac{\Delta C_{z_a}}{\Delta\alpha} \tag{2.22}$$

After striking the gust, the airplane normal force coefficient increases by an amount determined from Eqs. (2.21) and (2.22):

$$\Delta C_{z_a} = \frac{\beta KU}{V} \tag{2.23}$$

The increase in the airplane load factor Δn may be obtained by substituting the value of ΔC_{z_a} from Eq. (2.23) into Eq. (2.20):

$$\Delta n = \frac{\Delta C_{z_a} \rho S V^2}{2W}$$

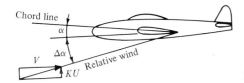

Figure 2.14

or
$$\Delta n = \frac{\rho S \beta K U V}{2W} \tag{2.24}$$

where ρ = standard sea-level air density, 0.002378 slug/ft^3
    S = wing area, ft^2
    β = slope of the curve of C_{z_a} versus α, rad
  KU = effective gust velocity, ft/s
    V = indicated airspeed, ft/s
    W = gross weight of airplane, lb

For purposes of calculation, it is more convenient to determine the slope β per degree and the airspeed V in miles per hour. Introducing the necessary constants in Eq. (2.24) yields

$$\Delta n = 0.1 \frac{\beta K U V}{W/S} \tag{2.25}$$

where β is the slope of C_{z_a} versus α per degree, V is the indicated airspeed in miles per hour, and other terms correspond to those in Eq. (2.24).

When the airplane is in level flight, the load factor is unity before the plane strikes the gust. The change in load factor Δn from Eq. (2.25) must be combined with the unit-load factor in order to obtain the total gust load factor:

$$n = 1 \pm 0.1 \frac{\beta K U V}{W/S} \tag{2.26}$$

Equation (2.26) may be plotted on the V-n diagram, as shown by the inclined straight lines through points F and H of Fig. 2.13. These lines represent load factors obtained when the airplane is in a horizontal attitude and strikes positive or negative gusts. Equation (2.25) is similarly plotted, as shown by the inclined lines through points G and E of Fig. 2.13. These lines represent load factors obtained when the airplane is in a vertical attitude and strikes positive or negative gusts in directions normal to the thrust line.

The gust load factor represented by point F of Fig. 2.13 may be more severe than the maneuvering load factor represented by point C. In the case shown, however, the maneuvering load factor is obviously greater and will represent the PLAA condition. The negative gust load factor represented by point E is greater than the negative maneuvering load factor represented by point D and will determine the NLAA condition. It might seem that the gust load factors should be added to the maneuvering load factors, in order to provide for the possibility of the airplane's striking a severe gust during a violent maneuver. While this condition is possible, it is improbable because the maneuvering load factors are under the pilot's control, and the pilot will restrict maneuvers in gusty weather. Both the maneuvering and gust load factors correspond to the most severe conditions expected during the life of the airplane, and there is little probability of a combined gust and maneuver producing a condition which would exceed the limit-load factor for the design condition.

The "effective, sharp-edged gust" velocity KU is the velocity of a theoretical gust which, if encountered instantaneously, would give the same load factor as the actual gust. Actually, it is impossible for the upward air velocity to change suddenly from zero to its maximum value. There is always a finite distance in which the air velocity changes gradually from zero to the maximum gust velocity, and a short time is required for the airplane to move through this transition region. Most specifications require that the airplane be designed for a gust velocity U of 30 ft/s with the gust effectiveness factor K of 0.8 to 1.2, depending on the wing loading W/S. Airplanes with higher wing loadings usually are faster and pass through the transition region from calm air to air with the maximum gust velocity in a shorter time, and hence they must be designed for larger values of K. The design values of KU are obtained from accelerometer readings for airplanes flying in turbulent air and represent the maximum effective gust velocities which will ever be encountered during the service life of the airplane. Some specifications require gust velocities of 50 ft/s, with corresponding gust reduction factors K of about 0.6. Since the values of KU in this case are also about 30 ft/s, the net effect is equivalent to a gust velocity U of 30 ft/s with a K of 1.0. The actual maximum vertical air velocities probably exceed 50 ft/s, but the transition is gradual, corresponding to the values of $K = 0.6$. High gust load factors exist for only a fraction of a second, and the airplane cannot move far in this time.

In order to understand the effect of gusts, it is necessary to study the motion of the airplane after it encounters a gust. If the gust is encountered instantaneously, the factor K is 1.0, and the effective gust velocity is U. The airplane is accelerated upward with an initial acceleration a_0 and attains a variable vertical velocity v. The gust angle of attack ($\Delta\alpha$ of Fig. 2.14) has a maximum value of U/V at the time the gust is encountered ($t = 0$), but this angle of attack is decreased to $(U - v)/V$ after the airplane attains an upward velocity. When the upward velocity v is equal to U, the relative wind is again horizontal, and the airplane is no longer accelerated. The variable vertical acceleration a is therefore

$$a = \frac{dv}{dt} = a_0 \frac{U - v}{U} \tag{2.27}$$

Separating the variables and integrating, we have

$$\int_0^v \frac{dv}{U - v} = \frac{a_0}{U} \int_0^t dt$$

or

$$\log_e \frac{U - v}{U} = -\frac{a_0 t}{U}$$

By using the exponential form and substituting the value of a from Eq. (2.27), the following expression for the acceleration a at time t is obtained:

$$\frac{a}{a_0} = e^{-a_0 t/U} \tag{2.28}$$

As a numerical example, consider a gust velocity U of 30 ft/s and an initial

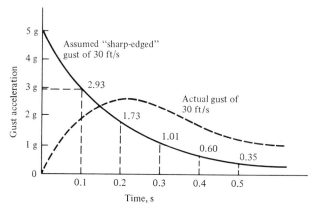

Figure 2.15

acceleration a_0 of 5g, corresponding to a gust load factor of 6.0. By substituting these values into Eq. (2.28) and plotting a versus t, the curve of Fig. 2.15 is obtained. The gust acceleration is seen to approach zero asymptotically in an infinite time, but it decreases greatly in the first 0.1 s. Thus, if the airplane had a forward speed of 500 ft/s (340 mi/h) it would move forward only 50 ft in 0.1 s. It seems logical to expect that atmospheric conditions are such that it is more than 50 ft from any region of calm air to a region in which the gust velocity is 30 ft/s. The actual gust acceleration probably is represented more accurately by the dotted line of Fig. 2.15, which would indicate an effectiveness factor K of about 0.6. However, since airplane accelerometer readings have shown effective gust velocities KU of 30 ft/s, the true conditions probably are represented by gust velocities U of more than 50 ft/s with effectiveness factors K less than 0.6.

2.9 EXAMPLES

Example 2.1 When landing on a carrier, a 10,000-lb airplane is given a deceleration of 3g (96.6 ft/s²) by means of a cable engaged by an arresting hook as shown in Fig. 2.16.

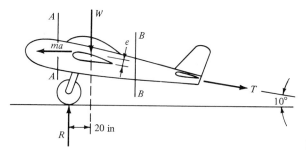

Figure 2.16

(a) Find the tension in the cable, the wheel reaction R, and the distance e from the center of gravity to the line of action of the cable.

(b) Find the tension in the fuselage at vertical sections AA and BB if the portion of the airplane forward section AA weighs 3000 lb and the portion aft of section BB weighs 1000 lb.

(c) Find the landing run if the landing speed is 80 ft/s.

SOLUTION

(a) First consider the entire airplane as a free body.

$$Ma = \frac{W}{g} a = \frac{10,000}{g} 3g = 30,000 \text{ lb}$$

$$\Sigma F_x = T \cos 10° - 30,000 = 0$$

$$T = 30,500 \text{ lb}$$

$$\Sigma F_y = R - 10,000 - 30,500 \sin 10° = 0$$

$$R = 15,300 \text{ lb}$$

$$\Sigma M_{cg} = 20 \times 15,300 - 30,500e = 0$$

$$e = 10 \text{ in}$$

(b) Consider the aft section of the fuselage as a free body, as shown in Fig. 2.17. It is acted on by an inertia force of

$$Ma = \frac{1000}{g} 3g = 3000 \text{ lb}$$

The tension on section BB is found as follows:

$$\Sigma F_x = 30,000 - 3000 - T_1 = 0$$

$$T_1 = 27,000 \text{ lb}$$

Since there is no vertical acceleration, there is no vertical inertia force. Section BB has a shear force V_1 of 6300 lb, which is equal to the sum of the weight and the vertical component of the cable force.

Consider the portion of the airplane forward of section AA as a free

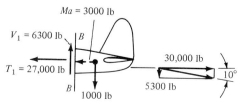

Figure 2.17

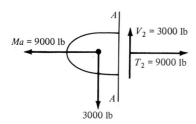

$V_2 = 3000$ lb

$Ma = 9000$ lb

$T_2 = 9000$ lb

3000 lb

Figure 2.18

body, as shown in Fig. 2.18. The inertia force is the following.

$$Ma = \frac{3000}{g} 3g = 9000 \text{ lb}$$

$$\Sigma F_x = T_2 - 9000 = 0$$

$$T_2 = 9000 \text{ lb}$$

The section AA must also resist a shearing force V_2 of 3000 lb and a bending moment obtained by taking moments of the forces shown in Fig. 2.18.

The forces T_1, T_2, V_1, and V_2 may be checked by considering the equilibrium of the center portion of the airplane, as shown in Fig. 2.19.

$$Ma = \frac{6000}{g} 3g = 18,000 \text{ lb}$$

$$\Sigma F_x = 27,000 - 18,000 - 9000 = 0$$

$$\Sigma F_y = 15,300 - 3000 - 6000 - 6300 = 0$$

(c) From elementary dynamics, the landing run s is obtained as follows:

$$v^2 - v_0^2 = 2as$$

$$0 - (80^2) = 2(-96.6)s$$

$$s = 33 \text{ ft}$$

Example 2.2 A 30,000-lb airplane is shown in Fig. 2.20a at the time of landing impact, when the ground reaction on each main wheel is 45,000 lb.
(a) If one wheel and tire weighs 500 lb, find the compression C and bending

6000 lb

$T_2 = 9000$ lb

$V_2 = 3000$ lb

$T_1 = 27,000$ lb

$V_1 = 6300$ lb

15,300 lb

$Ma = 18,000$ lb

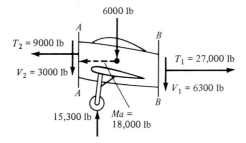

Figure 2.19

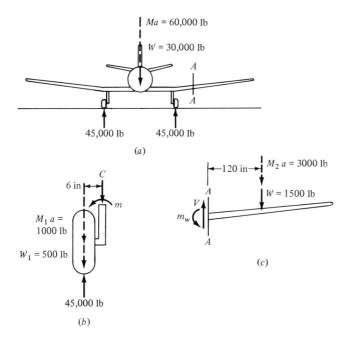

Figure 2.20

moment m in the oleo strut if the strut is vertical and is 6 in from the centerline of the wheel, as shown in Fig. 2.20b.

(b) Find the shear and bending moment at section AA of the wing if the wing outboard of this section weighs 1500 lb and has its center of gravity 120 in outboard of section AA.

(c) Find the required shock strut deflection if the airplane strikes the ground with a vertical velocity of 12 ft/s and has a constant vertical deceleration until the vertical velocity is zero. This neglects the energy absorbed by the tire deflection, which may be large in some cases.

(d) Find the time required for the vertical velocity to become zero.

SOLUTION

(a) Considering the entire airplane as a free body and taking a summation of vertical forces yield

$$\Sigma F_y = 45{,}000 + 45{,}000 - 30{,}000 - Ma = 0$$

$$Ma = 60{,}000 \text{ lb}$$

$$a = \frac{60{,}000}{M} = \frac{60{,}000g}{30{,}000} = 2g$$

Consider the landing gear as a free body, as shown in Fig. 2.20b. The inertia force is

$$M_1 a = \frac{w_1}{g} a = \frac{500}{g} 2g = 1000 \text{ lb}$$

The compression load C in the oleo strut is found from a summation of vertical forces:

$$\Sigma F_y = 45,000 - 500 - 1000 - C = 0$$

$$C = 43,500 \text{ lb}$$

The bending moment m is found as follows:

$$m = 45,000 \times 6 - 1000 \times 6 - 500 \times 6 = 261,000 \text{ in} \cdot \text{lb}$$

(b) The inertia force acting on the portion of the wing shown in Fig. 2.20c is

$$M_2 a = \frac{w_2}{g} a = \frac{1500}{g} 2g = 3000 \text{ lb}$$

The wing shear at section AA is found from a summation of vertical forces.

$$\Sigma F_y = V - 3000 - 1500 = 0$$

$$V = 4500 \text{ lb}$$

The wing bending moment is found by taking moments about section AA.

$$m_w = 3000 \times 120 + 1500 \times 120 = 540,000 \text{ in} \cdot \text{lb}$$

(c) The shock strut deflection is found by assuming a constant vertical acceleration of $-2g$, or -64.4 ft/s², from an initial vertical velocity of 12 ft/s to a final zero vertical velocity.

$$v^2 - v_0^2 = 2as$$

$$0 - (12^2) = 2(-64.4)s$$

$$s = 1.12 \text{ ft}$$

(d) The time required to absorb the landing shock is found based on elementary dynamics.

$$v - v_0 = at$$

$$0 - 12 = -64.4t$$

$$t = 0.186 \text{ s}$$

Since the landing shock occurs for such a short time, it may be less injurious to the structure and less disagreeable to the passengers than would a sustained load.

Example 2-3 A 60,000-lb airplane with a tricycle landing gear makes a hard two-wheel landing in soft ground so that the vertical ground reaction is

270,000 lb and the horizontal ground reaction is 90,000 lb. The moment of inertia about the center of gravity is 5,000,000 lb · s² · in, and the dimensions are shown in Fig. 2.21.

(a) Find the inertia forces on the airplane.
(b) Find the inertia forces on a 400-lb gun turret in the tail which is 500 in aft of the center of gravity. Neglect the moment of inertia of the turret about its own center of gravity.
(c) If the nose wheel is 40 in from the ground when the main wheels touch the ground, find the angular velocity of the airplane and the vertical velocity of the nose wheel when the nose wheel reaches the ground, assuming no appreciable change in the moment arms. The airplane's center of gravity has a vertical velocity of 12 ft/s at the moment of impact, and the ground reactions are assumed constant until the vertical velocity reaches zero, at which time the vertical ground reaction becomes 60,000 lb and the horizontal ground reaction becomes 20,000 lb.

SOLUTION

(a) The inertia forces on the entire airplane may be considered as horizontal and vertical forces Ma_x and Ma_y, respectively, at the center of gravity and a couple $I_c \alpha$, as shown in Fig. 2.21. These correspond to the inertia forces shown on the mass of Fig. 2.7, since the forces at the center of gravity represent the product of the mass and the acceleration components of the center of gravity.

$$\Sigma F_x = 90,000 - Ma_x = 0$$

$$Ma_x = 90,000 \text{ lb}$$

$$\Sigma F_y = 270,000 - 60,000 - Ma_y = 0$$

$$Ma_y = 210,000 \text{ lb}$$

$$\Sigma M_{cg} = -270,000 \times 40 - 90,000 \times 100 + I_c \alpha = 0$$

$$I_c \alpha = 19,800,000 \text{ in} \cdot \text{lb}$$

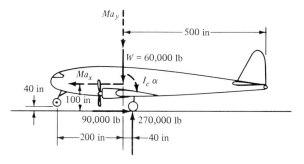

Figure 2.21

$$a_x = \frac{90{,}000}{M} = \frac{90{,}000}{60{,}000} \, g = 1.5g$$

$$a_y = \frac{210{,}000}{M} = \frac{210{,}000}{60{,}000} \, g = 3.5g$$

$$\alpha = \frac{I_c \alpha}{I_c} = \frac{19{,}800{,}000}{5{,}000{,}000} = 3.96 \text{ rad/s}^2$$

(b) The acceleration of the center of gravity of the airplane is now known, and the acceleration and inertia forces for the turret can be obtained by the method shown in Fig. 2.8, where the center of gravity of the airplane corresponds to point O of Fig. 2.8 and the center of gravity of the turret corresponds to point C. These forces are shown in Fig. 2.22 and have the following values:

$$Ma_x = \frac{400}{g} \, 1.5g = 600 \text{ lb}$$

$$Ma_y = \frac{400}{g} \, 3.5g = 1400 \text{ lb}$$

$$\alpha \bar{x} M = 3.96 \times 500 \times \frac{400}{386} = 2050 \text{ lb}$$

In calculating the term $\alpha \bar{x} M$, $\bar{x}$ is in inches and g is used as 386 in/s². If $\bar{x}$ is in feet, g will be 32.2 ft/s². The total force on the turret is 600 lb forward and 3850 lb down. This total force is seen to be almost 10 times the weight of the turret.

(c) The center of gravity of the airplane is decelerated vertically at 3.5g, or 112.7 ft/s². The time of deceleration from an initial velocity of 12 ft/s to a zero vertical velocity is found from the following.

$$v = v_0 = at$$

$$0 - 12 = 112.7t$$

$$t = 0.106 \text{ s}$$

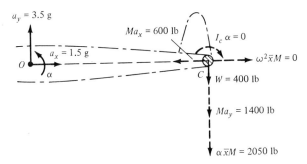

Figure 2.22

During this time, the center of gravity moves through a distance found from

$$s = v_0 t + \tfrac{1}{2}at^2$$
$$= 12 \times 0.106 - \tfrac{1}{2} \times 112.7 \times 0.106^2$$
$$= 0.636 \text{ ft, or } 7.64 \text{ in}$$

The angular velocity of the airplane at the end of 0.106 s after the landing is found from

$$\omega - \omega_0 = \alpha t$$
$$\omega - 0 = 3.96 \times 0.106$$
$$\omega = 0.42 \text{ rad/s}$$

The angle of rotation during this time is found from

$$\theta_1 = \omega_0 t + \tfrac{1}{2}\alpha t^2$$
$$\theta_1 = 0 + \tfrac{1}{2}(3.96)(0.106^2) = 0.0222 \text{ rad}$$

The vertical motion of the nose wheel resulting from this rotation, shown in Fig. 2.23, is

$$s_1 = \theta_1 x = 0.0222(200) = 4.44 \text{ in}$$

The distance of the nose wheel from the ground, after the vertical velocity of the center of gravity of the airplane has become zero, is

$$s_2 = 40 - 7.64 - 4.44 = 27.92 \text{ in}$$

The remaining angle of rotation θ_2, shown in Fig. 2.23, is

$$\theta_2 = \frac{s_2}{x} = \frac{27.92}{200} = 0.1396 \text{ rad}$$

Since the ground reaction decreases by the ratio of $\frac{60,000}{270,000}$ after the vertical acceleration of the airplane becomes zero, the angular acceleration decreases in the same proportion, as found by equating moments

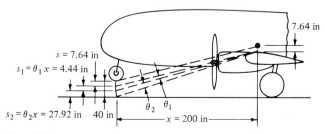

Figure 2.23

about the center of gravity:

$$\alpha_2 = \frac{60,000}{270,000}\, 3.96 = 0.88 \text{ rad/s}^2$$

The angular velocity of the airplane at the time the nose wheel strikes the ground is found from the following equation.

$$\omega^2 - \omega_0^2 = 2\alpha_2\theta_2$$

$$\omega^2 - (0.42^2) = 2 \times 0.88 \times 0.1396$$

$$\omega = 0.65 \text{ rad/s}$$

Since at this time the motion is rotation, with no vertical motion of the center of gravity, the vertical velocity of the nose wheel is found as follows:

$$v = \omega x$$

$$v = 0.65 \times \tfrac{200}{12} = 10.8 \text{ ft/s}$$

This velocity is smaller than the initial sinking velocity of the airplane. Consequently, the nose wheel would strike the ground with a higher velocity in a three-wheel level landing.

It is of interest to find the centrifugal force on the turret $\omega^2 \bar{x} M$ at the time the nose wheel strikes the ground. This force was zero when the main wheels hit because the angular velocity ω was zero. For the final value of ω, the following value is obtained:

$$\omega^2 \bar{x} M = (0.65^2) \times 500 \times \tfrac{400}{386} = 219 \text{ lb}$$

This force is much smaller than other forces acting on the turret, and usually it is neglected. In part c, certain simplifying assumptions are made which do not quite correspond with actual landing conditions. Aerodynamic forces are neglected, and the ground reactions on the landing gear are assumed constant while the landing gear is a combination of the tire deflection, in which the load is approximately proportional to the deformation, and the oleo strut deflection, in which the load is almost constant during the entire deformation, as assumed. The tire deflection may be as much as one-third to one-half the total deflection. The aerodynamic forces, which have been neglected, would probably reduce the maximum angular velocity of the airplane, since the horizontal tail moves upward as the airplane pitches, and the combination of upward and forward motions would give a downward aerodynamic force on the tail, tending to reduce the pitching acceleration.

The aerodynamic effects of the lift on the wing and tail surfaces are not shown in Fig. 2.21, but they will not affect the pitching acceleration appreciably if the ground reactions remain the same. Just before the airplane strikes the ground, the lift forces on the wing and tail are in equilibrium with the gravity force of 60,000 lb. Since the horizontal velocity of the airplane and the angle of attack are not appreciably changed ($\theta_1 = 0.0222$ rad $= 1.27°$), the lift forces continue to balance the weight of the airplane when the center of gravity is being decelerated. Instead of the weight of 60,000 lb shown in Fig. 2.21, there should be an additional inertia force of 60,000 lb down at the center of gravity. The moments about the center of gravity and the pitching acceleration are not changed, but the vertical deceleration a_y is increased. At the end of the deceleration of the center of gravity, the ground reactions are almost zero, since most of the airplane weight is carried by the lift on the wings. The airplane then pitches forward through the angle θ_2, which appreciably changes the angle of attack ($\theta_2 = 0.1396$ rad $= 8°$). The wing lift is then decreased, and most of the weight is supported by the ground reactions on the wheels. For the structural design of the airplane, usually only the loads during the initial impact are significant.

Example 2-4 Construct the V-n diagram and determine the wing internal load resulting from aerodynamic forces for the airplane (Fig. 2.24) whose wing planform is shown in Fig. 2.25. The following conditions are specified:

$$W = \text{airplane gross weight} = 8000 \text{ lb}$$

$$S = \text{airplane wing area} = 266 \text{ ft}^2$$

$$KU = \text{effective gust velocity} = 34 \text{ ft/s}$$

$$V_d = \text{design diving speed} = 400 \text{ mi/h}$$

$$n = \text{limit-load factor} = +6.0 \text{ and } -3.0$$

The aerodynamic characteristics of the airplane with the horizontal tail removed have been obtained from corrected wind tunnel data and are given in Table 2.1. The moment coefficient C_M is about the center of gravity of the airplane and is expressed in terms of the wing area and the mean aerodynamic chord for the wing, $\bar{c} = 86$ in. The stalling angle of the wing is $20°$, corresponding to a maximum lift coefficient of 1.67. The aerodynamic data

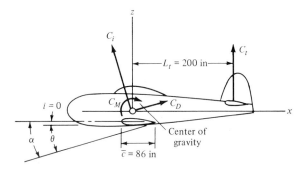

Figure 2.24

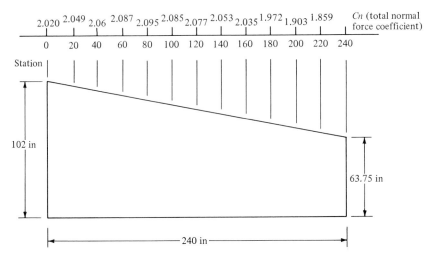

Figure 2.25

are extrapolated to the angle of attack of 26°. The negative stalling angle is −17°.

The force coefficients acting normal to the thrust line are calculated in Table 2.2. The components of C_L and C_D are calculated in columns 2 and 3. The tail load coefficient C_t is calculated in column 4 by means of Eq. (2.3). The final values of C_{z_a}, the normal force coefficient for the entire airplane, are obtained in column 5 as the sum of values from columns 2, 3, and 4.

The V-n diagram is constructed from the calculated data for C_{z_a}. For the OA portion of the curve of Fig. 2.13, the value of C_{z_a} is assumed to be 1.25 times the value at the stalling angle for the wing, or

$$C_{z_a} = 1.25(1.656) = 2.070$$

This corresponds with the angle of attack of 26°, within the accuracy of the data, and this angle is assumed. The equation for the curve OA of Fig. 2.13 is

Table 2.1

$\alpha = \theta$, deg	C_L	C_D	C_M
26	2.132	0.324	0.0400
20	1.670	0.207	0.0350
15	1.285	0.131	0.0280
10	0.900	0.076	0.0185
5	0.515	0.040	0.0070
0	0.130	0.023	−0.0105
−5	−0.255	0.026	−0.0316
−10	−0.640	0.049	−0.0525
−15	−1.025	0.092	−0.0770
−17	−1.180	0.115	−0.0860

Table 2.2

θ, deg (1)	$C_D \sin \theta$ (2)	$C_L \cos \theta$ (3)	C_t (4)	C_{z_a} (5)
26	0.143	1.918	0.017	2.078
20	0.071	1.570	0.015	1.656
15	0.034	1.240	0.012	1.286
10	0.013	0.887	0.008	0.908
5	0.004	0.512	0.003	0.519
0	0	0.130	-0.004	0.126
-5	-0.002	-0.254	-0.013	-0.269
-10	-0.008	-0.630	-0.022	-0.660
-15	-0.024	-0.990	-0.032	-1.046
-17	-0.034	-1.130	-0.036	-1.200

found as follows:

$$n = 2.078 \frac{\rho S V^2}{2W} = 2.078 \times 0.00256 \left(\frac{266}{8000} V^2\right)$$

$$= 0.0001772 V^2$$

For point A, $n = 6$ and $V = 184$ mi/h. The equation for the curve OB of Fig. 2.13 is found as follows:

$$n = -1.200 \frac{\rho S V^2}{2W} = -0.0001024 V^2$$

For point B, $n = -3$ and $V = 172$ mi/h. Points C and D are plotted with coordinates (400, 6), and (400, -3). The diagram is shown in Fig. 2.26.

The gust load factors are now obtained from Eqs. (2.25) and (2.26). The slope β may be obtained from the extreme coordinates of the curve for C_{z_a} if we assume a straight-line variation:

$$\beta = \frac{2.078 + 1.200}{26 + 17} = 0.0763 \text{ per degree}$$

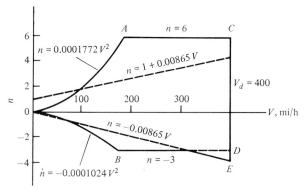

Figure 2.26

Table 2.3

Station no.	Y_i	Chord length C_i	Force coefficient C_{ni}	Shear V_i	Bending moment $M_i/1000$
0	240		0	0	0
1	220	50	1.859	560	5
2	200	66	1.903	1,880	29
3	180	73.2	1.972	3,510	83
4	160	76.4	2.035	5,310	171
5	140	79.6	2.053	7,230	296
6	120	82.8	2.077	9,250	461
7	100	86.0	2.085	11,370	667
8	80	89.2	2.095	13,580	917
9	60	92.4	2.087	15,870	1,212
10	40	95.6	2.06	18,220	1,553
11	20	98.8	2.049	20,630	1,942
12	0	102	2.02	23,090	2,379

From Eq. (2.26),

$$\Delta n = 0.1 \frac{\beta K U V}{W/S} = \frac{0.1(0.0763)(34)}{30} V$$

$$= 0.00865 V$$

For $V = 400$ mi/h, $\Delta n = 3.46$. Points F and E represent gust load factors of 4.46 and -3.46, respectively.

The wing internal bending and shear loads are now calculated for the PHAA condition, which is represented by point A on the V-n diagram. The wing has an angle of attack of $26°$ at an indicated airspeed of 184 mi/h. The total force coefficients normal to the wing chord are given in Fig. 2.25.

The shear and bending loads are calculated based on

$$V_i = V_{i-1} + \frac{q}{144} \frac{C_i C_{ni} + C_{i-1} C_{n,\,i-1}}{2} (Y_{i-1} - Y_i)$$

and
$$M_i = M_{i-1} + \frac{V_i + V_{i-1}}{2} (Y_{i-1} - Y_i)$$

where i indicates the station number and q is the dynamic pressure in pounds per square foot. All results are summarized in Table 2.3.

PROBLEMS

2.1 An airplane weighing 5000 lb strikes an upward gust of air which produces a wing lift of 25,000 lb (see Fig. P2.1). What tail load P is required to prevent a pitching acceleration if the dimensions are as shown? What will be the vertical acceleration of the airplane? If this lift force acts until the airplane obtains a vertical velocity of 20 ft/s, how much time is required?

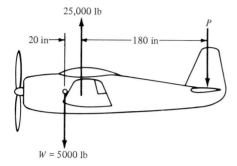

25,000 lb

20 in—

—180 in—

P

W = 5000 lb

Figure P2.1

2.2 An airplane weighing 8000 lb has an upward acceleration of 3g when landing. If the dimensions are as shown in Fig. P2.2, what are the wheel reactions R_1 and R_2? What time is required to decelerate the airplane from a vertical velocity of 12 ft/s? What is the vertical compression of the landing gear during this deceleration? What is the shear and bending moment on a vertical section AA if the weight forward of this section is 2000 lb and has a center of gravity 40 in from this cross section?

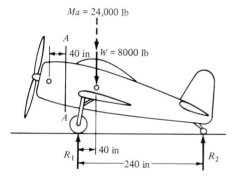

Ma = 24,000 lb

A

40 in W = 8000 lb

A

40 in

R_1

R_2

—240 in—

Figure P2.2

2.3 The airplane shown in Fig. P2.3 is making an arrested landing on a carrier deck. Find the load factors n and n_x, perpendicular and parallel to the deck, for a point at the center of gravity, a point 200 in aft of the center of gravity, and a point 100 in forward of the center of gravity. Find the relative

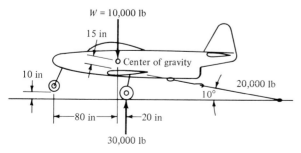

W = 10,000 lb

15 in

10 in

Center of gravity

20,000 lb

10°

—80 in—   —20 in—

30,000 lb

Figure P2.3

vertical velocity with which the nose wheel strikes the deck if the vertical velocity of the center of gravity is 12 ft/s and the angular velocity is 0.5 rad/s counterclockwise for the position shown. The radius of gyration for the mass of the airplane about the center of gravity is 60 in. Assume no change in the dimensions or loads shown.

2.4 An airplane is flying at 550 mi/h in level flight when it is suddenly pulled upward into a curved path of 2000-ft radius. (See Fig. P2.4.) Find the load factor of the airplane.

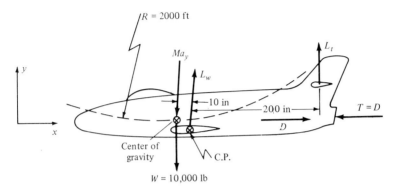

Figure P2.4

2.5 If the airplane in Prob. 2.4 is given a pitching acceleration of 2 rad/s^2, find its load factor, assuming that the change in lift due to pitching may be neglected.

2.6 A large transport aircraft is making a level landing, as shown in Fig. P2.6. The gross weight of the aircraft is 150,000 lb, and its pitching mass moment of inertia is 50×10^6 lb · in · s^2 about the center of gravity. The landing rear-wheel reaction is 350,000 lb at an angle of 15° with the vertical. Determine whether passenger A or B will receive the most load. Assume that each passenger weights 170 lb and neglect the airplane lift.

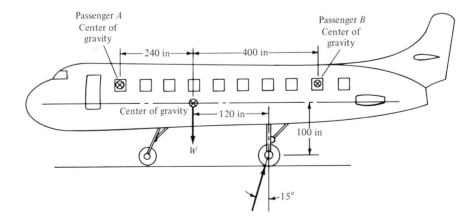

Figure P2.6

2.7 Assume that the center of gravity of the airplane in Example 2.4 is moved forward 8 in without changing the external aerodynamic configuration. The distance L_t is now 208 in, and the values of the aerodynamic pitching moments about the center of gravity are $C_M - 8C_z/86$, where values of C_M are given in Table 2.1.

(a) Calculate curves for C_t and C_{z_a}.

(b) Construct a V-n diagram, using the conditions specified in Sec. 2.7.

(c) Calculate the wing bending-moment diagram for air loads normal to the wing chord for the PHAA condition.

(d) Calculate the wing bending-moment diagram for chordwise air loads.

(e) Calculate the air-load torsional moments about the wing's leading edge if the leading edge is straight and perpendicular to the plane of symmetry of the airplane. Assume the airfoil at any section to have an aerodynamic center at the quarter-chord point and to have a negligible pitching moment about this point.

2.8 Calculate the wing normal and chordwise bending-moment diagrams for the PLAA condition for the airplane analyzed in Sec. 2.7.

2.9 If the airplane wing of Example 2.4 weighs $4.0\,\text{lb/ft}^2$, which is assumed distributed uniformly over the area, calculate the wing bending moments resulting from gravity and inertia forces normal to the wing chord for the four primary loading conditions.

THREE

ELASTICITY OF STRUCTURES

3.1 INTRODUCTION

This chapter defines stresses and strains and their fundamental relationships. The stress behavior of structures undergoing elastic deformation that is due to the action of external applied loads is also discussed. The term *elasticity* or *elastic behavior* is used here to imply a recovery property of an original size and shape.

3.2 STRESSES

Consider the solid body shown in Fig. 3.1 which is acted on by a set of external forces Q_i, as indicated. If we assume that rigid-body motion is prevented, the solid will deform in accordance with the external applied forces; as a result, internal loads between all parts of the body will be produced.

If the solid is separated into two parts by passing a hypothetical plane, as shown in Fig. 3.1b, then there exist internal forces whose resultants are indicated by Q_I and Q_{II} acting on parts I and II, respectively.

The forces which hold together the two parts of the body are normally distributed over the entire surface of the cut plane. If we consider only an infinitesimal area δA acted on by a resultant force δQ, then an average force per unit may be expressed as

$$\sigma_{\text{av}} = \frac{\delta Q}{\delta A} \tag{3.1}$$

In the limit as δA approaches zero, Eq. (3.1) becomes

$$\sigma = \frac{dQ}{dA} \tag{3.2}$$

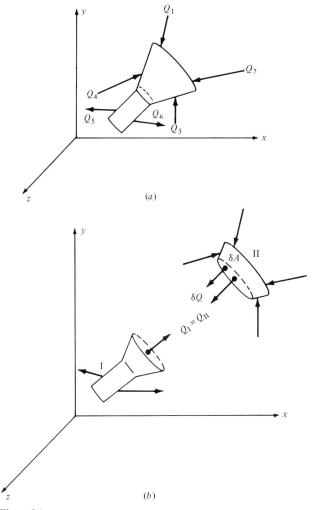

Figure 3.1

where σ now is the limiting value of the average force per unit area and, by definition, the *stress* at that point. A stress is completely defined if its magnitude and direction and the plane on which it acts are all known. For instance, it is not appropriate to ask about the stress at point 0 of the solid shown in Fig. 3.2 unless the plane on which the stress is acting is specified. An infinite number of planes may be passed through point 0, thus resulting in an infinite number of different stresses.

In the most general three-dimensional state of stress, nine stress components may exist:

$$[\sigma] = \begin{bmatrix} \sigma_{xx} & \sigma_{xy} & \sigma_{xz} \\ \sigma_{yx} & \sigma_{yy} & \sigma_{yz} \\ \sigma_{zx} & \sigma_{zy} & \sigma_{zz} \end{bmatrix} \tag{3.3}$$

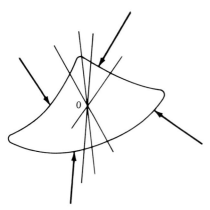

Figure 3.2

where σ_{ii} ($i = x$, y, z) are the normal stresses and $\sigma_{ij} = \sigma_{ji}$ ($i \neq j = x$, y, z) are the shearing stresses. The first subscript on σ_{ij} (i, $j = x$, y, z) denotes the plane at a constant i on which the stress is acting, and the second subscript denotes the positive direction of the stress. Figure 3.3 illustrates the stress notation.

In the case of the two-dimensional state of stress, or what is commonly referred to as the plane stress problem ($\sigma_{zz} = \sigma_{zx} = \sigma_{zy} = 0$), Eq. (3.3) becomes

$$[\sigma] = \begin{bmatrix} \sigma_{xx} & \sigma_{xy} \\ \sigma_{yx} & \sigma_{yy} \end{bmatrix} \quad (\sigma_{xy} = \sigma_{yx}) \tag{3.4}$$

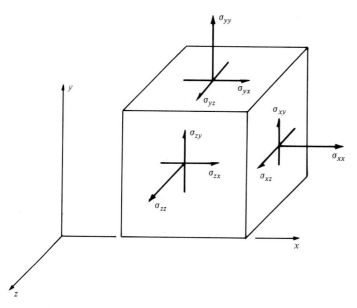

Figure 3.3

3.3 STRESS EQUILIBRIUM EQUATIONS IN A NONUNIFORM STRESS FIELD

In general, a solid which is acted on by a set of external applied loads experiences a state of stress that is not uniform throughout the body. This condition gives rise to a set of equations which are referred to as the equations of equilibrium. Consider the three-dimensional solid shown in Fig. 3.4.

Using the equilibrium equations of statics yields the following:

$$\Sigma F_x = 0$$

$$X \, dx \, dy \, dz + (\sigma_{xx} + \sigma_{xx, x} \, dx) dy \, dz - \sigma_{xx} \, dy \, dz + (\sigma_{xy} + \sigma_{xy, y} \, dy) dx \, dz$$

$$- \sigma_{xy} \, dx \, dz + (\sigma_{zx} + \sigma_{zx, z} \, dz) dx \, dy - \sigma_{zx} \, dx \, dy = 0$$

or
$$\sigma_{xx, x} + \sigma_{xy, y} + \sigma_{zx, z} + X = 0 \tag{3.5}$$

Similarly,

$$\Sigma F_y = 0 \qquad \sigma_{xy, x} + \sigma_{yy, y} + \sigma_{zy, z} + Y = 0 \tag{3.6}$$

$$\Sigma F_z = 0 \qquad \sigma_{yx, x} + \sigma_{zy, y} + \sigma_{zz, z} + Z = 0 \tag{3.7}$$

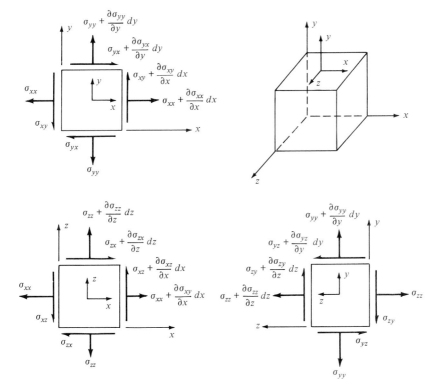

Figure 3.4

where X, Y, and Z are unit body forces. The comma denotes partial differentiation with respect to the following subscript:

$$\sigma_{zx,\,x} = \frac{\partial \sigma_{zx}}{\partial x} \qquad \sigma_{yy,\,y} = \frac{\partial \sigma_{yy}}{\partial y} \qquad \text{etc.}$$

In a cylindrical coordinate set of axes, the equilibrium equations may be easily derived from Fig. 3.5:

$$\sigma_{rr,\,r} + \frac{1}{r}\,\sigma_{r\theta,\,\theta} + \frac{\sigma_{rr} - \sigma_{\theta\theta}}{r} + \sigma_{rz,\,z} + R = 0 \qquad (3.8)$$

$$\frac{1}{r}\,\sigma_{\theta\theta,\,\theta} + \sigma_{r\theta,\,r} + \frac{2\sigma_{r\theta}}{r} + \sigma_{z\theta,\,z} + \Theta = 0 \qquad (3.9)$$

$$\sigma_{zz,\,z} + \frac{1}{r}\,\sigma_{\theta z,\,\theta} + \sigma_{rz,\,r} + \frac{\sigma_{rz}}{r} + Z = 0 \qquad (3.10)$$

where again R, Θ, and Z are unit body forces.

For plane stress problems, the equilibrium equations simplify to the following:

$$\sigma_{xx,\,x} + \sigma_{xy,\,y} + X = 0$$
$$\sigma_{xy,\,x} + \sigma_{yy,\,y} + Y = 0 \qquad (3.11)$$

or, in cylindrical coordinates,

$$\sigma_{rr,\,r} + \frac{1}{r}\,\sigma_{r\theta,\,\theta} + \frac{\sigma_{rr} - \sigma_{\theta\theta}}{r} + R = 0$$

$$\frac{1}{r}\,\sigma_{\theta\theta,\,\theta} + \sigma_{r\theta,\,r} + \frac{2\sigma_{r\theta}}{r} + \Theta = 0 \qquad (3.12)$$

3.4 STRAINS AND STRAIN-DISPLACEMENT RELATIONSHIPS

Strains are nondimensional quantities associated with the deformations (displacements) of an element in a solid body under the action of external applied loads.

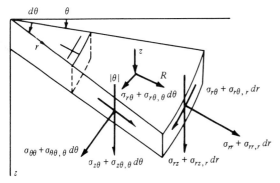

Figure 3.5

To arrive at a mathematical definition of the strain components, take the solid body shown in Fig. 3.6 and consider only the infinitesimal elements OA, OB, and OC.

If, after deformations take place, the displacements of point O are denoted by q_x, q_y, and q_z in the x, y, and z directions, respectively, then the displacements of points A, B, and C which are dx, dy, and dz away from point O will be $q_x + q_{x,x} \, dx$, $q_y + q_{y,y} \, dy$, and $q_z + q_{z,z} \, dz$, respectively. Figure 3.7 shows all the displacements resulting from the application of external applied loads. When these relative displacements occur in a solid body, the body is said to be in a state of *strain*. The strains associated with the relative change in length are referred to as *normal strains*, and those related to relative change in angles are called *shearing strains*.

The normal strain is defined as

$$\epsilon = \lim_{L \to 0} \frac{\Delta L}{L} \tag{3.13}$$

where ΔL is the change in length of an element whose original length was L before deformation took place. On the basis of Eq. (3.13), the normal strain in the x direction, for instance, is obtained as follows:

$$\epsilon_{xx} = \frac{\Delta(OA)}{OA} = \frac{O'A' - OA}{OA}$$

$$OA = dx$$

$$O'A' = [(dx + q_{x,x} \, dx)^2 + (q_{y,x} \, dx)^2 + (q_{z,x} \, dx)^2]^{1/2}$$

The normal strain in the x direction is then

$$\epsilon_{xx} = [(1 + q_{x,x})^2 + (q_{y,x})^2 + (q_{z,x})^2]^{1/2} - 1$$

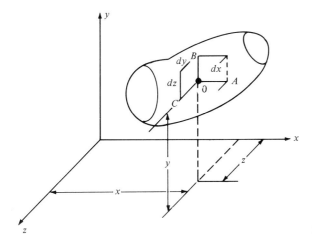

Figure 3.6

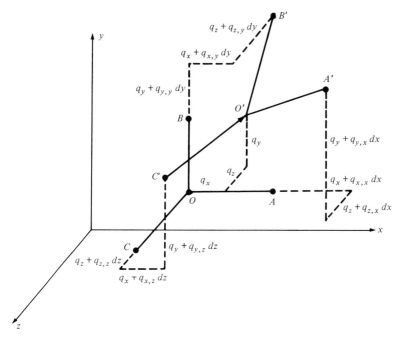

Figure 3.7

which reduces to the following by using the binomial expansion technique:

$$\epsilon_{xx} = q_{x,x} + \tfrac{1}{2}[(q_{x,x})^2 + (q_{y,x})^2 + (q_{z,x})^2] + \cdots$$

For small displacements, the terms involving the squares of derivatives may be neglected in comparison with the derivative in the first term. Therefore, the x-direction linear normal strain becomes

$$\epsilon_{xx} = q_{x,x} \qquad (3.14a)$$

In similar manner, the linearized normal strains in the y and z directions may be derived and are given by

$$\epsilon_{yy} = q_{y,y} \qquad (3.14b)$$

and

$$\epsilon_{zz} = q_{z,z} \qquad (3.14c)$$

The shearing strains may be derived by finding the relative change in the angle between a given pair of the three line segments shown in Fig. 3.7.

$$\epsilon_{xy} = q_{x,y} + q_{y,x}$$

$$\epsilon_{xz} = q_{x,z} + q_{z,x} \qquad (3.15)$$

$$\epsilon_{yz} = q_{y,z} + q_{z,y}$$

The derivation of Eq. (3.15) may be found in Refs. 11, 12, and 13.

Thus, in a three-dimensional state of strain, the strain field components may be compacted in a matrix form:

$$[\epsilon] = \begin{bmatrix} \epsilon_{xx} & \epsilon_{xy} & \epsilon_{xz} \\ \epsilon_{yx} & \epsilon_{yy} & \epsilon_{yz} \\ \epsilon_{zx} & \epsilon_{zy} & \epsilon_{zz} \end{bmatrix} \tag{3.16}$$

where $\epsilon_{ij} = \epsilon_{ji}$ for $i \neq j$ is a shearing strain and for $i = j$ is the normal strain.

In the case of two-dimensional state of strain (plane-strain problem), $\epsilon_{zz} = \epsilon_{zx} = \epsilon_{zy} = 0$, Eq. (3.16) reduces to

$$[\epsilon] = \begin{bmatrix} \epsilon_{xx} & \epsilon_{xy} \\ \epsilon_{yx} & \epsilon_{yy} \end{bmatrix} \tag{3.17}$$

In cylindrical coordinates, Eqs. (3.14) and (3.15) may be written as

$$\epsilon_{rr} = q_{r,\,r} \qquad \epsilon_{\theta z} = \frac{q_{z,\,\theta}}{r} + q_{\theta,\,z}$$

$$\epsilon_{\theta\theta} = \frac{q_{\theta,\,\theta}}{r} + \frac{q_r}{r} \qquad \epsilon_{zr} = q_{z,\,r} + q_{r,\,z} \tag{3.18}$$

$$\epsilon_{zz} = q_{z,\,z} \qquad \epsilon_{r\theta} = \frac{q_{r,\,\theta}}{r} + q_{\theta,\,r} - \frac{q_\theta}{r}$$

where q_r, q_θ, and q_z are the displacements in the r, θ, and z directions, respectively.

3.5 COMPATIBILITY EQUATIONS FOR PLANE-STRESS AND PLANE-STRAIN PROBLEMS

The strain-displacement relationships for plane-strain problems are given by the following.

$$\epsilon_{xx} = q_{x,\,x}$$

$$\epsilon_{yy} = q_{y,\,y} \tag{3.19}$$

$$\epsilon_{xy} = q_{x,\,y} + q_{y,\,x}$$

By examining Eq. (3.19), it is apparent that there exist three components of strain which are expressed in terms of only two components of displacements. Thus, it may be concluded that not all these strain components are independent of one another. This may be easily verified by differentiating twice the first and the second equations of (3.19) with respect to y and x, respectively, and the last equation with respect to x and y.

$$\epsilon_{xx,\,yy} = q_{x,\,xyy}$$

$$\epsilon_{yy,\,xx} = q_{y,\,yxx}$$

$$\epsilon_{xy,\,xy} = q_{x,\,xyy} + q_{y,\,xxy}$$

Substituting the first two equations into the third yields

$$\epsilon_{xy, xy} = \epsilon_{xx, yy} + \epsilon_{yy, xx} \tag{3.20}$$

which is the compatibility equation of deformation.

Thus, the choice of the three strain components cannot be arbitrary, but must be such that the compatibility equation is satisfied. This compatibility equation ensures the existence of single-valued displacement functions q_x and q_y for a solid. In fact, when Eq. (3.20) is expressed in terms of stresses, it ensures the existence of a unique solution for a stress problem, as is illustrated in later chapters.

For three-dimensional state of stress, the compatibility equations may be derived in similar manner to that of Eq. (3.20) and are given by the following.

$$
\begin{aligned}
\epsilon_{xy, xy} &= \epsilon_{xx, yy} + \epsilon_{yy, xx} \\[2mm]
\epsilon_{xz, xz} &= \epsilon_{xx, zz} + \epsilon_{zz, xx} \\[2mm]
\epsilon_{yz, yz} &= \epsilon_{yy, zz} + \epsilon_{zz, yy} \\[2mm]
2\epsilon_{xx, yz} &= \epsilon_{xy, xz} + \epsilon_{xz, xy} - \epsilon_{yz, xx} \\[2mm]
2\epsilon_{yy, xz} &= \epsilon_{xy, yz} - \epsilon_{xz, yy} + \epsilon_{yz, xy} \\[2mm]
2\epsilon_{zz, xy} &= -\epsilon_{xy, zz} + \epsilon_{xz, yz} + \epsilon_{yz, xz}
\end{aligned}
\tag{3.21}
$$

3.6 BOUNDARY CONDITIONS

Boundary conditions are those conditions for which the displacements and/or the surface forces are prescribed at the boundary of a given solid. For instance, if the displacements q_x, q_y, and q_z are prescribed at the boundary, then conditions are referred to as *displacement boundary conditions* and may be written as

$$q_x = \bar{q}_x(\eta) \qquad q_y = \bar{q}_y(\eta) \qquad q_z = \bar{q}_z(\eta) \tag{3.22}$$

where $\bar{q}_x$, $\bar{q}_y$, and $\bar{q}_z$ are known functions of displacements at the boundary.

On the other hand, if σ_{xx}, σ_{yy}, and σ_{xy} are prescribed at the boundary, then the conditions are called *force boundary conditions* and are normally expressed as

$$
\begin{bmatrix} \bar{N}_{xx} \\ \bar{N}_{yy} \\ \bar{N}_{zz} \end{bmatrix} =
\begin{bmatrix} \bar{\sigma}_{xx} & \bar{\sigma}_{xy} & \bar{\sigma}_{zx} \\ \bar{\sigma}_{xy} & \bar{\sigma}_{yy} & \bar{\sigma}_{yz} \\ \bar{\sigma}_{zx} & \bar{\sigma}_{yz} & \bar{\sigma}_{zz} \end{bmatrix}
\begin{bmatrix} \eta_x \\ \eta_y \\ \eta_z \end{bmatrix}
\tag{3.23}
$$

where the $\bar{N}_{ii}$ ($i = x, y, z$) are the surface boundary forces, the $\bar{\sigma}_{ij}$ ($i, j = x, y, z$) are the prescribed stresses at the boundary, and η_x, η_y, η_z are the direction cosines. For the derivation of Eq. (3.23) see Ref. 14.

It is appropriate to note at this point that in order to obtain the exact stress field in any given solid under the action of external loads, the equations of equilibrium, compatibility equations, and the boundary conditions must all be satisfied.

3.7 STRESS-STRAIN RELATIONSHIPS

Structural behavior may be classified into three basic categories based on the functional relationship between stresses and strains:

1. Inelastic nonlinear
2. Elastic nonlinear
3. Linear elastic

Consider a steadily loaded bar as shown in Fig. 3.8. If upon loading the functional relationship between the stress and its corresponding strain takes on a curved path, and upon unloading it takes a different curved path, as in Fig. 3.9a, then the structural behavior is referred to as *inelastic nonlinear* behavior. If such relationship follows the same curved path upon both loading and unloading the bar as in Fig. 2.9b, then the behavior is said to be *elastic nonlinear* behavior. Last, if such relationship takes on the same straight path as in Fig. 2.9c, then the behavior is termed *linear elastic* behavior.

In the most general case for a linear elastic anisotropic solid, Hooke's law, which relates stresses to strains, may be written in a matrix form as follows, in which $\epsilon =$ strain, $\sigma =$ stress, and a_{ij} $(i, j = x, y, z)$ are the material elastic constants $(a_{ij} = a_{ji})$.

$$
\begin{bmatrix} \epsilon_{xx} \\ \epsilon_{yy} \\ \epsilon_{zz} \\ \epsilon_{yz} \\ \epsilon_{xz} \\ \epsilon_{xy} \end{bmatrix} = \begin{bmatrix} a_{11} & a_{12} & a_{13} & a_{14} & a_{15} & a_{16} \\ a_{21} & a_{22} & a_{23} & a_{24} & a_{25} & a_{26} \\ a_{31} & a_{32} & a_{33} & a_{34} & a_{35} & a_{36} \\ a_{41} & a_{42} & a_{43} & a_{44} & a_{45} & a_{46} \\ a_{51} & a_{52} & a_{53} & a_{54} & a_{55} & a_{56} \\ a_{61} & a_{62} & a_{63} & a_{64} & a_{65} & a_{66} \end{bmatrix} \begin{bmatrix} \sigma_{xx} \\ \sigma_{yy} \\ \sigma_{zz} \\ \sigma_{yz} \\ \sigma_{xz} \\ \sigma_{xy} \end{bmatrix} \tag{3.24}
$$

If there exist three orthogonal planes of elastic symmetry through every point of the solid body, then Eq. (3.24) becomes

$$
\begin{bmatrix} \epsilon_{xx} \\ \epsilon_{yy} \\ \epsilon_{zz} \\ \epsilon_{yz} \\ \epsilon_{xz} \\ \epsilon_{xy} \end{bmatrix} = \begin{bmatrix} a_{11} & a_{12} & a_{13} & 0 & 0 & 0 \\ a_{21} & a_{22} & a_{23} & 0 & 0 & 0 \\ a_{31} & a_{32} & a_{33} & 0 & 0 & 0 \\ 0 & 0 & 0 & a_{44} & 0 & 0 \\ 0 & 0 & 0 & 0 & a_{55} & 0 \\ 0 & 0 & 0 & 0 & 0 & a_{66} \end{bmatrix} \begin{bmatrix} \sigma_{xx} \\ \sigma_{yy} \\ \sigma_{zz} \\ \sigma_{yz} \\ \sigma_{xz} \\ \sigma_{xy} \end{bmatrix} \tag{3.25}
$$

The elastic constants a_{ij} in Eq. (3.25) may be defined in terms of the engineering-material constants as follows:

Figure 3.8

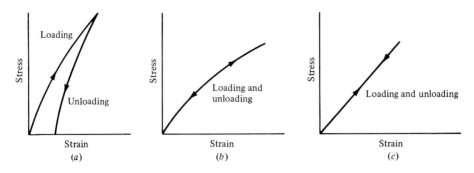

Figure 3.9 (a) Inelastic nonlinear; (b) elastic nonlinear; (c) linear elastic.

$$a_{11} = \frac{1}{E_{xx}} \qquad a_{22} = \frac{1}{E_{yy}} \qquad a_{33} = \frac{1}{E_{zz}}$$

$$a_{12} = a_{21} = -\frac{v_{yx}}{E_{yy}} = -\frac{v_{xy}}{E_{xx}}$$

$$a_{23} = a_{32} = -\frac{v_{zy}}{E_{zz}} = -\frac{v_{yz}}{E_{yy}} \qquad\qquad (3.26)$$

$$a_{13} = a_{31} = -\frac{v_{xz}}{E_{xx}} = -\frac{v_{zx}}{E_{zz}}$$

$$a_{44} = \frac{1}{G_{yz}} \qquad a_{55} = \frac{1}{G_{xz}} \qquad a_{66} = \frac{1}{G_{xy}}$$

where E_{ij}, G_{ij}, and v_{ij} $(i, j = x, y, z)$ are the modulus of elasticity, shear modulus, and Poisson's ratio, respectively.

A body which obeys Eq. (3.25)—i.e., at each point there exist three mutually perpendicular planes of elastic symmetry—is commonly referred to as an *orthotropic* body. Plywood, for instance, and most reinforced plastics may be designated as orthotropic materials.

If all directions in a solid are elastically equivalent and any plane which passes through any point of the body is a plane of elastic symmetry, then it is called an *isotropic* body. In this case, the elastic constants in Eq. (3.25) simplify to

$$a_{11} = a_{22} = a_{33} = \frac{1}{E}$$

$$a_{44} = a_{55} = a_{66} = \frac{1}{G}$$

$$\qquad\qquad (3.27)$$

$$a_{12} = a_{21} = a_{23} = a_{32} = a_{13} = a_{31} = -\frac{v}{E}$$

$$G = \frac{E}{2(1 + v)}$$

where E = Young's modulus, G = shear modulus, and v = Poisson's ratio.

Most metals, such as aluminum, steel, and titanium, are isotropic materials.

It is important to note here that in an anistropic body, normal strains will induce not only normal stresses but also shearing stresses; likewise, shearing strains will produce normal stresses, as may easily be seen from Eq. (3.24). However, normal strains in an isotropic or orthotropic body will cause only normal stresses, while shearing strains will cause shearing stresses. This may be verified by examining Eq. (3.25).

For plane-stress problems where σ_{zz}, σ_{yz}, and σ_{xz} are zero, if we assume isotropic material, Eq. (3.25) becomes

$$
\begin{bmatrix} \epsilon_{xx} \\ \epsilon_{yy} \\ \epsilon_{xy} \end{bmatrix} = \frac{1}{E} \begin{bmatrix} 1 & -v & 0 \\ -v & 1 & 0 \\ 0 & 0 & \dfrac{1}{2(1+v)} \end{bmatrix} \begin{bmatrix} \sigma_{xx} \\ \sigma_{yy} \\ \sigma_{xy} \end{bmatrix} \tag{3.28}
$$

By substituting Eq. (3.28) into Eq. (3.20) and utilizing Eq. (3.11), the compatibility equation in terms of stresses alone becomes

$$
\sigma_{xx,\, xx} + 2\sigma_{xy,\, xy} + \sigma_{yy,\, yy} = 0 \tag{3.29}
$$

The body forces are assumed to be zero.

3.8 TRANSFORMATION OF STRESSES AND STRAINS

A stress field $(\sigma_{xx}, \sigma_{yy}, \sigma_{xy})$, such as shown in Fig. 3.10, which is known in one set of systems axes may be transformed to any other arbitrary set of axes such as $\eta\beta$.

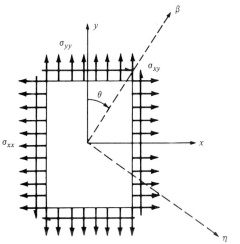

Figure 3.10

For instance, assume that at a given point in a solid, the stresses (σ_{xx}, σ_{yy}, and σ_{xy}) are known in reference to the x and y axes, as shown in Fig. 3.11. The object is to find the set of stresses in reference to a new set of axes, $\eta\beta$, which are rotated through an angle θ as shown. The stress $\sigma_{\eta\eta}$ may be found by considering the free-body diagram which is cut by a plane along the β axis at an angle θ from the vertical, as shown in Fig. 3.11b. If side OB is assumed to have area A, then sides OC and CB will each have an area of $A\cos\theta$ and $A\sin\theta$, respectively. Since equilibrium conditions prevail, the summation of forces along each of the η and β directions must be zero, or

$$\Sigma F_\eta = 0 \overset{+}{\searrow}$$

$$\sigma_{\eta\eta} A - \sigma_{xx} A \cos\theta \cos\theta - \sigma_{yy} A \sin\theta \sin\theta + \sigma_{xy} A \cos\theta \sin\theta$$
$$+ \sigma_{yx} A \sin\theta \cos\theta = 0$$

By simplifying and noting that $\sigma_{xy} = \sigma_{yx}$, the following is obtained:

$$\sigma_{\eta\eta} = \sigma_{xx} \cos^2\theta + \sigma_{yy} \sin^2\theta - \sigma_{xy} \sin 2\theta$$

$$\Sigma F_\beta = 0 \overset{+}{\swarrow}$$

$$\sigma_{\eta\beta} A + \sigma_{xx} A \cos\theta(\sin\theta) - \sigma_{yy} A \sin\theta(\cos\theta) + \sigma_{xy} A \cos\theta(\cos\theta)$$
$$- \sigma_{yx} A \sin\theta(\sin\theta) = 0 \qquad (3.30a)$$

or

$$\sigma_{\eta\beta} = -\frac{\sigma_{xx} \sin 2\theta}{2} + \frac{\sigma_{yy} \sin 2\theta}{2} + (1 - 2\cos^2\theta)\sigma_{xy} \qquad (3.30b)$$

In a similar manner, $\sigma_{\beta\beta}$ may be obtained and is given by

$$\sigma_{\beta\beta} = \sigma_{xx} \sin^2\theta + \sigma_{yy} \cos^2\theta + \sigma_{xy} \sin 2\theta \qquad (3.30c)$$

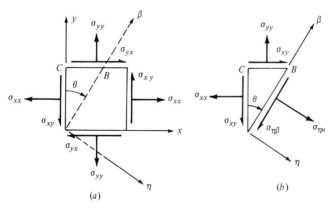

Figure 3.11

Equations (3.30a), (3.30b), and (3.30c) may be assembled in a matrix form as

$$
\begin{bmatrix} \sigma_{\eta\eta} \\ \sigma_{\beta\beta} \\ \sigma_{\eta\beta} \end{bmatrix} = \begin{bmatrix} \cos^2\theta & \sin^2\theta & -\sin 2\theta \\ \sin^2\theta & \cos^2\theta & \sin 2\theta \\ -\dfrac{\sin 2\theta}{2} & \dfrac{\sin 2\theta}{2} & 1 - 2\cos^2\theta \end{bmatrix} \begin{bmatrix} \sigma_{xx} \\ \sigma_{yy} \\ \sigma_{xy} \end{bmatrix} \tag{3.31a}
$$

where θ is the angle of rotation and is positive in the clockwise direction.

In compact matrix form, Eq. (3.31a) becomes

$$\{\Sigma\}_{\eta,\,\beta} = [T]\{\Sigma\}_{x,\,y} \tag{3.31b}$$

where $[T]$ is referred to as the transformation matrix.

In a similar manner, the strains may be transformed as follows:

$$
\begin{bmatrix} \epsilon_{\eta\eta} \\ \epsilon_{\beta\beta} \\ \gamma_{\eta\beta} \end{bmatrix} = [T] \begin{bmatrix} \epsilon_{xy} \\ \epsilon_{yy} \\ \gamma_{xy} \end{bmatrix} \tag{3.32}
$$

where γ_{ij} (tensor shearing strain) $= \frac{1}{2}\epsilon_{ij}$ (engineering shearing strain) and the matrix $[T]$ is the same as that in Eq. (3.31).

PROBLEMS

3.1 Derive the compatibility equation in cylindrical coordinates for a two-dimensional state of stress.

3.2 Find the direction along which a hole may be drilled in the solid shown in Fig. P3.2 such that no shearing stresses exist along the hole direction.

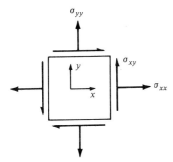

σ_{yy}

σ_{xy}

y

σ_{xx}

x

Figure P3.2

3.3 Find the maximum and minimum normal stresses and the planes on which they act for the block shown in Fig. P3.2. Also, determine the maximum shear stress and the plane on which it acts in Prob. 3.2.

3.4 Derive Eq. (3.30c).

3.5 For a two-dimensional state of stress, show that the stress-strain relationship for an orthotropic material is

$$
\begin{bmatrix} \sigma_{xx} \\ \sigma_{yy} \\ \sigma_{xy} \end{bmatrix} = \begin{bmatrix} S_{11} & S_{12} & 0 \\ S_{12} & S_{22} & 0 \\ 0 & 0 & S_{66} \end{bmatrix} \begin{bmatrix} \epsilon_{xx} \\ \epsilon_{yy} \\ \epsilon_{xy} \end{bmatrix}
$$

where S_{ij} are the stiffness constants, defined as follows.

$$S_{11} = \frac{a_{22}}{\Delta} \qquad S_{12} = \frac{a_{12}}{\Delta} \qquad S_{22} = \frac{a_{11}}{\Delta}$$

$$S_{66} = \frac{1}{a_{66}} \qquad \Delta = a_{11}a_{22} - a_{12}^2$$

3.6 Find the matrix of the stiffness constants if a new, arbitrary set of chosen axes is taken as shown in Fig. P3.6. Assume that the stiffness constants with respect to the x and y axes are as given in Prob. 3.5.

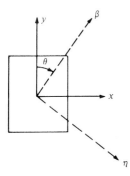

η **Figure P3.6**

3.7 A narrow cantilever beam has unit width and is loaded with a force Q, as shown in Fig. P3.7. The deformations in the x and y directions are

$$q_x = \frac{1}{2EI}\left(-Qx^2y - \tfrac{1}{3}vQy^3 + QL^2y\right) + \frac{1}{2GI}\left(\tfrac{1}{3}Qy^3 - Qc^2y\right)$$

$$q_y = \frac{1}{2EI}\left(vQxy^2 + \tfrac{1}{3}Qx^3 - QL^2x + \tfrac{2}{3}QL^3\right)$$

(a) If $\sigma_{yy} = 0$, find the stress fields for the beam under consideration at any point (x, y).

(b) Show that the stress fields in part (a) are the stress fields for the given beam.

(c) If $Q = 1000$ lb, $c = 2$ in, $L = 100$ in, $EI = 10^8$, and $v = 0.25$ and by utilizing the stress fields in part (a), find the principal normal stresses and the maximum shear stress at a point on the beam given by $x = L/2$, $y = -c/2$.

(d) Could the following stress fields be possible stress fields for the given beam?

$$\sigma_{xx} = \frac{Q}{EI}\left[y^2 + v(x^2 - y^2)\right]$$

$$\sigma_{yy} = \frac{Q}{EI}\left[x^2 + v(y^2 - x^2)\right]$$

$$\sigma_{xy} = -2\frac{Q}{EI}vxy$$

Figure P3.7

3.8 You are given a rectangular plate with positive applied stresses (σ_{xx} and σ_{yy}). What must be the magnitude of σ_{xx} in order for the contraction in the x direction to be prevented? If this plate is subjected to the positive stresses σ_{xx}, σ_{yy}, and σ_{xy}, in what direction should the plate be drawn in order to preserve its angles during stretching?

3.9 A plate of unit thickness is subjected to a set of loads P_η and P_β uniformly distributed over the sides a and b, respectively. (See Fig. P3.9.) What must be the ratio of magnitudes of P_η and P_β in order for the contraction of the plate in the x direction to be prevented?

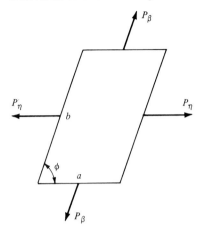

Figure P3.9

FOUR

BEHAVIOR AND EVALUATION
OF VEHICLE MATERIAL

4.1 INTRODUCTION

It is of utmost importance for the structural analyst to have a full understanding
of the behavior of vehicle materials and be able to intelligently evaluate and
select the material best suited to the constraints and operational requirements of
the design.

The materials used in various parts of vehicle structures generally are selected by different criteria. The criteria are predicated on the constraints and operational requirements of the vehicle and its various structural members. Some of
these more important requirements involve

1. Environment
2. Fatigue
3. Temperature
4. Corrosion
5. Creep
6. Strength and stiffness
7. Weight limitation
8. Cost
9. Human factor

This chapter familiarizes the reader with the roles that some of these requirements play in the final selection and evaluation of materials to be used in vehicle
structures.

4.2 MECHANICAL PROPERTIES OF MATERIALS

Materials, in general, may be classified according to their constituent composition as single-phase or multiphase. All metals, such as aluminum, steel, and titanium, are referred to as *single-phase materials*. All composites, which are made out of filaments (fibers) embedded in a matrix (binder), are called *multiphase materials*. Plywood and reinforced fiber glass are examples of multiphase materials.

Almost all important structural properties of single- or multiphase materials are obtained by three basic tests: tensile test, compression test, and shear test. The American Society for Testing Materials (ASTM) sets all the specifications and test procedures for materials testing.

Tensile Test

Figure 4.1 shows the basic configuration of a tensile test specimen. The load P is applied gradually through the use of a tensile testing machine. The normal strain ϵ_n usually is measured either by utilizing electrical strain gage techniques or by measuring the total elongation δ in an effective gage length L for various values of the tension load P. For small loads, the elongation is assumed to be uniform over the entire gage length L and therefore the normal strain may be mathematically expressed in the form

$$\epsilon_n = \frac{\delta}{L} \tag{4.1}$$

where δ and L are both measured in the same units of length. The corresponding normal stress σ_n is also assumed to be uniformly distributed over the cross-sectional area A of the test specimen and is obtained as follows:

$$\sigma_n = \frac{P}{A} \tag{4.2}$$

For common engineering units, the load P is in pounds, the area A is in square inches, and the stress σ is in pounds per square inch. The stress-strain diagram for a material is obtained by plotting values of the stress σ against corresponding values of the strain ϵ, as shown in Fig. 4.2. For small values of the

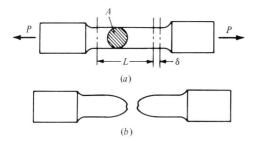

(a)

(b)

Figure 4.1

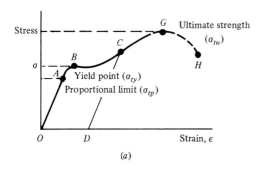

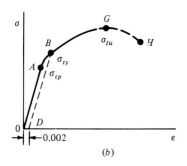

Figure 4.2

stress, the stress-strain curve is a straight line, as shown by line OA of Fig. 4.2. The constant ratio of stress to strain for this portion of the curve is called the *modulus of elasticity E*, as defined in the following equation:

$$E = \frac{\sigma_n}{\epsilon_n} \tag{4.3}$$

where E has units of pounds per square inch.

A material such as plain low-carbon steel, which is commonly used for bridge and building structures, has a stress-strain diagram such as that shown in Fig. 4.2a. At point B, the elongation increases with no increase in load. This stress at this point is called the *yield point*, or *yield stress*, σ_{ty} and is very easy to detect when such materials are tested.

The stress at point A, where the stress-strain curve first deviates from a straight line, is called the *proportional limit* σ_{tp} and is much more difficult to measure while a test is being conducted. Specifications for structural steel usually are based on the yield stress rather than the proportional limit, because of the ease in obtaining this value.

Flight-vehicle structures are made of materials such as aluminum alloys, high-carbon steels, and composites which do not have a definite yield point, but which do exhibit stress-strain behavior similar to that shown in Fig. 4.2b. It is convenient to specify arbitrarily the yield stress for such materials as the stress at which a permanent strain of 0.002 in/in is obtained. Point B of Fig. 4.2b represents this yield stress and is obtained by drawing line BD parallel to OA

through point D, representing zero stress and 0.002-in/in strain, as shown. When the load is removed from a test specimen which has passed the proportional limit, the specimen does not return to its original length, but retains a permanent strain. For the material represented by Fig. 4.2a, the load might be removed gradually at point C. The stress-strain curve would then follow line CD, parallel to OA, until at point D a permanent strain equal to OD were obtained for no stress. Upon a subsequent application of load, the stress-strain curve would follow lines DC and CG. Similarly, if the specimen represented by Fig. 4.2b were unloaded at point B, the stress-strain curve would follow line BD until a permanent strain of 0.002 in/in were obtained for no stress.

It is customary to use the initial area A of the tension test specimen rather than the actual area of the necked-down specimen in computing the unit stress σ. While the true stress, calculated from the reduced area, continues to increase until failure occurs, the apparent stress, calculated from the initial area, decreases, as shown by the dotted lines GH of Fig. 4.2. The actual failure occurs at point H, but the maximum apparent stress, represented by point G, is the more important stress to use in design calculations. This value is defined as the *ultimate strength* σ_{tu}. In the design of tension members for vehicle structures, it is accurate to employ the initial area of the member and the apparent ultimate tensile strength σ_{tu}. In using the stress-strain curve to calculate the ultimate bending strength of beams, as shown in a later chapter, the results are slightly conservative because the beams do not neck down in the same manner as tension members.

Compression Test

The compressive strength of materials is more difficult to identify from stress-strain curves than the corresponding tensile strength. Compressive failures for most structural designs in engineering applications are associated with instabilities which are related to yield stress rather than ultimate stress. Yield-stress values which are obtained based on the 0.2 percent offset method (0.002-in/in permanent strain) have been proved to be relatively successful for correlating instabilities in most metals; however, the correlation is less satisfactory for nonmetals, specifically composites. Specimen geometry and means of supports seem to have considerable effect on compression test results. The compressive stress-strain diagrams for most materials are similar to the tensile stress-strain curves.

Shear Test

In-plane shear properties are the most difficult to obtain and have the least standardized testing procedures of all major mechanical properties. The design of a shear-test specimen having a test section subjected to a uniform shearing stress is impossible. The closest practical approach is probably a thin-walled circular cylinder loaded in torsion, as shown in Fig. 4.3. For small displacements, the shear strain is expressed as

$$\epsilon_s = \frac{\Delta S}{L} \tag{4.4}$$

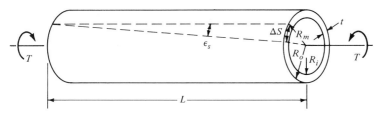

Figure 4.3

where ϵ_s = shear strain
ΔS = change in arc length
L = effective gage length

The shear stress may be calculated from the well-known strength-of-material torsional equation as

$$\sigma_s = \frac{Tr}{J} \tag{4.5}$$

where σ_s = shear stress
r = cylinder radius
J = cross-sectional polar moment of inertia

For very thin-walled cylindrical test specimens, Eq. (4.5) may be written as

$$\sigma_s = \frac{TR_m}{J_m} \tag{4.6}$$

where R_m = mean radius = $(R_o + R_i)/2$ and $J_m = 2\pi R_m^3 t$.

The shear modulus of elasticity may be obtained from the shear stress–shear strain diagram as

$$G = \frac{\sigma_s}{\epsilon_s} \tag{4.7}$$

4.3 EQUATIONS FOR STRESS-STRAIN CURVE IDEALIZATION

In the design of vehicle structural members, it is necessary to consider the properties of the stress-strain curve at stresses higher than the elastic limit. In other types of structural and machine designs, it is customary to consider only stresses below the elastic limit; but weight considerations are so important in flight vehicle design that it is necessary to calculate the ultimate strength of each member and to provide the same factor of safety against failure for each part of the entire structure. The ultimate bending or compressive strengths of many members are difficult to calculate, and it is necessary to obtain information from destruction tests of complete members. In order to apply the results on tests of members of one material to similar members of another material, it is desirable to obtain an idealized analytical expression for the stress-strain diagrams of various materials.

Ramberg and Osgood[19] have developed a method of expressing any stress-strain curve in terms of the modulus of elasticity E, a stress σ_1 (which is approximately equal to the yield stress), and a material shape factor n. The equation for the stress-strain diagram is

$$\bar{\epsilon} = \bar{\sigma} + \tfrac{3}{7}\bar{\sigma}^n \tag{4.8}$$

where $\bar{\epsilon}$ and $\bar{\sigma}$ are dimensionless terms defined as follows:

$$\bar{\epsilon} = \frac{E\epsilon}{\sigma_1} \tag{4.9}$$

and

$$\bar{\sigma} = \frac{\sigma}{\sigma_1} \tag{4.10}$$

The curves expressed by Eq. (4.8) are plotted in Fig. 4.4 for various values of n. A material such as mild steel, in which the stress remains almost constant above the yield point, is represented by the curve for $n = \infty$. Other materials, with various types of stress-strain diagrams, may be represented by the curves for other values of n. In order to represent the stress-strain diagrams for all materials by the single equation, it is necessary to use the reference stress value of σ_1 rather than the yield stress. The value of σ_1 is obtained as shown in Fig. 4.5 by drawing the line $\sigma = 0.7E\epsilon$ from the origin to the stress-strain curve and obtaining the stress coordinate σ_1 of this point of intersection. The stress σ_1 is approximately equal to the yield stress for typical flight vehicle materials. The value of n may be determined so that Eq. (4.8) fits the experimental stress-strain curve in the desired region. Ramberg and Osgood show that for most materials the value of n may be accurately determined from the stress σ_1 and a similar stress σ_2 on the line $\sigma = 0.85E\epsilon$.

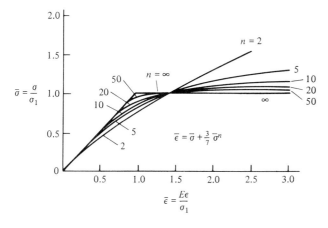

Figure 4.4

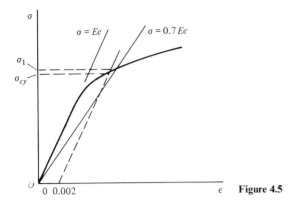

Figure 4.5

It is appropriate to note here that similar equations for idealizing shear stress–shear strain curves may be formulated and fitted into various materials' shear test data.

4.4 FATIGUE

Fatigue is a dynamic phenomenon which may be defined as the initiation and propagation of microcracks into macrocracks as a result of repeated applications of stresses. It is a process of localized progressive structural fracture in material under the action of dynamic stresses. A structure which may not ever fail under a single application of load may very easily fail under the same load if it is applied repeatedly. This failure under repeated application of loads is termed *fatigue failure*.

In spite of the many studies and vast amount of experimental data accumulated over the years, fatigue is still the most common cause of failure in machinery and various structures as well as the least understood of all other structural behavior. This lack of understanding is attributed to the fact that the initiation and the propagation of microscopic cracking are inherently statistical in nature. In fact, the analyst often is confronted with the wide variations in the statistics of what may be estimated as (1) type of service and environment, (2) magnitude of service loads and frequency of occurrence, (3) the quality control during the fabrication operations, (4) the extent and accuracy of the analyses in determining stresses, and (5) the applicability of the material strength data.

The main objective of all fatigue analyses and testing is the prediction of fatigue life of a given structure or machine part subjected to repeated loading. Such loads may have constant amplitude, as indicated in Fig. 4.6; however, in flight vehicle structures, the load history is usually random in nature, as shown in Fig. 4.7.

Fatigue-life prediction Most of the fatigue-life prediction methods used in the design of structures have been based on fatigue allowable data generated by sine

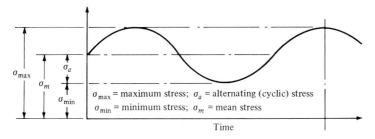

σ_{max} = maximum stress; σ_a = alternating (cyclic) stress
σ_{min} = minimum stress; σ_m = mean stress

Figure 4.6 Sine loading.

wave excitation. Only recently closed-loop, servo-controlled hydraulic machines have become available for true random loading testing. Fatigue test data usually are presented graphically (Fig. 4.8) and the curves are referred to as the allowable *S-N curves*. The curves in conjunction with the "cumulative damage" concept form the basis for most of the methods used in the prediction of fatigue life.

Among the several theories proposed for fatigue-life prediction, the Palmgren-Miner theory,[26] because of its simplicity, seems to be the most widely used. The method hypothesizes that the useful life expended may be expressed as the ratio of the number of applied cycles η_i to the number of cycles N_i to failure at a given constant stress level σ_i. When the sum of all the fractions reaches 1, failure should occur. Mathematically this failure criterion is written as

$$\left(\frac{\eta_1}{N_1}\right)_{\sigma_1 = \text{const}} + \left(\frac{\eta_2}{N_2}\right)_{\sigma_2 = \text{const}} + \cdots + \left(\frac{\eta_r}{N_r}\right)_{\sigma_r = \text{const}} = 1$$

or

$$\sum_{i=1}^{r} \left(\frac{\eta_i}{N_i}\right)_{\sigma_i = \text{const}} = 1 \qquad (4.11)$$

It is important to note that in Palmgren-Miner theory no provisions are made to take into account the various effects on fatigue life, such as notch sensitivity effect, loading sequence effect (high-low or low-high), and the consequences of

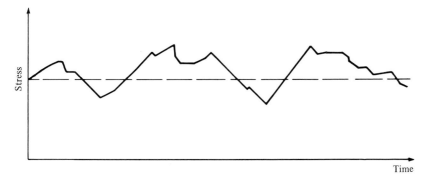

Figure 4.7 Random stress loading.

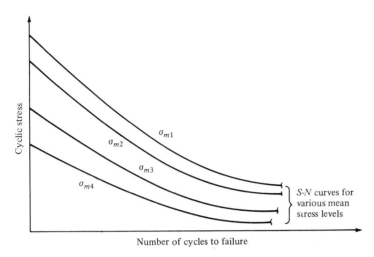

Figure 4.8 *S-N* fatigue curves.

different levels of mean stress. In fact, the theory has been shown to yield unconservative results in some test case studies conducted by Gassner,[20] Kowalewski,[21] and Corten and Dolan.[22]

To illustrate the use of Eq. (4.11), consider a bracket which supports an electronic box in the aircraft cockpit. In a typical mission, the bracket encounters a stress history spectrum idealized as shown in Fig. 4.9. The fatigue allowable of the bracket material is given in Fig. 4.10. The problem is to find the number of missions the aircraft may accomplish before the bracket fails.

From Figs. 4.9 and 4.10, Table 4.1 may be easily constructed. Therefore, in one mission 0.433 percent of the useful life of the bracket is expended. This means

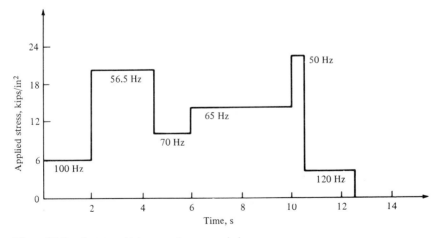

Figure 4.9 Bracket stress history spectrum per mission.

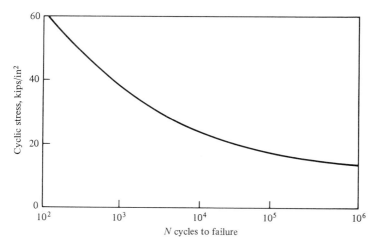

Figure 4.10 *S-N* curves.

that the aircraft might accomplish about 200 missions without bracket failure, based on Eq. (4.6).

Fatigue test data: *S-N* **curves** Fatigue tests are conducted for a wide variety of reasons, one of which is to establish materials' fatigue allowables, or what is commonly referred to as the *S-N* curves. Contrary to results of static tests, it has been observed that the scatter in fatigue test results can be quite large. This inherent scatter characteristic leaves no choice but to treat the results statistically. One of the more widely used statistical distribution functions is the log-normal distribution, whose mean value is taken as

$$M = \frac{1}{n} \sum_{i=1}^{n} \log N_i \qquad (4.12)$$

where n = total number of specimens tested at the same stress level and N_i = number of cycles to failure for specimen i.

Table 4.1

η_i, cycles	σ_i, kips/in^2	N_i, cycles	$\dfrac{\eta_i}{N_i}$
200	6	∞	0
141	20	50,000	0.00282
105	10	∞	0
260	14	10^6	0.00026
25	22	2×10^4	0.00125
240	4	∞	0
$\sum \dfrac{\eta_i}{N_i}$			0.00433

The unbiased standard deviation of M is defined by

$$\delta = \left[\frac{\sum\limits_{i=1}^{n} (\log N_i - M)^2}{n - 1} \right]^{1/2} \qquad (4.13)$$

In order to calculate the number of cycles to failure, based on some confidence level, the standard variable ζ is taken as

$$\zeta = \frac{\log N - M}{\delta} \qquad (4.14)$$

or

$$\log N = M + \zeta\delta \qquad (4.15)$$

The probability of surviving $\log N$ cycles is

$$\text{Probability}(\log N) = P(\log N) = \frac{1}{\sqrt{2\pi}} \int_{\zeta}^{\infty} e^{-\zeta^2/2} \, d\zeta \qquad (4.16)$$

Equation (4.16) may be used to tabulate the probabilities of survival for various values of the standard variable ζ, as shown in Table 4.2.

To illustrate the procedure, consider the following actual fatigue test results for seven tested specimens:

Specimen no., i	Cycles to failure N_i
1	61,318
2	39,695
3	62,803
4	51,039
5	83,910
6	35,631
7	96,500

From Eq. (4.12) the mean value is

$$M = \frac{1}{n} \sum_{i=1}^{n} \log N_i = \frac{1}{7} \sum_{i=1}^{7} \log N_i$$

$$= \tfrac{1}{7}(\log 61{,}318 + \log 39{,}695 + \cdots + \log 96{,}500)$$

$$= 4.76463$$

Table 4.2

ζ	Probability of survival, %
−1.280	90.0
−1.645	95.0
−2.330	99.0

From Eq. (4.13) the unbiased standard deviation is ·

$$\delta = \left[\sum_{i=1}^{7} \frac{(\log N_i - M)^2}{6} \right]^{1/2}$$

$$= \left[\frac{(\log 61{,}318 - 4.76463)^2}{6} + \cdots + \frac{(\log 96{,}500 - 4.76463)^2}{6} \right]^{1/2}$$

$$= 0.1588$$

From Eq. (4.15), if we assume a probability of survival of 95 percent.

$$\log N = M + \zeta\delta = 4.76463 + (-1.645)(0.1588) = 4{,}503$$

Therefore the number of cycles to failure for a 95 percent probability of survival is

$$N = \text{antilog } 4.68189 = 31{,}800 \text{ cycles}$$

4.5 STRENGTH-WEIGHT COMPARISONS OF MATERIALS

The criterion commonly used in the selection of structural materials for aerospace vehicle application is that which yields *minimum weight*. This involves selecting the proper combination of material and structural proportions with the weight as the objective function to be minimized to yield an optimum design. Although weight comparisons of materials may be based on several factors, such as resistance to corrosion, fatigue behavior, creep characteristics, strength, and so on, the only treatment given here is with respect to strength.

As an illustration, let us consider the three loaded members shown in Fig. 4.11; for simplicity, it is assumed that there exists only one free variable (the thickness *t*, in this case) to be chosen in the design. The criteria which govern the design of members in Fig. 4.11*a*, *b*, and *c* are ultimate uniaxial tension, ultimate

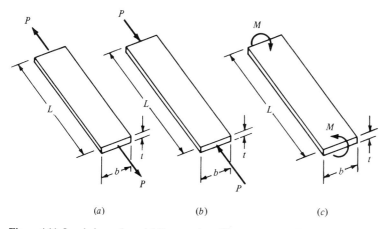

Figure 4.11 Loaded members. (*a*) Pure tension; (*b*) pure compression; (*c*) pure bending.

uniaxial compression or buckling, and ultimate bending. The expressions relating the applied external loads to the induced actual stresses are

Tension:
$$\sigma_t = \frac{P}{A} \quad (A = bt) \tag{4.17}$$

Compression (buckling):
$$\sigma_c = \frac{\pi^2 EI}{AL^2} \quad \left(I = \frac{bt^3}{12}\right) \tag{4.18}$$

Bending:
$$\sigma_b = \frac{Mt}{2I} \tag{4.19}$$

where σ_t, σ_c, σ_b = ultimate tensile, compression, and flexural stresses, respectively
A = cross-sectional area
I = moment of inertia of member

The weight of the member may be expressed in terms of the material density (ρ pounds per cubic inch) and the fixed and free geometric dimensions as

$$W = Lbt\rho \tag{4.20}$$

Solving for the free variable t from Eqs. (4.17) through (4.19) and substituting into Eq. (4.20) yields the material weight required to meet each specified design criterion. Thus

$$W = \frac{PL\rho}{\sigma_t} \qquad \text{(tension)} \tag{4.21}$$

$$W = \frac{L^2 b\rho}{\pi} \left(\frac{12\sigma_c}{E}\right)^{1/2} \qquad \text{(compression)} \tag{4.22}$$

$$W = L\rho \left(\frac{6Mb}{\sigma_b}\right)^{1/2} \qquad \text{(bending)} \tag{4.23}$$

With Eqs. (4.21) to (4.23) available, weight comparisons of different materials may be conducted. Thus the weights of two different materials required to carry the axial load P may be readily obtained from Eq. (4.21) as

$$W_1 = \frac{PL\rho_1}{\sigma_{t1}} \qquad W_2 = \frac{PL\rho_2}{\sigma_{t2}}$$

where the subscripts 1 and 2 refer to materials 1 and 2, respectively, and σ_{t1} and σ_{t2} are the ultimate tensile stresses of materials 1 and 2, respectively.

$$\frac{W_1}{W_2} = \frac{\rho_1}{\rho_2} \frac{\sigma_{t2}}{\sigma_{t1}} \tag{4.24}$$

Similarly, the ratio of weights of two members of two different materials resisting the same bending moment may be easily obtained by utilizing Eq. (4.23):

$$\frac{W_1}{W_2} = \frac{\rho_1}{\rho_2} \left(\frac{\sigma_{b2}}{\sigma_{b1}}\right)^{1/2} \tag{4.25}$$

Likewise, the ratio of weights for two members of two different materials resisting the same compressive (buckling) load may be written immediately by using Eq. (4.22):

$$\frac{W_1}{W_2} = \frac{\rho_1}{\rho_2}\left(\frac{E_2}{E_1}\right)^{1/3} \tag{4.26}$$

Typical aerospace vehicle sheet materials are compared in Table 4.3 by means of Eqs. (4.24) through (4.26). The weights of the various materials are compared with the aluminum alloy 2024-T3. The weight ratios for tension members, shown in column 5, do not vary greatly for the different materials. For members in bending, however, the lower-density materials have a distinct advantage, as shown in column 6. Similarly, the lower-density materials have an even greater advantage in compression buckling, as indicated in column 7. Values of σ vary with sheet thickness, and those shown are used only for comparison.

The computations of Table 4.3 indicate that the last three materials, having lower densities, are superior to the aluminum alloys. However, it is important to note that magnesium alloys are more subject to corrosion than aluminum alloys, while wood and plastic materials are less ductile. Brittle materials are undesirable for structures with numerous bolted connections and cutouts which produce local high-stress concentrations. Ductile materials, which have a large unit elongation at the ultimate tensile strength, will yield slightly at points of high local stress and will thus relieve the stress, whereas brittle materials may fail under the same conditions. Fiber-reinforced plastics have been used successfully for aerospace vehicle structures as long ago as the late 1940s and early 1950s. In those days, the main reinforcement was glass fiber in fabric form with polyester resin as the bonding agent. Since then and primarily in the last few years, development of new high-modulus fibers (such as boron, silicon carbide, graphite, and beryllium) in combination with high-modulus, high-temperature-resistant resins (such as cycloaliphatic epoxies, polymeric and polybenzimidazole resins) has added a new dimension to materials for applications in aerospace and marine- and land-based structures. These new fibers and resins are being combined in a unidirectional,

Table 4.3 Strength-weight comparisons of materials

Sheet material (1)	σ, kips/in^2 average (2)	ρ, lb/in^3 (3)	$E(10^3)$, kips/in^2 (4)	Tension: $\frac{\rho_1}{\rho_2}\frac{\sigma_{t2}}{\sigma_{t1}}$ (5)	Bending: $\frac{\rho_1}{\rho_2}\sqrt{\frac{\sigma_{b2}}{\sigma_{b1}}}$ (6)	Buckling: $\frac{\rho_1}{\rho_2}\sqrt[3]{\frac{E_2}{E_1}}$ (7)
Stainless steel	185	0.286	26	1.23	1.72	2.12
Aluminum alloy 2024-T3	66	0.100	10.5	1.00	1.00	1.00
Aluminum alloy 7075-T6	77	0.101	10.4	0.87	0.93	1.01
Magnesium alloy	40	0.065	.5	1.07	0.83	0.77
Laminated plastic	30	0.050	2.5	1.10	0.74	0.83
Spruce wood	9.4	0.0156	1.3	1.09	0.42	0.31

The "Ratio of weight to weight of 2024-T3 aluminum alloy" header spans columns 5, 6, and 7 (Tension, Bending, Buckling).

preimpregnated form which gives the analyst complete freedom to tailor the composite (a composite is made of a certain number of unidirectional plies) to meet the imposed load requirements in both magnitude and direction. Studies have indicated that through the use of composite materials, the total weight of an aerospace vehicle could be reduced by more than 35 percent.

4.6 SANDWICH CONSTRUCTION

The problem of increasing weight which accompanies increasing material thickness is being met frequently by the use of sandwich construction in application to aerospace vehicles. This type of construction consists of thin, outer- and inner-facing layers of high-density material separated by a low-density, thick core material.

In aerospace applications, depending on the specific mission requirements of the vehicle, the material of the sandwich facings may be reinforced composites, titanium, aluminum, steel, etc. Several types of core shapes and core materials may be utilized in the construction of the sandwich. The most popular core has been the "honeycomb" core, which consists of very thin foils in the form of hexagonal cells perpendicular to the facings.

Although the concept of sandwich construction is not new, only in the last decade has it gained great prominence in the construction of practically all aerospace vehicles, including missiles, boosters, and spacecraft. This is primarily a result of the high structural efficiency that can be developed with sandwich construction. Other advantages offered by sandwich construction are its excellent vibration and flutter characteristics, superior insulating qualities, and design versatility.

An element of a sandwich beam is shown in Fig. 4.12. For simplicity, the facings are assumed to have equal thickness t_f, and the core thickness is t_c. It is also assumed that the core carries no longitudinal normal stress σ. Let us consider the case where it is desired to find the optimum facing thickness which results in a minimum weight of the sandwich beam carrying a bending moment M. Without any loss in generality, b and L may be taken as unit values, and thus

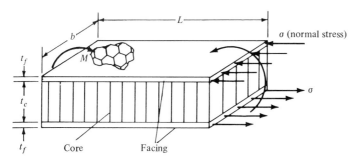

Figure 4.12 Sandwich beam element.

the resisting bending moment can be expressed as

$$M = \sigma t_f (t_f + t_c)$$

If we assume that t_f is small compared to t_c, which is normally the case, then the above equation reduces to

$$M = \sigma t_f t_c$$

or

$$M = \sigma \beta t_c^2 \tag{4.27}$$

where t_f is expressed in terms of t_c by the equation $t_f = \beta t_c$.

The weight of a unit element of the beam is approximately

$$W = \rho_c t_c + 2\rho_f \beta t_c \tag{4.28}$$

where ρ_c and ρ_f equal the core and facing densities, respectively. Eliminating the variable t_c from Eqs. (4.27) and (4.28) yields

$$W = (\rho_c + 2\rho_f \beta) \sqrt{\frac{M}{\beta \sigma}} \tag{4.29}$$

The value of β for the minimum weight may be obtained by differentiating Eq. (4.29) with respect to β and equating the derivative to zero. Performing the differentiation and solving for β yield

$$\beta = \frac{\rho_c}{2\rho_f} \tag{4.30}$$

Equation (4.30) indicates that for a sandwich material resisting bending moment, the minimum weight is obtained when the two layers of the face material have approximately the same total weight as the core. Note that this condition does not yield minimum weight if the beam element is under the action of compressive (buckling) load.

It is now possible to compare the weight of a sandwich-construction beam element with that of a solid element corresponding to the sandwich face material, if we assume that they both resist the same loads. A sandwich element designed to resist bending moments will have a total weight equal to twice the total weight of the facings, if the face and core materials have equal weights. Thus, from Eq. (4.28)

$$W = 4\rho\beta t \tag{4.31}$$

By solving for t from Eq. (4.27) and substituting into Eq. (4.31), the following weight for the sandwich element is obtained:

$$W = 4\rho\beta \sqrt{\frac{M}{\beta \sigma}} \tag{4.32}$$

The weight W_s of a solid beam element from Eq. (4.23) is

$$W_s = \rho \sqrt{\frac{6M}{\sigma}} \tag{4.33}$$

Hence, the ratio of the weight of a sandwich beam element to that of a solid element of the corresponding sandwich face material is

$$\frac{W}{W_s} = \frac{4\rho\beta \sqrt{M/(\beta\sigma)}}{\rho \sqrt{6M/\sigma}} = 1.63 \sqrt{\beta} \tag{4.34}$$

It is important to note that Eq. (4.34) is valid only for a sandwich in which the total weight of the facings is equal to the weight of the core. In order to compare the weights of a sandwich with solid elements studied in Table 4.3, consider a sandwich whose facings are made of 2024-T3 aluminum alloy and a core material whose density is 0.01 lb/in^3. From Eq. (4.30)

$$\beta = \frac{\rho_c}{2\rho_f} = \frac{0.01}{2(0.1)} = 0.05$$

From Eq. (4.34)

$$\frac{W}{W_s} = 1.63 \sqrt{0.05} = 0.37$$

Thus, the sandwich has only 37 percent of the weight of a solid element resisting the same bending moment. Also note that the value of 0.37 is less than any of the other values in column 6 of Table 4.3.

In the preceding discussion, it was assumed that the proportions for the sandwich element were limited only by theoretical considerations. In actual structures, practical considerations are much more important. The thickness of the face material, for example, usually is greater than the theoretical value, because it might not be feasible to manufacture and form very thin sheets. Likewise, the core was assumed to support the facings sufficiently to develop the same unit stress as in a solid element, whereas the actual low-density materials might not provide such support.

4.7 TYPICAL DESIGN DATA FOR MATERIALS

In the manufacture of materials, it is not possible to obtain exactly the same structural properties for all specimens of a material. In a large number of tested specimens of the same material, the ultimate strength may vary as much as 10 percent. In the design of an aerospace vehicle structure, therefore, it is necessary to use stresses which are the minimum values that may be obtained in any specimen of the material. These values are termed the *minimum guaranteed values* of the manufacturer. The licensing and procuring agencies specify the minimum values to be used in the design of aerospace vehicles. These values are contained

Table 4.4 Typical mechanical data for materials

Tension	
σ_{tu}	Ultimate stress
σ_{ty}	Yield stress
σ_{tp}	Proportional limit
E	Modulus of elasticity
e	Elongation
Compression	
σ_{cu}	Ultimate (block) stress
σ_{cy}	Yield stress
σ_{cp}	Proportional limit
σ_{co}	Column yield stress
E_c	Modulus of elasticity
Shear	
σ_{su}	Ultimate stress
σ_{st}	Torsional modulus of rupture
σ_{sp}	Proportional limit (torsion)
G	Modulus of rigidity (torsion)
Bearing	
σ_{bru}	Ultimate stress
σ_{bry}	Yield stress

in Military Handbooks, such as MIL-HDBK-5A, MIL-HDBK-5, MIL-HDBK-17, MIL-HDBK-23. Table 4.4 shows the typical mechanical data required in the design of aerospace structures. Normally these data are presented in accordance with one of the following bases:

A basis: At least 99 percent of all mechanical property values are expected to fall above the specified property values with a confidence of 95 percent.
B basis: At least 90 percent of all mechanical property values are expected to fall above the specified property values with a confidence of 95 percent.
S basis: Minimum mechanical property values as specified by various agencies.

PROBLEMS

4.1 The buckling load for a sandwich column is approximately given by $P = \pi^2 EI/L^2$. Find the thickness ratio of facing to the core which results in an optimum design (minimum weight) for the column. See Fig. P4.1.

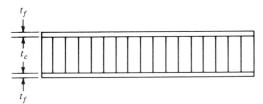

Figure P4.1

4.2 Find the weight ratio of a sandwich column to that of a solid column whose material is the same as that of the sandwich facings.

4.3 Work Prob. 4.2 for a core density of 0.015 lb/in³ and the following specific cases of facing materials:

    (*a*) 2024-T3 aluminum alloy (density = 0.1 lb/in³)
    (*b*) 6A1-4V titanium (density = 0.16 lb/in³)
    (*c*) 321 stainless steel (density = 0.286 lb/in³)
    (*d*) Inconel (density = 0.3 lb/in³)
    (*e*) Beryllium (density = 0.069 lb/in³)
    (*f*) Reinforced composite (unidirection)
        (1) Glass fiber (density = 0.09 lb/in³)
        (2) Boron fiber (density = 0.095 lb/in³)
        (3) Graphite (density = 1.053 lb/in³)

4.4 A missile-holding fixture on an aircraft is subject during each flight to the stress-load history shown in Fig. P4.4. After how many flights will the fixture fail if the material fatigue allowable is that shown in Fig. 4.10?

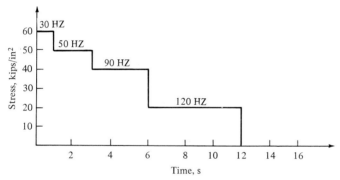

Figure P4.4

4.5 Ten parts were fatigue-tested at the same stress level, and the following failure cycles were reported:

Specimen no.	Cycles to failure
1	300,000
2	190,000
3	225,000
4	350,000
5	260,000
6	280,000
7	490,000
8	310,000
9	360,000
10	390,000

After how many cycles should the part be replaced so that only the following percentage of the parts in service fails before replacement?

    (*a*) 1
    (*b*) 5
    (*c*) 10

FIVE

STRESS ANALYSIS

5.1 INTRODUCTION

In order to select sizes of structural members to meet the design load requirements on a specific aerospace vehicle, it is necessary to find the unit stresses acting on the cross section of each structural element. The unit stress referred to here is the force intensity at any point, and it has units of force per unit area, or pounds per square inch in common engineering units.

It was shown in Chap. 3 that there exist two distinct components of stress, normal and shear stress. A normal stress is a unit stress which acts normal to the cross section of the structural element, while the shear stress is parallel and in the plane of the cross section. A normal stress is induced by bending moments and axial forces. A shear stress, however, is caused by torsional moments and shear forces. This chapter discusses the theory and the application of these two fundamental stress components.

5.2 FORCE-STRESS RELATIONSHIPS

The stress field at any chosen point in a solid beam may be entirely defined by the components of force resultants or stresses acting along the directions of some "gaussian" coordinate system, as shown in Fig. 5.1. The forces and stresses are taken to be positive if they act in the positive direction of the corresponding coordinate axis.

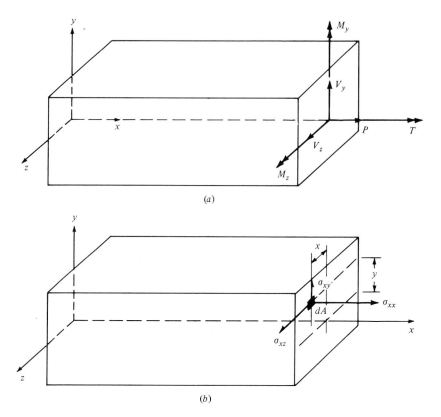

Figure 5.1 Stresses and force resultants.

From Fig. 5.1 the force resultants may be related to the stresses as follows:

$$P = \int_A \sigma_{xx}\, dA \qquad M_z = -\int_A y\sigma_{xx}\, dA$$

$$V_y = \int_A \sigma_{xy}\, dA \qquad M_y = \int_A z\sigma_{xx}\, dA \tag{5.1}$$

$$V_z = \int_A \sigma_{xz}\, dA \qquad T = \int_A (y\sigma_{xz} - z\sigma_{xy})\, dA$$

where P = axial force
V_y, V_z = shear forces
M_z, M_y = bending moments
T = torque
σ_{xx} = normal stress
σ_{xy}, σ_{xz} = shearing stresses

5.3 NORMAL STRESSES IN BEAMS

The normal stresses in beam elements are induced by bending and/or extensional actions. Two approaches may be used to determine stresses; the first is based on the theory of elasticity, and the second is based on strength-of-materials theory. The latter, which is used here, assumes that plane sections remain plane after extensional-bending deformation takes place. This assumption implies that the deformations due to transverse shear forces (V_z and V_y) are very small and therefore may be neglected. In addition, this assumption allows the displacements (deflections) of any point in the beam to be expressed in terms of the displacements of points located on the beam axis.

Assume that the displacement in the x direction of any point in the beam is represented by $q_x(x, y, z)$. If we take $u_x(s)$ to be the extensional displacement of any point on the beam axis ($y = z = 0$) and ψ_z and ψ_y to be the rotational displacements of the beam cross section, then

$$q_x(x, y, z) = u_x(x) - y\,\psi_z(x) + z\psi_y(x) \tag{5.2}$$

The axial strain from Eq. (3.14a) is defined by

$$\epsilon_{xx} = q_{x,\,x}$$

Hence, from Eq. (5.2)

$$\epsilon_{xx} = u_{x,\,x} - y\psi_{z,\,x} + z\psi_{y,\,x} \tag{5.3}$$

At any given cross section $x = x_0$,

$$\frac{du_x(x_0)}{dx} = \text{const} = B_1$$

$$\frac{d\psi_z(x_0)}{dx} = \text{const} = B_2$$

$$\frac{d\psi_y(x_0)}{dx} = \text{const} = B_3$$

where Eq. (5.3) becomes

$$\epsilon_{xx} = B_1 + B_2\,y + B_3\,z \tag{5.4}$$

In order to determine the stresses which correspond to the strains in Eq. (5.4), the stress-strain relationship in Chap. 3 is utilized. By assuming that the stresses σ_{zz} and σ_{yy} are negligible compared to σ_{xx}, the following relationship for an isotropic material may be obtained easily from Eq. (3.25):

$$\sigma_{xx} = E\epsilon_{xx} \tag{5.5}$$

where E = modulus of elasticity of the material.

Substituting Eq. (5.4) into Eq. (5.5) yields

$$\sigma_{xx} = E(B_1 + B_2\,y + B_3\,z) \tag{5.6}$$

Constants B_1, B_2, and B_3 may now be determined through the use of Eq. (5.1), or or

$$P = \int_A E(B_1 + B_2\, y + B_3\, z)\, dA$$

$$M_z = -\int Ey(B_1 + B_2\, y + B_3\, z)\, dA$$

$$M_y = \int Ez(B_1 + B_2\, y + B_3\, z)\, dA$$

Carrying out the integrations yields

$$\frac{P}{E} = B_1 A + B_2\, A\bar{y} + B_3\, A\bar{z}$$

$$\frac{-M_z}{E} = B_1 A\bar{y} + I_z\, B_2 + I_{yz}\, B_3 \tag{5.7}$$

$$\frac{M_y}{E} = B_1 A\bar{z} + I_{yz}\, B_2 + I_y\, B_3$$

where A = cross-sectional area

$$I_z = \int_A y^2\, dA = \text{moment of inertia of cross section about } t \text{ axis} \tag{5.8a}$$

$$I_y = \int_A z^2\, dA = \text{moment of inertia of cross section about } y \text{ axis} \tag{5.8b}$$

$$I_{yz} = \int_A yz\, dA = \text{product moment of inertia of cross section} \tag{5.8c}$$

$$\bar{y} = \int_A y\, dA \bigg/ \int_A dA$$

$$\bar{z} = \int_A z\, dA \bigg/ \int_A dA \tag{5.9}$$

If the z and y axes are taken through the geometric centroid of the cross section, then $\bar{y}$ and $\bar{z}$ become identically zero. Hence Eq. (5.7) reduces to

$$\frac{P}{E} = B_1 A$$

$$\frac{-M_z}{E} = I_z\, B_2 + I_{yz}\, B_3 \tag{5.10}$$

$$\frac{M_y}{E} = I_{yz}\, B_2 + I_y\, B_3$$

Solving Eqs. (5.10) for the unknown constants yields

$$B_1 = \frac{P}{AE}$$

$$B_2 = -\frac{I_y M_z + I_{yz} M_y}{E(I_y I_z - I_{yz}^2)}$$

(5.11)

$$B_3 = \frac{I_z M_y + I_{yz} M_z}{E(I_y I_z - I_{yz}^2)}$$

Substituting Eqs. (5.11) into Eq. (5.6) yields the general expression of the normal stress:

$$\sigma_{xx} = \frac{P}{A} - \frac{I_y M_z + I_{yz} M_y}{I_y I_z - I_{yz}^2} y + \frac{I_z M_y + I_{yz} M_z}{I_y I_z - I_{yz}^2} z$$

(5.12)

When Eq. (5.12) is used, it is important to observe the sign convention used in the derivation. See Fig. 5.1. In cases where y and z axes are principal axes of the cross-sectional area, the product of the moment of inertia I_{yz} about these axes is zero. For this condition, Eq. (5.12) reduces to

$$\sigma_{xx} = \frac{P}{A} - \frac{M_z}{I_z} y + \frac{M_y}{I_y} z$$

(5.13)

If there is no axial force acting on the beam and bending occurs about the z axis only, then Eq. (5.13) reduces to the familiar strength-of-material pure bending equation

$$\sigma_{xx} = -\frac{M_z}{I_z} y$$

(5.14)

5.4 SHEAR STRESSES IN BEAMS

The shear stresses in beams are induced by pure shear force action and/or torsional action. In this section, only shear stresses due to shear forces are considered, while the latter are dealt with in a separate section.

Consider a small section of a beam, as shown in Fig. 5.2. For simplicity, assume that the beam cross section is symmetrical and the theory of strength of materials holds. The shear force V_y parallel to the beam cross section produces shear stresses σ_{xy} of varying intensity over the cross-sectional area. Corresponding to the vertical shear stress σ_{xy} there exists a shear stress σ_{yx} in the xz plane which is equal to σ_{xy} at the points of intersection of the two planes. Thus, the expression of the vertical shear stress σ_{xy} at any point in the cross section is obtained by determining the shear stress σ_{yx} on a horizontal plane through the point.

The bending stresses on the left and right sections of the beam element (Fig.

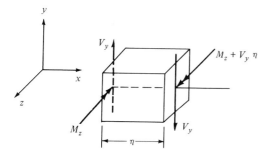

Figure 5.2 Beam element.

5.2) are shown in Fig. 5.3. At any point a distance y from the neutral axis, the bending stress will be $M_z y / I_z$ on the left face and $M_z y / I_z + V_y \eta y / I_z$ on the right face. In order to obtain the shear stress at a distance $y = y_1$ above the neutral axis, the portion of the beam above that point is considered as a free body, as shown in Fig. 5.3c. For equilibrium of the horizontal forces, the force produced by the shear stress σ_{yx} on the horizontal area of width t and length δx must be equal to the difference in the normal forces on the two cross sections. Summing forces in the horizontal direction yields

$$\sigma_{yx} t\eta = \int_{y_1}^{c} \frac{V_y \, y\eta}{I_z} \, dA \tag{5.15}$$

Equation (5.15) may be written in standard form as

$$\sigma_{yx} = \frac{V_y}{I_z t} \int_{y_1}^{c} y \, dA \tag{5.16}$$

where the integral represents the moment of the area of the cross section above the point where the shear stress is being determined, with the moment arms measured from the neutral axis. The cross-sectional area considered is shown by the shaded portion in Fig. 5.3a. It is important to note that Eq. (5.16) is appli-

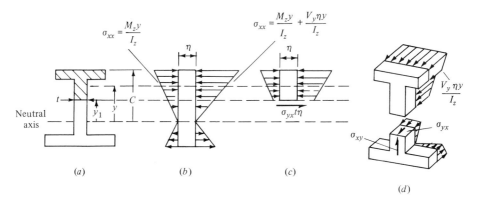

(a)   (b)   (c)

(d)

Figure 5.3

cable only to beams of uniform, symmetrical cross sections. Tapered beams and beams of unsymmetrical cross section are considered later.

Example 5.1 Find the maximum normal stress in the beam in Fig. 5.4 and the shear stress distribution over the cross section.

SOLUTION The maximum normal stress due to bending will occur at the point of maximum bending moment, or at the fixed end of the beam. Since the shear force is constant throughout the beam span, the shear stress distribution will be the same at any cross section. The moment of inertia for the cross is obtained as follows:

$$I_z = 2\left[\left(3 \times \frac{1^3}{12}\right) + 3 \times 2.5^2\right] + \left(1 \times \frac{4^3}{12}\right) = 43.3 \text{ in}^4$$

$$I_{yz} = P = M_y = 0$$

The maximum normal stress is

$$\sigma_{xx} = -\frac{M_z y}{I_z} = -\frac{40 \times 20(\pm 3)}{43.3} = \mp 55.4 \text{ kips/in}^2$$

For a point 1 in below the top of the beam, the integral of Eq. (5.16) is equal to the moment of the area of the upper rectangle about the neutral axis:

$$\int_{y_1}^{c} y \, dA = 2.5(3) = 7.5 \text{ in}^3$$

The average shear stress just above this point, where $t = 3$ in is

$$\sigma_{yx} = \frac{V_y}{I_t t} \int_{y_1}^{c} y \, dA = \frac{40,000}{43.3 \times 3} 7.5 = 2310 \text{ lb/in}^2$$

The average shear stress just below this point, where $t = 1$ in is

$$\sigma_{yx} = \frac{V_y}{I_z t} \int_{y_1}^{c} y \, dA = \frac{40,000}{43.3 \times 1} 7.5 = 6930 \text{ lb/in}^2$$

For a point 2 in below the top of the beam, the integral of Eq. (5.16) is

$$\int_{y_1}^{c} y \, dA = 2.5 \times 3 + 1.5 \times 1 = 9.0$$

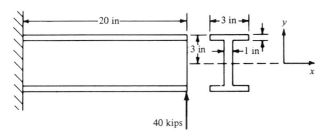

Figure 5.4

This shear stress at this point is

$$\sigma_{yx} = \frac{V_y}{I_z} \int_{y_1}^{c} y\, dA = \frac{40,000}{43.3 \times 1}\, 9.0 = 8320 \text{ lb/in}^2$$

At a point on the neutral axis of the beam, the shear stess is

$$\sigma_{yx} = \frac{V_y}{I_z} \int_{y_1}^{c} y\, dA = \frac{40,000}{43.3 \times 1}\, (2.5 \times 3 + 1 \times 2) = 8780 \text{ lb/in}^2$$

The distribution of shear stress over the cross section is shown in Fig. 5.5. The stress distribution over the lower half of the beam is similar to the distribution over the upper half because of the symmetry of the cross section about the neutral axis.

Alternative Solutions for Shear Stresses

In some problems it is more convenient to find shear stresses by obtaining the forces resulting from the change in bending stresses between two cross sections than it is to use Eq. (5.16). Portions of the beam between two cross sections a unit distance apart are shown in Fig. 5.6. The bending moment increases by V_y in this unit distance, and the bending stresses on the left face of the beam are larger than those on the right face by an amount $V_y \eta y / I_t$, where $\eta = 1$. At the top of the beam, this difference is

$$\frac{V_y y}{I_z} = \frac{40(3)}{43.3} = 2.77 \text{ kips/in}^2$$

The differences in bending stresses at other points of the cross section are obtained by substituting various values of y and are shown in Fig. 5.6b. Cutting sections and utilizing the equations of static equilibrium in each case (Fig. 5.6c, d, and e) yield the shearing stresses at these various points:

$$\sigma_{yx} = 6930/3 = 2310 \text{ lb/in}^2 \text{ at 1 in below top of beam}$$

$$\sigma_{yx} = 8320/1 = 8320 \text{ lb/in}^2 \text{ at 2 in below top of beam}$$

$$\sigma_{yx} = 8780/1 = 8780 \text{ lb/in}^2 \text{ at neutral axis}$$

Note that these shear stress values are the same as shown in Fig. 5.5.

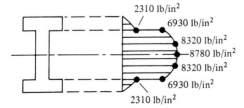

Figure 5.5

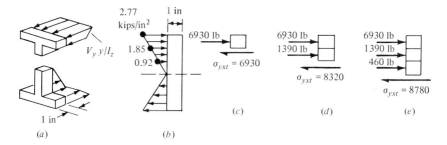

Figure 5.6

Example 5.2 In the beam cross section shown in Fig. 5.7, the webs are considered to be ineffective in resisting normal stresses but capable of transmitting shear. Each stringer area of 0.5 in^2 is assumed to be lumped at a point. Find the shear stress distribution in the webs.

SOLUTION If we neglect the moments of inertia of the webs and of the stringers about their own centroids, the cross-sectional moment of inertia about the neutral axis is

$$I_z = 2(0.5)(6^2) + 2(0.5)(2^2) = 40 \text{ in}^4$$

If two cross sections 1 in apart are considered, the difference in bending stresses $V_y y/I_z$ on the two cross sections will be $8(\frac{6}{40}) = 1.2$ kips/in^2 on the outside stringers and $8(\frac{2}{40}) = 0.4$ kips/in^2 on the inside stringers. The differences in axial loads on the stringers at the two cross sections are found as the product of these stresses and the stringer areas and are shown in Fig. 5.7c.

The shear stress in the web at a point between the upper two stringers is found from the equilibrium of spanwise forces on the upper stringer.

$$\sigma_{yx}(0.04)(1) = 600$$

or

$$\sigma_{yx} = 15{,}000 \text{ lb/in}^2$$

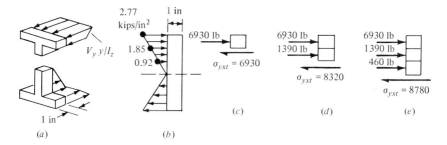

(a)   (b)   (c)   (d)

Figure 5.7

If the webs resist no bending stress, the shear stress will be constant along each web, as shown in Fig. 5.7d. If the webs resist bending stresses, the shear stress in each web will vary along the length of the web and will be greater at the end nearer the neutral axis. The shear stress in the web between the two middle stringers is found by considering spanwise forces on the two upper stringers:

$$\sigma_{yx}(0.04)(1) = 600 + 200$$

or

$$\sigma_{yx} = 20,000 \ \text{lb/in}^2$$

In problems involving shear stresses in thin webs, the shear force per inch length of web often is obtained rather than the shear stress. The shear per inch, or *shear flow*, is equal to the product of the shear stress and the web thickness. The shear flow for each web, shown in Fig. 5.7c, is equal to the sum of the longitudinal loads above the web.

The shear stresses may also be obtained by using Eq. (5.16). For a point between the two upper stringers,

$$\sigma_{yx} = \frac{V_y}{I_z t} \int_{y_1}^{c} y \, dA = \frac{8000}{40 \times 0.040} (0.5 \times 6) = 15,000 \ \text{lb/in}^2$$

For a point between the two middle stringers,

$$\sigma_{yx} = \frac{V_y}{I_z t} \int_{y_1}^{c} y \, dA = \frac{8000}{40 \times 0.040} (0.5 \times 6 + 0.5 \times 2) = 20,000 \ \text{lb/in}^2$$

Example 5.3 Find expressions for the normal stress for all beams whose unsymmetrical cross sections are given in Fig. 5.8a and b.

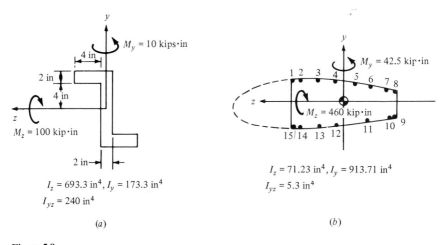

$$I_z = 693.3 \ \text{in}^4, \ I_y = 173.3 \ \text{in}^4$$
$$I_{yz} = 240 \ \text{in}^4$$

(a)

$$I_z = 71.23 \ \text{in}^4, \ I_y = 913.71 \ \text{in}^4$$
$$I_{yz} = 5.3 \ \text{in}^4$$

(b)

Figure 5.8

SOLUTION From Eq. (5.12) with P set to zero, the normal stress for the beam in Fig. 5.8a is

$$\sigma_{xx} = -\frac{I_y M_z + I_{yz} M_y}{I_y I_z - I_{yz}^2} y + \frac{I_z M_y + I_{yz} M_z}{I_y I_z - I_{yz}^2} z$$

$$= -\frac{173.3(100,000) + 240(10,000)}{(693.3)(173.3) - 240^2} y + \frac{693.3(10,000) + 240(100,000)}{(693.3)(173.3) - 240^2} z$$

$$= -315y + 494z$$

Similarly, for the beam in Fig. 5.8b, the normal stress expression is

$$\sigma_{xx} = -6457y + 9.06z$$

5.5 SHEAR FLOW IN THIN WEBS

Shear flow is defined as the product of the shear stress and the thickness of the web. For all practical purposes, it is sufficiently accurate to assume that shear stresses in thin webs are always parallel to the surfaces for the entire thickness of the web. In Fig. 5.9, a curved web representing the leading edge of a wing is shown, and the shear stresses are parallel to the surfaces of the web at all points. Air loads normal to the surface must, of course, be resisted by shear stresses perpendicular to the web, but these stresses usually are negligible and are not considered here. It might appear that a thin, curved web is not an efficient structure for resisting shearing stresses, but this is not the case. The diagonal tensile and compressive stresses σ_t and σ_c are shown in Fig. 5.9 on principal planes at 45° from the planes of maximum shear σ_s. From Mohr's circle for a condition of pure shear, it may be shown that the diagonal compressive stress σ_c and the diagonal tensile stress σ_t are both equal to the maximum shear stress σ_s. If the diagonal compression alone were acting on the curved web, it would bend the web to an increased curvature. The diagonal tensile stress, however, tends to decrease the curvature, and the two effects counteract each other. Consequently, the curved web will resist high shear stress without deforming from its original curvature.

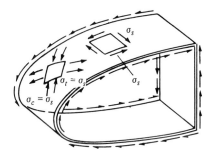

Figure 5.9

The shear flow q, which is the product of the shear stress σ_s and the web thickness t, usually is more convenient to use than the shear stress. The shear flow may be obtained before the web thickness is determined, but the shear stress depends on the web thickness. Often it is necessary to obtain the resultant force on a curved web in which the shear flow q is constant for the length of the web. The element of the web shown in Fig. 5.10 has length ds, and the horizontal and vertical components of this length are dz and dy, respectively. The force on this element of length is $q\,ds$, and the components of the force are $q\,dz$ horizontally and $q\,dy$ vertically. The total horizontal force is

$$F_z = \int_0^z q\,dz = qz \tag{5.17}$$

where z is the horizontal distance between the ends of the web. The total vertical force on the web is

$$F_y = \int_0^y q\,dy = qy \tag{5.18}$$

where y is the vertical distance between the ends of the web. The resultant force is qL, where L is the length of the straight line joining the ends of the web, and the resultant force is parallel to this line. Equations (5.17) and (5.18) are independent of the shape of the web, but depend on the components of the distance between the ends of the web. The induced torsional moment of the resultant force depends on the shape of the web. The torque induced at any point such as O, shown in Fig. 5.11a, is equal to $rq\,ds$. The area dA of the triangle formed by joining point O and the extremities of the element of length ds is $r\,ds/2$. Then the torque induced by the shear flow along the entire web may be obtained as follows:

$$T = \int_s qr\,ds = \int_A 2q\,dA = 2q \int_A dA$$

or

$$T = 2Aq \tag{5.19}$$

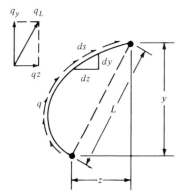

Figure 5.10

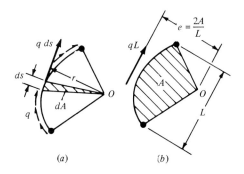

(a) (b) **Figure 5.11**

where A is the area enclosed by the web and the lines joining the ends of the web with point O, as shown in Fig. 5.11b. The distance e, shown in Fig. 5.11b, of the resultant force from point O may be obtained by dividing the torque by the force:

$$e = \frac{2Aq}{qL} = \frac{2A}{L} \tag{5.20}$$

It is important to note that the shear flow q is assumed to be constant in the derivation of Eqs. (5.19) and (5.20).

5.6 SHEAR CENTER

Open-section thin web beams, such as in Fig. 5.12, are unstable in carrying torsional loads. Thus if a beam cross section is symmetrical about a vertical axis, then the vertical loads must be applied in the plane of symmetry in order to produce no torsion on the cross section. However, if the beam cross section is not symmetrical, then the loads must be applied at a point such that they produce no torsion. This point is called the *shear center* and may be obtained by finding the position of the resultant of the shear stresses on any cross section. The simplest type of beam for which the shear center may be calculated is made of two concentrated flange areas joined by a curved shear web, as shown in Fig. 5.12.

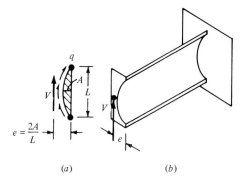

(a) (b) **Figure 5.12**

The two flanges must lie in the same vertical plane if the beam carries a vertical load. If the web resists no bending, the shear flow in the web will have a constant value q. The resultant of the shear flow will be $qL = V$, and the position of this resultant from Eq. (5.20) will be a distance $e = 2A/L$ to the left of the flanges, as shown in Fig. 5.12a. Therefore, all loads must be applied in a vertical plane which is a distance e from the plane of the flanges.

A beam with only two flanges that are in a vertical plane is not stable for horizontal loads. The vertical location of the shear center would have no significance for this beam. For beams which resist horizontal loads as well as vertical loads, it is necessary to determine the vertical location of the shear center. If the cross section is symmetrical about a horizontal axis, the shear center must lie on the axis of symmetry. If the cross section is not symmetrical about a horizontal axis, the vertical position of the shear center may be calculated by taking moments of the shear forces produced by horizontal loads. The method of calculating the shear center of a beam can be illustrated best by numerical examples.

Example 5.4 Find the shear flows in the webs of the beam shown in Fig. 5.13a. Each of the four flange members has an area of 0.5 in². The webs are assumed to carry no bending stress. Find the shear center for the area.

SOLUTION Two cross sections 1 in apart are shown in Fig. 5.13b. The increase in bending moment in the 1-in length is equal to the shear V. The increase of bending stress on the flanges in the 1-in length is

$$\frac{V_y y}{I_z} = \frac{10,000 \times 5}{50} = 1000 \text{ lb/in}^2$$

The load on each 0.5-in² area resulting from this stress is 500 lb and is shown in Fig. 5.13b. The actual magnitude of the bending stress is not needed in the shear-flow analysis, since the shear flow depends on only the change in bending moment or the shear. If each web is cut in the spanwise direction, as shown, the shear forces on the cut webs must balance the loads on the

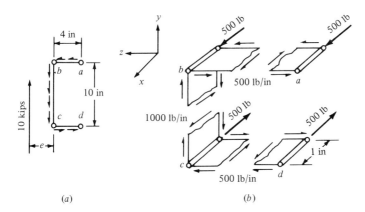

(a)   (b)

Figure 5.13

flanges. The force in web *ab* must balance the 500-lb force on flange *a*, and since this spanwise force acts on a 1-in length, the shear flow in the web will be 500 lb/in in the direction shown. The shear flow in web *bc* must balance the 500-lb force on flange *b* as well as the 500-lb spanwise force in web *ab*, and consequently the shear flow has a value of 1000 lb/in. The shear flow in web *cd* must balance the 1000-lb spanwise force in web *bc* as well as the 500-lb force on flange *c*, which is in the opposite direction. The shear flow in web *cd* is therefore 500 lb/in and is checked by the equilibrium of flange *d*.

The directions of the shear flow on the vertical beam cross section are obtained from the directions of the spanwise forces. Since each web has a constant thickness, the shear flow, like shear stresses, must be equal on perpendicular planes. The shear flow on a rectangular element must form two equal and opposite couples. The directions of all shear flows are shown in Fig. 5.13b and the back section in Fig. 5.13a. The shear center is found by taking moments about point *c*:

$$\sum T_c = 0^{+\rangle}$$

$$-10,000e + 500(4)(10) = 0$$

or
$$e = 2 \text{ in}$$

The shear center will be on a horizontal axis of symmetry, since a horizontal force along this axis will produce no twisting of the beam.

Example 5.5 Find the shear flows in the webs of the beam shown in Fig. 5.14a. Each of the four flanges has an area of 1.0 in². Find the shear center for the area.

SOLUTION The moment of inertia of the area about the horizontal centroidal axis is

$$I_z = 4(1 \times 4^2) = 64 \text{ in}^4$$

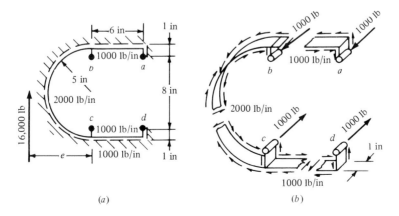

(a)

(b)

Figure 5.14

The change in axial load in each flange between the two cross sections 1 in apart is

$$\frac{V_y}{I_z} yA = \frac{16,000}{64} \times 4 \times 1 = 1000 \text{ lb}$$

The axial loads and shear flows are shown in Fig. 5.14b. The shear flows in the webs are obtained by a summation of the spanwise forces on the elements, as in Example 5.4.

The distance e to shear center is found by taking moments about a point below c, on the juncture of the webs. The shear flow in the nose skin produces a moment equal to the product of the shear flow and twice the area enclosed by the semicircle. The shear flow in the upper horizontal web has a resultant force of 6000 lb and a moment arm of 10 in. The short vertical webs at a and d each resist forces of 1000 lb with a moment arm of 6 in. The resultant forces on the other webs pass through the centers of moment:

$$\sum T_c = 0^{+\curvearrowright}$$

$$-16,000e + 2(39.27)(2000) + 6000(10) + 2(1000)(6) = 0$$

or
$$e = 14.32 \text{ in}$$

5.7 TORSION OF CLOSED-SECTION BOX BEAMS

The thin-web, open-section box beams previously considered are capable of resisting loads which are applied at the shear center but become unstable under torsional loads. In many structures, especially in aerospace vehicles, the resultant load takes on different positions for different loading conditions and consequently may produce torsion. On an aircraft wing, for example, the resultant aerodynamic load is farther forward on the wing at high angles of attack than at low angles of attack. The position of this load also changes when the ailerons or wing flaps are deflected. Thus a closed-section box beam, which is capable of resisting torsion, is used for aircraft wings and similar structures. Typical types of wing construction are shown in Fig. 5.15. The wing section of Fig. 5.15a has only

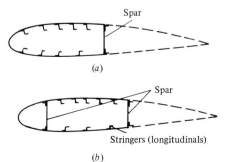

Spar

(a)

Spar

Stringers (longitudinals)

(b)

Figure 5.15 Typical wing construction.

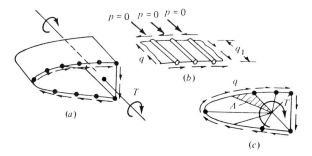

Figure 5.16 Box beam loaded in torsion.

one spar, and the skin forward of this spar forms a closed section which is designed to resist the wing torsion, whereas the portion aft of the spar is lighter and is designed not to resist any loads on the wing but to act as an aerodynamic surface. The wing section shown in Fig. 5.15b has two spars which form a closed-section box beam. In some wings, two or more closed boxes may act together in resisting torsion, but such sections are statically indeterminate and are considered in a later chapter.

The box section shown in Fig. 5.16 is loaded only by a torsional moment T. Since the axial loads in the stringers are produced by wing bending, they are zero for the condition of pure torsion. If the upper stringers are considered as a free body, as shown in Fig. 5.16b, the spanwise forces must be in equilibrium; that is, $qa = q_1 a$ or $q = q_1$. If similar sections containing other flanges are considered, it becomes obvious that the shear flow at any point must be equal to q. The constant shear flow q around the circumference has no resultant horizontal or vertical force, since in the application of Eqs. (5.17) and (5.18) the horizontal and vertical distances between the endpoints of the closed web are zero. The resultant of the shear flow is thus a torque equal to the applied external torque T, taken about any axis perpendicular to the cross section. If we take point O in Fig. 5.16c as a reference, the following may be immediately written from Eq. (5.19):

$$T = \sum 2(\Delta A)q = 2Aq \qquad (5.21)$$

where A is the sum of the triangular areas ΔA and is equal to the total area enclosed by the box section. The area A is the same regardless of the position of point O, since the moment of a couple is the same about any point. If point O is chosen outside the section, some of the triangular areas ΔA will be negative, corresponding to the direction of the moment of the shear flow about point O, but the algebraic sum of all areas ΔA will be equal to the enclosed area A.

5.8 SHEAR FLOW IN CLOSED-SECTION BOX BEAMS

Consider a box beam containing only two stringers, as shown in Fig. 5.17. Since this section is stable under the action of torsional loads, the vertical shear force V

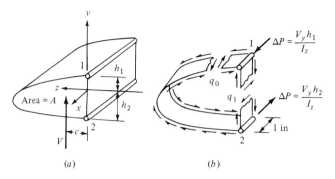

Figure 5.17 One-cell–two-stringers box beam.

may be applied at any point in the cross section. Note that this beam is unstable under the action of a horizontal load, since the two stringers in the same vertical plane cannot resist a bending moment about a vertical axis. If two cross sections 1 in apart are considered, as shown in Fig. 5.17b, the difference in axial load on the stringers, ΔP, between the two cross sections may be found from the difference in the bending stress $\sigma_{xx} = -M_z y/I_z = V_y(1)y/I_z$, or $|\Delta P| = \sigma_{xx} A_f = V_y A_1 h_1/I_z = V_y A_2 h_2/I_z$, where A_1 and A_2 are stringer areas. These loads must be balanced by the shear flow shown in Fig. 5.17b. If we consider equilibrium, the summation of forces on the upper stringer in the spanwise direction must be zero:

$$q_1(1 \text{ in}) + q_0(1 \text{ in}) - \frac{V_y A_1 h_1}{I_z} = 0$$

or
$$q_1 = \frac{V_y A_1 h_1}{I_z} - q_0 \tag{5.22}$$

The shear flow q_0 may be found by summing torsional moments for the back section about a perpendicular axis through the lower stringer:

$$V_y c - 2A q_0 = 0$$

or
$$q_0 = \frac{V_y c}{2A}$$

where A is the total area enclosed by the box.
Substituting this value in Eq. (5.22) yields

$$q_1 = \frac{V_y h_1 A_1}{I_z} - \frac{V_y c}{2A} \tag{5.23}$$

The shear flows in box beams with several stringers may be obtained by a method similar to that previously used. From a summation of spanwise loads on various stringers the shear flows may all be expressed in terms of one unknown shear flow. Then this shear flow may be obtained by equating the moments of the shear flows to the external torsional moment about a spanwise axis. For the box beam shown in Fig. 5.18, all the shear flows $q_1, q_2, \ldots, q_n$ may be expressed in

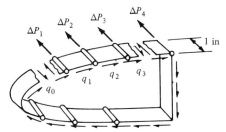

Figure 5.18

terms of the shear flow q_0 by considering the spanwise equilibrium of the stringers between web 0 and the web under consideration:

$$q_1 = q_0 + \Delta P_1$$

$$q_2 = q_0 + \Delta P_1 + \Delta P_2$$

$$\cdots \cdots \cdots \cdots \cdots \cdots \qquad (5.24)$$

or
$$q_n = q_0 + \sum_0^n \Delta P_n$$

where $\sum_0^n \Delta P_n$ represents the summation of loads ΔP between 0 and any web n. After all the shear flows are expressed in terms of the unknown q_0, the value of q_0 may be obtained from the summation of the torsional moments. Note that the shear flow in any other web could have been considered as the unknown q_0. For the case of general bending, the difference in axial load on the stringers ΔP between two sections 1 in apart may be found from Eq. (5.12). Making the substitutions $M_z = V_y(1 \text{ in})$, $M_y = V_z(1 \text{ in})$, and $P = 0$ yields

$$\Delta P_f = \left(-\frac{I_y V_y + I_{yz} V_z}{I_y I_z - I_{yz}^2} y + \frac{I_z V_z + I_{yz} V_y}{I_y I_z - I_{yz}^2} z \right) A_f \qquad (5.25)$$

where y and z are the coordinates of the stringer area A_f.

Example 5.6 Find the shear flow in all webs of the box beam shown in Fig. 5.19a.

SOLUTION The moment of inertia of the beam cross section about the neutral axis is $I = (4 \times 0.5 + 2 \times 1)(5^2) = 100 \text{ in}^4$. The difference in bending stress between the two cross sections 1 in apart is $V(1)y/I = 10,000 \ (1)(5)/100 = 500$ lb/in^2. This produces compressive loads ΔP of 500 lb on the 1-in^2 upper stringer areas and 250 lb on the 0.5 in^2 stringer areas, as shown in Fig. 5.19b. The shear flow in the leading-edge skin is considered as the unknown q_0, although the shear flow in any other web could have been considered as the unknown. Now the shear flow in all other webs may be obtained in terms of q_0 by considering the equilibrium of the spanwise forces on the stringers, as shown in Fig. 5.19b.

The value of q_0 is obtained by considering the equilibrium of torsional

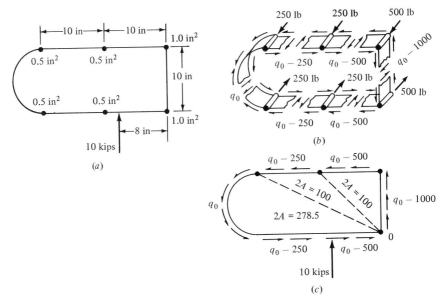

Figure 5.19

moments due to the shear flow and the external applied loads in reference to any axis normal to the back cross section of the beam. Taking the axis through point O, for example, as shown in Fig. 5.19c, and summing torques to zero yield

$$\sum T_0 = 0^+ \circlearrowleft$$

$$-10,000(8) + (q_0 - 500)(100) + (q_0 - 250)(100) + q_0(278.5) = 0$$

or
$$q_0 = 324 \text{ lb/in}$$

The shear flow in the rest of the webs may be computed easily from Fig. 5.19c.

Example 5.7 Find the shear flow in the webs of the box beam shown in Fig. 5.20.

SOLUTION The change in bending stress between two cross sections is obtained from Eq. (5.25). The terms to be used in this equation are obtained as follows:

$$I_z = (2 \times 3 + 2 \times 1)(5^2) = 200 \text{ in}^4$$
$$I_y = (2 \times 3 + 2 \times 1)(10^2) = 800 \text{ in}^4$$
$$I_{yt} = 1(5)(-10) + 3(-5)(-10) + 1(10)(-5) + 3(5)(10) = 200 \text{ in}^4$$
$$V_z = 4 \text{ kips}$$
$$V_y = 10 \text{ kips}$$

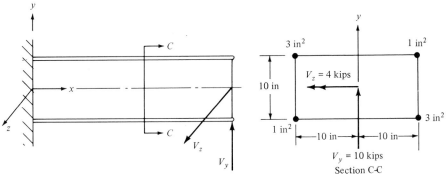

Figure 5.20 Box beam.

The substitution in Eq. (5.25) yields

$$\Delta P = (23.33z - 73.33y)A_f$$

With the above equation, the ΔP on each stringer may be obtained easily by making the proper substitution for the flange area and its corresponding coordinates. The results are shown in Fig. 5.21a. Now the shear flow in each web can be obtained from the increments of the flange loads, as was done in Example 5.6 for the symmetrical box beam. The shear flow in the left-hand web is designated q_0. The shear flows in the rest of the webs are obtained by considering the equilibrium of forces in the spanwise direction and are given in Fig. 5.21a. Now the unknown shear flow q_0 is obtained from the equilibrium of torsional moments. Taking point O as a reference point and summing moments about the x axis through O yield

$$(q_0 - 400)(100) + q_0(100) + (q_0 - 600)(100) + (q_0 - 1000)(100) = 0$$

or

$$400q_0 - 200{,}000 = 0$$

$$q_0 = 500 \text{ lb/in}$$

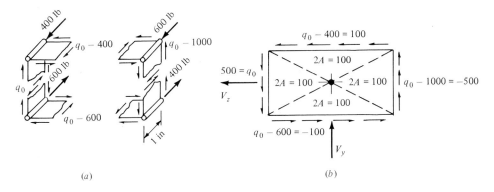

(a)

(b)

Figure 5.21

The final shear-flow results are indicated on Fig. 5.21*b*. The minus sign implies that the wrong direction of shear flow has been assumed.

5.9 SPANWISE TAPER EFFECT

In the preceding analysis of shear stresses in beams, we assumed that the cross section of the beam remained constant. Since in aerospace vehicle structures a minimum weight is always sought, usually the beams are tapered in order to achieve maximum structural efficiency. While this variation in cross section may not cause appreciable errors in the application of the flexure formula for bending stresses, often it causes large errors in the shear stresses determined from Eq. (5.16).

As an illustration, consider the beam shown in Fig. 5.22 which, for simplicity, is assumed to consist of two stringers joined by a vertical web that resists no bending. The resultant axial loads in the stringers must be in the direction of the stringers and must have horizontal components $P_x = M_z/h$. The vertical components of this load which act on the stringers, $P \tan \alpha_1$ and $P \tan \alpha_2$, as shown in Fig. 5.22*b*, resist part of the external applied shear V_y. By designating the shear

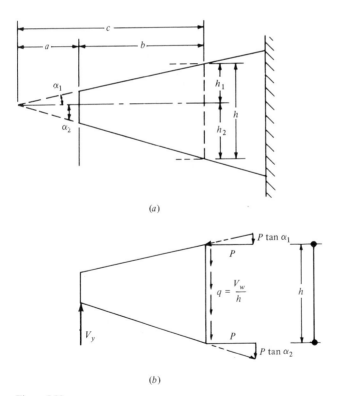

(a)

(b)

Figure 5.22

resisted by the stringers as V_f and that resisted by the webs as V_w,

$$V_y = V_f + V_w \tag{5.26a}$$

$$V_f = P(\tan \alpha_1 + \tan \alpha_2) \tag{5.26b}$$

From the geometry of the beam, $\tan \alpha_1 = h_1/c$, $\tan \alpha_2 = h_2/c$, and $\tan \alpha_1 + \tan \alpha_2 = (h_1 + h_2)/c = h/c$. Substituting this value into Eq. (5.26b) yields

$$V_f = P \frac{h}{c} \tag{5.27}$$

Equation (5.27) will apply for a beam with any system of vertical loads. For the present loading, the value of P is $V_y b/h$. Substituting this value for P into Eq. (5.27) yields

$$V_f = V \frac{b}{c} \tag{5.28}$$

From Eqs. (5.26a) and (5.28) and from the geometry,

$$V_w = V_y \frac{a}{c} \tag{5.29}$$

Equations (5.28) and (5.29) can be expressed in terms of the depths h_0 and h of the beam by making use of the proportion $a/c = h_0/h$:

$$V_w = \frac{V_y h_0}{h} \tag{5.30a}$$

and

$$V_f = V \frac{h - h_0}{h} \tag{5.30b}$$

The shear flow in the webs now can be found by using Eq. (5.16) in conjunction with the shear V_w in Eq. (5.30a). For instance, if we assume that the areas of both stringers in Fig. 5.22 are the same, the shear flow at a section where the distance between the stringers is h may be calculated as follows:

$$q = \frac{V_w}{I} \int y \, dA = \frac{V_w}{Ah^2/2} \frac{Ah}{2} = \frac{V_w}{h} \tag{5.31}$$

When the beam has several stringers, the shear flow may be obtained in a manner similar to that for the two-stringer beam as long as the stringer areas remain constant along the span. If the stringer areas vary along the span and not all vary in the same proportion, Eq. (5.16) cannot be applied.

Example 5.8 Find the shear flows in the web of the beam shown in Fig. 5.23 at 20-in intervals along the span.

SOLUTION The shear flows are obtained by the use of Eqs. (5.30a) and (5.31). The solution of these equations is shown in Table 5.1.

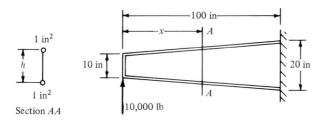

Section AA

Figure 5.23

While slide-rule accuracy is sufficient for shear-flow calculations, the values in Table 5.1 are computed to four significant figures for comparison with a method to be developed later.

Example 5.9 Find the shear flows at section AA of the box beam shown in Fig. 5.24.

SOLUTION The moment of inertia of the cross section at AA about the neutral axis is

$$I = 2(2 + 1 + 1)(5^2) = 200 \text{ in}^4$$

The bending stresses at section AA are

$$\sigma_{xx} = \frac{M_z y}{I} = \frac{400,000 \times 5}{200} = 10,000 \text{ lb/in}^2$$

The horizontal components of the forces acting on the 2-in^2 stringers are 20,000 lb, and the forces on the 1-in^2 stringers are 10,000 lb, as shown in Fig. 5.25a. The vertical components are obtained by multiplying the forces and the tangents of the angles between stringers and the horizontal. The sum of the vertical components of forces on all stringers V_f is 4000 lb, and the remaining shear V_w of 4000 lb is resisted by the shear flows in the webs. If one of the upper webs is cut, as shown in Fig. 5.25b, the shear flows in the webs may be obtained from

$$q = \frac{V_w}{I_z} \int y \, dA$$

Table 5.1

X	h	$\dfrac{h_0}{h}$	V_w	$q = \dfrac{V_w}{h}$
0	10	1	10,000	1,000
20	12	0.8333	8,333	694.4
40	14	0.7143	7,143	510.2
60	16	0.6250	6,250	390.6
80	18	0.5555	5,555	308.6
100	20	0.5	5,000	250.0

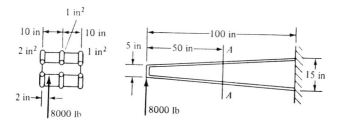

Figure 5.24

where the integral represents the moment of the areas between the cut web and the web under consideration. The change in bending stress on a stringer between the two cross sections 1 in apart is $V_w\,y/I$ when the effect of taper is considered, and the change in axial load on a stringer of area A_f is

$$\Delta P = \frac{V_w}{I}\,yA_f$$

These axial loads are shown in Fig. 5.25b in the same way as they were shown previously for beams with no taper. The equilibrium of forces in the spanwise direction yields the shear flow in terms of q_0 in all the webs, as shown in Fig. 5.25b. Now the shear flow q_0 can be found by summing torsional moments about the z axis through point O for the back section, as shown in Fig. 5.26a. The final shear flow in each web is shown in Fig. 5.26b.

$$\sum T_0 = 0^+ \circlearrowleft$$

$$8000(2) + (q_0 - 100)(200) + q_0(100) + (q_0 + 100)(100)$$

$$-2(500)(10) - 2(500)(20) = 0$$

$$q_0 = 60 \text{ lb/in}$$

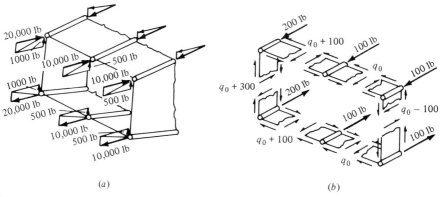

(a)   (b)

Figure 5.25

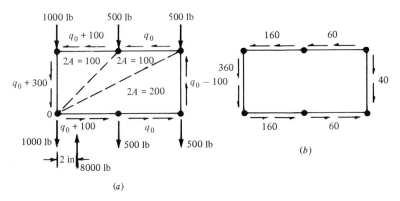

Figure 5.26

5.10 BEAMS WITH VARIABLE STRINGER AREAS

In Sec. 5.9, beams were considered which varied in depth but had stringers whose cross sections were constant. In many aerospace structural beams, the cross-sectional area of the stringer members varies as well as the depth of the beam. If the areas of all the stringer members are increased by a constant ratio, the method of Sec. 5.9 can be used; if the areas at one cross section are not proportional to the areas at another cross section, the method would be considerably in error. The airplane wing section in Fig. 5.27 represents a structure in which the variation in stringer areas must be considered. The stringer areas in this wing are designed in such a manner that the bending stresses are constant along the span. In order to resist the larger bending moments near the root of the wing, the bending strength is augmented by increasing the depth of the wing and the area of spar caps A and B. The stringers which resist the part of the bending moment not resisted by the spar caps have the same area for the entire span. Since the axial stresses on these stringers are the same at every point along the span, the increments of load increase ΔP will be zero except on spar caps A and B. It may be seen from Eq. (5.24) that the shear flow must be constant around the entire leading edge of the wing and changes only at the spar caps. Consequently, the methods of analysis previously used are not applicable to this problem.

The bending stresses and total stringer loads may be calculated for two cross sections of the beam. The actual dimensions and stringer areas for each cross section are used, so that any changes between the cross sections are taken into

Figure 5.27

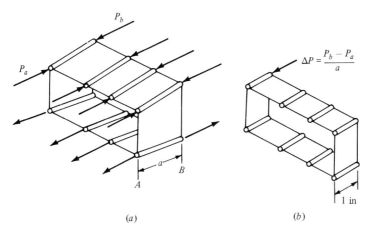

Figure 5.28

consideration. The stringer loads P_a and P_b are shown in Fig. 5.28 for two sections a distance apart. The increase in load in any stringer is assumed to be uniform in length a. The increase in stringer load per unit length along the span is

$$\Delta P = \frac{P_b - P_a}{a} \tag{5.32}$$

This typical force is shown in Fig. 5.28b. Now the shear flow can be obtained from these values of ΔP, as in the previous analysis.

It is seen that the shear force is not used in finding the values of ΔP; consequently, it is not necessary to calculate the vertical components of the stringer loads. The effect of beam taper and changes in stringer area are implemented automatically when the moments of inertia and bending stresses are calculated. Since it is necessary to determine the wing bending stresses at frequent stations along the wing span in order to design the stringers, the terms P_a and P_b can be obtained without too many additional calculations. Thus this method of analysis is often simpler and more accurate than the method which considered variations in depth but not variations in stringer area.

The distance a between two cross sections may be any convenient value. It is common practice to calculate wing bending stresses at intervals of 15 to 30 in along the span. The intervals are quite satisfactory for shear-flow calculations. Note that for very small values of a, small percentage errors in P_a and P_b result in large percentage errors in ΔP. However, if a is too large, the average shear flow obtained between two sections may not be quite the same as the shear flow midway between the sections.

Example 5.10 Find the shear flows in the beam of Fig. 5.23 by the method of using differences in bending stresses.

Table 5.2

x (1)	M (2)	h (3)	$P = \dfrac{M}{h}$ (4)	$P_b - P_a$ (5)	$q = \dfrac{P_b - P_a}{20}$ (6)	Percentage error (7)
10	100,000	11	9,091			
20				13,986	699.3	0.7
30	300,000	13	23,077			
40				10,256	512.8	0.5
50	500,000	15	33,333			
60				7,843	392.1	0.4
70	700,000	17	41,176			
80				6,192	309.6	0.3
90	900,000	19	47,368			

SOLUTION For this two-flange beam, the axial load in the flanges has a horizontal component $P = M/h$. The values of P for various sections are calculated in column 4 of Table 5.2. In computing the shear at any cross section, values of the axial loads at cross sections 10 in on either side are found. The free-body diagrams are shown in Fig. 5.29. The circled numbers represent stations, or the distance from the cross section to the left end of the beam. The difference in horizontal loads on the upper part of the beam between the cross sections 20 in apart must be balanced by the resultant of the horizontal shear flow, $20q$. The differences in axial loads are tabulated in column 5, and the shear flows $q = (P_b - P_a)/20$ are shown in column 6. The value of the shear flow at station 20 thus is assumed to be equal to the average horizontal shear between stations 10 and 30. Even though the shear

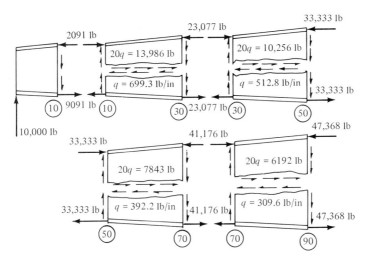

Figure 5.29

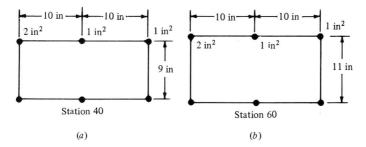

Figure 5.30

does not vary linearly along the span, the error in this assumption is only 0.7 percent, as found by comparison with the exact value obtained in Table 5.1. This error is even smaller at the other stations.

Example 5.11 Find the shear flows at cross section AA of the box beam shown in Fig. 5.24 by considering the difference in bending stresses at cross sections 10 in on either side of AA.

SOLUTION The moment of inertia at station 40 (40 in from the left end) is found from the dimensions shown in Fig. 5.30a. The bending stresses at station 40, resulting from the bending moment of 320,000 in · lb, are

$$\sigma_{xx} = \frac{M_z y}{I_z} = \frac{320,000(4.5)}{162} = 8888 \text{ lb/in}^2$$

The loads on the 1-in^2 areas are 8888 lb, and the loads on the 2-in^2 areas are 17,777 lb, as shown in Fig. 5.31a. The moment of inertia at station 60 is found from the dimensions shown in Fig. 5.31b:

$$I_z = 8(5.5^2) = 242 \text{ in}^4$$

The bending stresses resulting from the bending moment of 480,000 in · lb

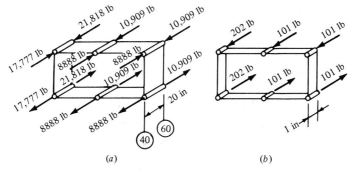

Figure 5.31

are

$$\sigma_{xx} = \frac{M_z y}{I_z} = \frac{480,000 \times 5.5}{242} = 10,909 \text{ lb/in}^2$$

The loads on the stringers are 10,909 and 21,818 lb, as shown in Fig. 5.31a. The increments of flange loads ΔP in a 1-in length are found from Eq. (5.32). For the area of 2 in^2,

$$\Delta P = \frac{21,818 - 17,777}{20} = 202 \text{ lb}$$

For the area of 1.0 in^2

$$\Delta P = \frac{10,909 - 8888}{20} = 101 \text{ lb}$$

The values of ΔP are shown in Fig. 5.31b. The remaining solution is identical to that of Example 5.9. The values of ΔP are 1 percent higher than the exact values shown in Fig. 5.25b. The reason for this small discrepancy is that the average shear flow between stations 40 and 60 is 1 percent higher than the shear flow at station 50. The other assumptions used in the two solutions are identical. The method of using differences in bending stresses automatically considers the effects of the shear carried by the stringers, and it is not necessary to calculate the angles of inclination of the stringers. It is, however, necessary to find the torsional moments about the proper axis if the stringer forces are omitted in the moment equation.

5.11 AIRY STRESS FUNCTION

It is shown in Chap. 3 that a stress field describes the exact state of stress in a solid if and only if it satisfies the conditions of equilibrium, compatibility, and prescribed boundary stresses. For two-dimensional stress problems in the absence of body forces, the equilibrium and compatibility equations are

$$\begin{aligned} \sigma_{xx, x} + \sigma_{xy, y} &= 0 \\ \sigma_{xy, x} + \sigma_{yy, y} &= 0 \end{aligned} \quad \text{(equilibrium)} \quad (3.11)$$

$$\sigma_{xx, xx} + 2\sigma_{xy, xy} + \sigma_{yy, yy} = 0 \quad \text{(compatibility)} \quad (3.29)$$

If a stress function $\Phi(x, y)$ is assumed such that the stresses in a solid are defined by

$$\begin{aligned} \sigma_{xx} &= \Phi_{,yy} \\ \sigma_{yy} &= \Phi_{,xx} \\ \sigma_{xy} &= -\Phi_{,xy} \end{aligned} \quad (5.33)$$

then Eqs. (3.11) are identically satisfied. Upon substituting Eqs. (5.33) into Eq. (3.29), the following is obtained:

$$\Phi_{,xxxx} + 2\Phi_{,xxyy} + \Phi_{,yyyy} = 0 \tag{5.34}$$

The solution of Eq. (5.34) satisfies both the equilibrium and the compatibility equations and therefore gives a possible stress field in an elastic solid. In order for the obtained stress field to describe the true state of stress for a specific problem, the prescribed boundary conditions must be satisfied also.

The solution of Eq. (5.34) may be obtained by two methods: the polynomial solution and the Fourier series solution.

Polynomial solution If the stress function $\Phi(x, y)$ is assumed to have a solution of the form

$$\Phi(x, y) = \Phi_1(x, y) + \Phi_2(x, y) + \cdots + \Phi_n(x, y) = \sum_{n=1}^{N} \Phi_n(x, y) \tag{5.35}$$

where

$$\Phi_n(x, y) = \sum_{i=0}^{n} A_{in} x^{n-i} y^i \tag{5.36}$$

then, by considering various degrees of polynomials and suitably adjusting their coefficients A_{in}, a number of practical problems may be solved. For instance, taking

$$\Phi(x, y) = \sum_{n=1}^{4} \Phi_n(x, y) = \sum_{n=1}^{4} \left(\sum_{i=0}^{n} A_{in} x^{n-i} y^i \right) \tag{5.37}$$

and assuming all coefficients to be zero except A_{22} and A_{02} yield

$$\sigma_{xx} = \Phi_{,yy} = 2A_{22}$$
$$\sigma_{yy} = \Phi_{,xx} = 2A_{02} \tag{5.38}$$
$$\sigma_{xy} = 0$$

By examining Eq. (5.37), it may be shown easily that this is the solution of a two-dimensional solid loaded as shown in Fig. 5.32a. If, in addition, A_{02} is taken as zero, then the solution corresponds to that of Fig. 5.32b. Figures 5.32c and d represent the conditions where A_{12} and A_{34}, respectively, are the only nonzero set of coefficients in Eq. (5.36).

To illustrate the use of the Airy stress function in the solution of practical problems, let us consider the beam problem shown in Fig. 5.33a. It is assumed that the external shear load V is distributed at the surface $x = x_0$ according to the strength-of-material parabolic shear-stress distribution, and the axial force S is applied uniformly over that surface.

Also, it is assumed that the bending stresses vary linearly, as shown in Fig.

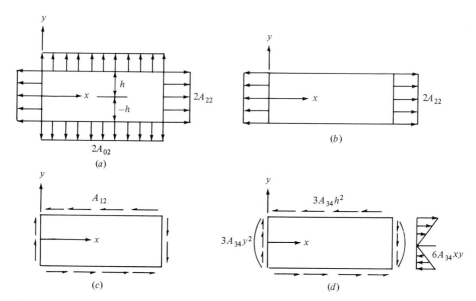

Figure 5.32

5.33*b*. Through comparison of Figs. 5.32*b*, *c*, and *d* and 5.33*b*, it can be seen that the loading of Fig. 5.33*b* may be obtained by superposing the loads shown in Fig. 5.32*b*, *c*, and *d*. Thus, the stress field for the beam in Fig. 5.33*a* is

$$\sigma_{xx} = 2A_{22} + 6A_{34}xy$$
$$\sigma_{yy} = 0 \tag{5.39}$$
$$\sigma_{xy} = A_{12} - 3A_{34}y^2$$

The boundary conditions which must be satisfied for the given beam in Fig. 5.33*a* are

$$\sigma_{xy} = 0 \quad \text{at} \quad y = \pm h$$
$$-\int_{-h}^{h} \sigma_{xy}\, dy = V \quad \text{at} \quad x = 0 \tag{5.40}$$
$$\int_{-h}^{h} \sigma_{xx}\, dy = S \quad \text{at} \quad x = 0$$

Utilizing Eqs. (5.39) in Eqs. (5.40) yields

$$-A_{12} - 3A_{34}h^2 = 0$$
$$\int_{-h}^{h} (-A_{12} - 3A_{34}y^2)\, dy = -V \tag{5.41}$$
$$\int_{-h}^{h} 2A_{22}\, dy = S$$

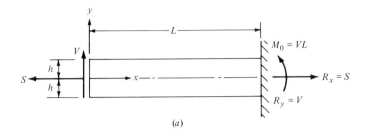

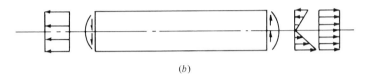

(b)

Figure 5.33

Solving Eq. (5.41) simultaneously gives

$$A_{12} = \frac{3}{4} \frac{V}{h}$$

$$A_{34} = -\frac{V}{4h^3}$$

$$A_{22} = \frac{S}{4h}$$

Hence the true beam stresses are

$$\sigma_{xx} = \frac{S}{2h} - \frac{3}{2} \frac{V}{h^3} xy = \frac{S}{2h} - \frac{V}{I} xy$$

$$\sigma_{yy} = 0 \tag{5.42}$$

$$\sigma_{xy} = -\frac{3}{4} \frac{V}{h} + \frac{3}{4} \frac{V}{h^3} y^2 = -\frac{V}{2I} (h^2 - y^2)$$

Fourier series solution The Fourier series technique is used whenever the load distribution is discontinuous over a portion of the solid, as shown in Fig. 5.34, for example. This method assumes that the solution of Eq. (5.34) may be expressed as

$$\Phi(x, y) = \Phi_x(x)\Phi_y(y) \tag{5.43}$$

where $\Phi_x(x)$ may be taken, in the form of Fourier series, as

$$\Phi_x(x) = \sin \frac{m\pi x}{L}$$

or $\tag{5.44}$

$$\Phi_x(x) = \cos \frac{m\pi x}{L}$$

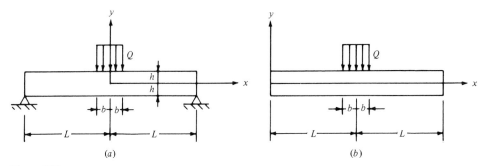

Figure 5.34

Utilizing Eq. (5.44a) or (5.44b) in Eq. (5.43) and then substituting into Eq. (5.34) yield

$$\frac{d^4\Phi_y}{dy^4} - 2\beta^2 \frac{d^2\Phi_y}{dy^2} + \beta^4\Phi_y = 0 \tag{5.45}$$

where $\beta = m\pi/L$.

Equation (5.45) is a fourth-order homogeneous differential equation with constant coefficients. Its solution can be obtained easily in terms of hyperbolic functions as

$$\Phi_y = C_1 \sinh \beta y + C_2 \cosh \beta y + C_3 y \sinh \beta y + C_4 y \cosh \beta y \tag{5.46}$$

Constants $C_1, C_2, \cdots$ are determined from the boundary conditions of the solid under consideration. The total solution of Eq. (5.34) thus becomes

$$\Phi(x, y) = (C_1 \sinh \beta y + C_2 \cosh \beta y + C_3 y \sinh \beta y$$
$$+ C_4 y \cosh \beta y) \sin \frac{m\pi x}{L} \tag{5.47a}$$

or

$$\Phi(x, y) = (C_1 \sinh \beta y + C_2 \cosh \beta y + C_3 y \sinh \beta y$$
$$+ C_4 y \cosh \beta y) \cos \frac{m\pi x}{L} \tag{5.47b}$$

The choice of the trigonometric function depends on the symmetry of the loading. For instance, if the loading is symmetrical about the chosen y axis of the beam, as shown in Fig. 5.34a, then the cosine function must be used. For the beam of Fig. 5.34b, the sine function must be assumed because of the antisymmetry of the loading about the chosen y axis.

PROBLEMS

5.1 Find the maximum tensile and maximum compressive stresses resulting from bending of the beam shown in Fig. P5.1. Find the distribution of shear stresses over the cross section at the section where the shear is a maximum, considering points in the cross section at vertical intervals of 1 in.

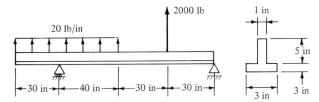

Figure P5.1

5.2 Find the maximum shear and bending stresses in the beam cross section shown in Fig. P5.2 if the shear V is 10,000 lb and the bending moment M is 400,000 in · lb. Both angles have the same cross section. Assume the web to be effective in resisting bending stresses.

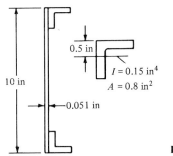

Figure P5.2

5.3 Find the shear stress and the shear-flow distribution over the cross section of the beam shown in Fig. P5.3. Assume the web to be ineffective in resisting bending and the stringer areas to be concentrated at points.

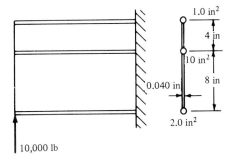

Figure P5.3

5.4 Each of the five upper stringers has an area of 0.4 in², and each of the five lower stringers hás an area of 0.8 in². Find the shear flows in all the webs if the vertical shearing force is 12,000 lb.

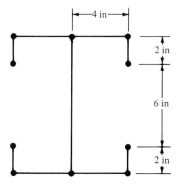

Figure P5.4

5.5 Each of the six stringers of the cross section shown in Fig. P5.5 and P5.6 has an area of 0.5 in². Find the shear flows in all webs and the location of the shear center for a vertical shearing force of 10,000 lb.

5.6 Find the shear flows in all webs in Fig. P5.5 and P5.6 for a horizontal shearing force of 3000 lb. Each stringer has an area of 0.5 in².

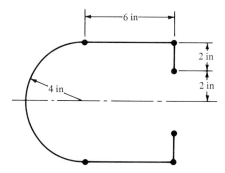

Figure P5.5 and P5.6

5.7 Find a general expression for the shear-flow distribution around the circular tube shown in Fig. P5.7. Assume the wall thickness t to be small compared with the radius R.

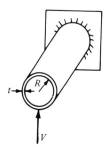

Figure P5.7

5.8 Use Eqs. (5.17) and (5.18) to find the shear flow in the webs of the two-stringer beam shown in Fig. P5.8 under the action of a vertical shear V_y.

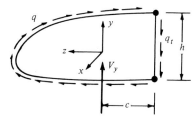

Figure P5.8

5.9 Find the shear-flow distribution for the section shown in Fig. P5.9. Each stringer has a cross-sectional area of 1.5 in².

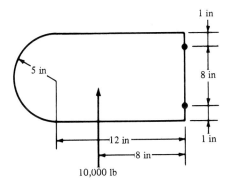

Figure P5.9

5.10 Find the shear flows in the webs of the box beam shown in Fig. P5.10 if the area is symmetrical about a horizontal centerline.

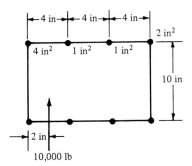

Figure P5.10

5.11 Find the shear flows in the webs of the beam shown in Fig. P5.11 and P5.12. All stringers have areas of 1.0 in².

5.12 Assume that the two right-hand stringers in Fig. P5.11 and P5.12 have areas of 3.0 in² and the other stringers have areas of 1.0 in². Find the shear flows in the webs by two methods.

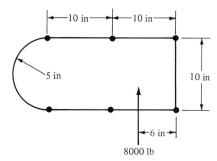

8000 lb   **Figure 5.11** and **P5.12**

5.13 Find the shear flows in all webs if the two right-hand stringers shown in Fig. P5.13 have areas of 1.5 in^2 and the other stringers have areas of 0.5 in^2.

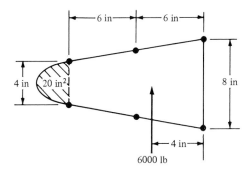

6000 lb   **Figure P5.13**

5.14 Find the shear-flow distribution in all webs shown in Fig. P5.14. All parts of the cross section resist bending stresses.

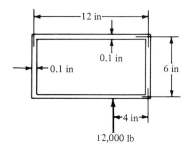

12,000 lb   **Figure P5.14**

5.15 Solve Example 5.8 if the beam depth varies from 5 in at the free end to 15 in at the support. (See Fig. P5.15.)

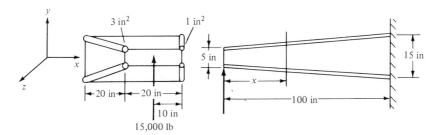

Figure P5.15

5.16 Find the shear flows for the cross section at $x = 50$ in. Consider only this one cross section, but calculate the torsional moments by two methods.

(a) Select the torsional axis arbitrarily, and calculate the in-plane components of the flange loads.

(b) Take moments about a torsional axis joining the centroids of the various cross sections.

5.17 Repeat Prob. 5.16 if there is an additional chordwise load of 6000 lb acting to the left at the center of the tip cross section.

5.18 A cantilever beam 30 in long carries a vertical load of 1000 lb at the free end. The cross section is rectangular and is 6 by 1 in. Find the maximum bending stress and the location of the neutral axis if (a) the 6-in side is vertical, (b) the 6-in side is tilted 5° from the vertical, and (c) the 6-in side is tilted 10° from the vertical.

5.19 A horizontal beam with a square cross section resists vertical loads. Find the angle of the neutral axis with the horizontal if one side of the beam makes an angle θ with the horizontal. At what angle should the beam be placed for the bending stress to have a minimum value?

5.20 Find the bending stresses and stringer loads for the box beam whose cross section is shown in Fig. P5.20 if $M_z = 100,000$ and $M_y = -40,000$ in · lb. Assume the areas of the stringer members are as follows:

(a) $a = b = c = d = 2$ in^2
(b) $a = b = 3$ in^2, $c = d = 1$ in^2
(c) $a = d = 3$ in^2 $b = c = 1$ in^2
(d) $a = c = 3$ in^2, $b = d = 1$ in^2
(e) $a = c = 1$ in^2, $b = d = 3$ in^2

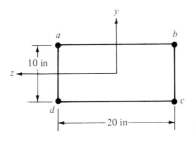

Figure P5.20

5.21 A beam with the cross section shown in Fig. P5.21 resists a bending moment $M_z = 100$ in · lb. Calculate the bending stresses at points A, B, and C.

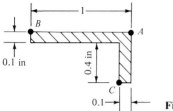

Figure P5.21

5.22 The box beam shown in Fig. P5.22 resists bending moments of $M_z = 1,000,000$ and $M_y = 120,000$ in · lb. Find the bending stress in each stringer member. Assume that the webs are ineffective in bending and the areas and coordinates of the stringers are as follows:

No.	Area, in²	z, in	y, in
1	1.8	2.62	8.3
2	0.4	− 10.81	9.12
3	0.8	− 24.70	9.75
4	2.3	− 24.70	− 1.3
5	1.0	2.62	− 1.2

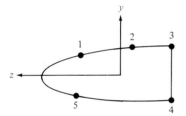

Figure 5.22

5.23 Find the shear flows at the cross section shown in Fig. P5.23 and P5.24 for $x = 50$ in. Consider only the one cross section, and calculate the in-plane components of the stringer loads.

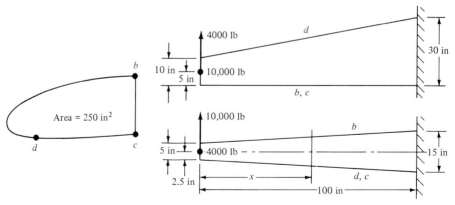

Figure P5.23 and P5.24

5.24 Repeat Prob. 5.23, using the differences in stringer loads shown in Fig. P5.23 and P5.24 at the cross section for $x = 40$ and $x = 60$ in.

5.25 Calculate the shear flows in the webs of the cross section shown in Fig. P5.25 and P5.26 at $x = 50$ in. Assume the flange areas as follows:

(a) $a = b = 3$ in^2, $c = d = 1$ in^2
(b) $a = c = 1$ in^2, $b = d = 3$ in^2
(c) $a = c = 3$ in^2, $b = d = 1$ in^2

Consider only the one cross section, and calculate the in-plane components of the flange loads.

5.26 Repeat Prob. 5.25, using the differences in flange loads at the cross section shown in Fig. P5.25 and P5.26 for $x = 40$ and 60 in. Use a torsional axis joining the centroids of the cross sections.

Figure P5.25 and **P5.26**

5.27 Find the shear flows at station 100 of the fuselage shown in Fig. P5.27. Assume all stringer areas to be 1 in^2.

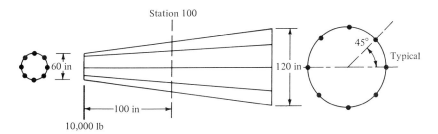

Figure P5.27

5.28 Using the Airy stress function, find the stresses in the beams shown in Fig. P5.28.

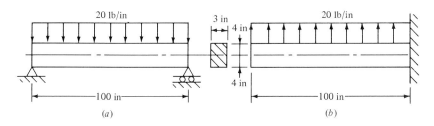

Figure P5.28

5.29 A circular, thin plate is under the action of uniformly distributed pressure applied around the outer edge. Find the stresses, using the Airy stress function. Assume the plate thickness is equal to t. (See Fig. P5.29.)

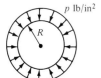

Figure P5.29

DEFLECTION ANALYSIS OF STRUCTURAL SYSTEMS

6.1 INTRODUCTION

The most important applications of the methods for calculating displacements (deflections) are in the analysis of redundant (indeterminate) structural systems, as is demonstrated here and in later chapters.

The deflections of most engineering structures are small and very seldom are used as an important design criterion. However, the relative rigidity of various elements in redundant structural systems affects the stress distribution in the structure; therefore, it is necessary to consider the deformations in the analysis of such structural systems.

The methods of deflection analyses presented in this chapter are Castigliano's method, the Rayleigh-Ritz method, and the finite difference method. The finite-element method is presented in a separate chapter. Also presented here is the unit-load method for the analysis of simple redundant structures. The energy methods treated are derived in accordance with the principles of virtual and complementary virtual work associated with virtual displacements and virtual forces, respectively. We assume small strains and corresponding small displacements in all the developments of this chapter. In addition, the material is assumed perfectly elastic.

6.2 WORK AND COMPLEMENTARY WORK: STRAIN AND COMPLEMENTARY STRAIN ENERGIES

Consider a structure to be acted on by a set of generalized forces $Q_i (i = 1, 2, \ldots, n)$ which result in a corresponding set of generalized displacements $q_i (i = 1, 2, \ldots, n)$. The force-displacement relationship for a typical force Q_n and its corresponding displacement q_n are shown graphically in Fig. 6.1a. The area under the curve represents the work W done by force Q_n in moving through the corresponding displacement q_n. The area above the curve is defined as the *complementary work* $\bar{W}$. For a system which is in a state of static equilibrium, and if heat dissipation is neglected, then based on the conservation of energy, it can be stated that the work done on the system is equal to the strain energy stored in the system, or

$$W = U \tag{6.1}$$

and
$$\bar{W} = \bar{U} \tag{6.2}$$

where U and $\bar{U}$ are the strain energy and complementary strain energy, respectively. Figure 6.1b shows a graphical representation of these quantities.

6.3 PRINCIPLE OF VIRTUAL DISPLACEMENTS AND RELATED THEOREMS

In variational mechanics, a structural system is imagined to have gone through a set of infinitesimal displacements consistent with the constraints, when, in reality, no such displacements exist. These fictitious movements of the structure are commonly referred to as the *virtual displacements*, and the corresponding work is called the *virtual work*.

Consider that a structure is given a small variation in virtual displacement δq_n, as shown in Fig. 6.1. This induces a variation in virtual work δW and a

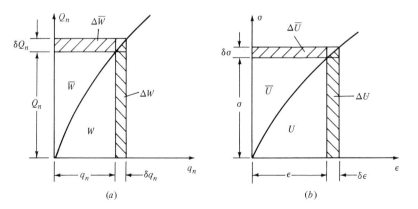

Figure 6.1 (a) work and complementary work; (b) strain and complementary strain energies.

corresponding variation in strain energy δU, as indicated by the vertical strips in Fig. 6.1. If we assume that the external applied forces $\{Q\}$ and the induced internal stresses $\{\Sigma\}$ remain constant during the virtual displacements, then the changes in virtual work and virtual strain energy can be obtained readily from Fig. 6.1 and are given by

$$\Delta W = \delta W = Q_n \, \delta q_n \qquad (6.3)$$

$$\Delta U^* = \delta U^* = \sigma \delta \epsilon \qquad (6.4)$$

where δ denotes the first variation, and it operates in the same manner as the differential operator d, and δU^* is the variation in strain energy per unit volume (energy density), i.e.,

$$\delta U = \int_V \delta U^* \, dV \qquad \text{(volume integral)} \qquad (6.5)$$

If the variations in δW and δU are considered for the variation of all displacements q_i ($i = 1, 2, \ldots, n$) consistent with the constraints, then

$$\delta W = \sum_{i=1}^{n} Q_i \, \delta q_i \qquad (6.6)$$

or, in matrix compact form, Eq. (6.6) becomes

$$\delta W = \lfloor \delta q \rfloor \{Q\} \qquad (6.7)$$

where the symbols $\lfloor \ \rfloor$ and $\{ \ \}$ indicate row and column matrices, respectively. Likewise, the corresponding variation in strain energy density can be expressed as

$$\delta U^* = \lfloor \delta E \rfloor \{\Sigma\} \qquad (6.8)$$

where the strain and stress fields, in general, are given by

$$\lfloor \delta E \rfloor = \delta [\epsilon_{xx} \quad \epsilon_{yy} \quad \epsilon_{zz} \quad \epsilon_{xy} \quad \epsilon_{xz} \quad \epsilon_{yz}] \qquad (6.9)$$

and

$$\{\Sigma\} = \begin{bmatrix} \sigma_{xx} \\ \sigma_{yy} \\ \sigma_{zz} \\ \sigma_{xy} \\ \sigma_{xz} \\ \sigma_{yz} \end{bmatrix} \qquad (6.10)$$

The principle of virtual displacements states that an elastic deformable structural system is in a state of equilibrium if the virtual work δW done by forces $\{Q\}$ is equal to the virtual strain energy δU for every arbitrary virtual displacement consistent with the constraints of the structure. Mathematically, this principle is expressed as

$$\delta W = \delta U \tag{6.11}$$

or

$$\lfloor \delta q \rfloor \{Q\} = \int_V \lfloor \delta E \rfloor \{\Sigma\} \, dV \tag{6.12}$$

where the external applied forces $\{Q\}$ and the internal stresses $\{\Sigma\}$ are assumed to be in equilibrium and the coordinate displacements $\{\delta q\}$ and the strains $\{\delta E\}$ satisfy the compatibility condition

$$\{\delta E\} = [\lambda]\{\delta q\} \tag{6.13}$$

Substituting Eq. (6.13) into Eq. (6.12) yields

$$\lfloor \delta q \rfloor \{Q\} = \int_V \lfloor \delta q \rfloor [\lambda]^T \{\Sigma\} \, dV \tag{6.14}$$

or, in expanded form,

$$\lfloor \delta q_1 \quad \delta q_2 \cdots \delta q_i \cdots \delta q_n \rfloor \begin{bmatrix} Q_i \\ Q_2 \\ \vdots \\ Q_i \\ \vdots \\ Q_n \end{bmatrix} = \int_V \lfloor \delta q_1 \quad \delta q_2 \cdots \delta q_i \cdots \delta q_n \rfloor$$

$$\times \begin{bmatrix} \lambda_{11} & \lambda_{12} & \lambda_{13} & \cdots & \lambda_{1n} \\ \lambda_{21} & \lambda_{22} & \lambda_{23} & \cdots & \lambda_{2n} \\ \lambda_{31} & \lambda_{32} & \lambda_{33} & \cdots & \lambda_{3n} \\ \cdots & \cdots & \cdots & \cdots & \cdots \\ \lambda_{61} & \lambda_{62} & \lambda_{63} & \cdots & \lambda_{6n} \end{bmatrix}^T \begin{bmatrix} \sigma_1 \\ \sigma_2 \\ \sigma_3 \\ \vdots \\ \sigma_6 \end{bmatrix} dV \tag{6.15}$$

where the superscript T denotes matrix transpose (i.e., rows and corresponding columns are exchanged).

If every virtual displacement δq_r $[(r = 1, 2, \ldots n), r \neq i]$ is set equal to zero with the ith displacement δq_i allowed to be a unit displacement, then it can be seen easily from Eq. (6.15) that Eq. (6.14) reduces to

$$1 \cdot Q_i = \int_V 1 \cdot \lfloor \lambda \rfloor \{\Sigma\} \, dV$$

or

$$Q_i = \int_V \lfloor \lambda_i \rfloor \{\Sigma\} \, dV \qquad (6.16)$$

where the row matrix $\lfloor \lambda_i \rfloor$ is defined by Eq. (6.13), which under the condition

$$\delta q_r = \begin{cases} 0 & \text{for } r \neq i \\ 1 & \text{for } r = i \end{cases}$$

becomes

$$\lfloor \lambda_i \rfloor = \lfloor \delta E_i \rfloor \qquad (6.17)$$

Equation (6.16) describes what is commonly referred to as the *unit-displacement method*. See Argyris and Kelsey[23] for more detailed information on this theorem.

For conservative structural systems, Eq. (6.11) may be expressed in the form

$$\delta(U + P) = \delta V = 0 \qquad (6.18)$$

where δV is the first variation of the total potential energy and

$$\delta W = -\delta P = \sum_{i=1}^{n} Q_i \, \delta q_i$$

Equation (6.18) is the principle of stationary total potential and may be shown readily to be equivalent to stating that

$$\frac{\partial V}{\partial q_i} = 0 \qquad (i = 1, 2, \ldots, n) \qquad (6.19)$$

which is the basis of the Rayleigh-Ritz method in structural analysis.

Another theorem that is based on the principle of virtual displacements is Castigliano's first theorem.

From Eqs. (6.3) and (6.11),

$$\delta U = \lfloor \delta q \rfloor \{Q\} \qquad (6.20)$$

By noting that δU is a function of the coordinate generalized displacements $[\delta q_i \ (i = 1, 2, \ldots, n)]$ and using the Taylor series expansion, Eq. (6.20) becomes

$$\delta U = \sum_{i=1}^{n} \frac{\partial U}{\partial q_i} \delta q_i + \frac{1}{2} \sum_{i=1}^{n} \sum_{j=1}^{n} \frac{\partial^2 U}{\partial q_i \, \partial q_j} \delta q_i \, \delta q_j + \cdots$$

or, in matrix form,

$$\delta U = \lfloor \delta q \rfloor \left\{ \frac{\partial U}{\partial q} \right\} + \tfrac{1}{2} \lfloor \delta q \rfloor [S] \{\delta q\} + \cdots \qquad (6.21)$$

where the sumbol ∂ denotes partial differentiation and the matrix $[S]$ is a stiffness matrix whose elements are defined by

$$S_{ij} = \frac{\partial^2 U}{\partial q_i \, \partial q_j} \qquad (i, j = 1, 2, \ldots, n) \qquad (6.22)$$

If only first-order variation in δq is retained in Eq. (6.21), then from Eq. (6.20),

$$\lfloor \delta q \rfloor \{Q\} = \lfloor q \rfloor \left\{ \frac{\partial U}{\partial q} \right\}$$

or

$$\lfloor \delta q \rfloor \left(\{Q\} - \left\{ \frac{\partial U}{\partial q} \right\} \right) = 0$$

Since $[\delta q]$ are independent displacements,

$$\{Q\} - \left\{ \frac{\partial U}{\partial q} \right\} = 0$$

or

$$\{Q\} = \left\{ \frac{\partial U}{\partial q} \right\} \tag{6.23}$$

For $i = r$, Eq. (6.23) states that

$$Q_r = \frac{\partial U}{\partial q_r} \tag{6.24}$$

Equation (6.24) is Castigliano's first theorem.

6.4 PRINCIPLE OF VIRTUAL FORCES AND RELATED THEOREMS

The treatment of virtual forces in variational mechanics is analogous to that of the virtual displacements presented in Sec. 6.3. Thus, from Fig. 6.1 it can be seen easily that for a given small variation in virtual force δQ_n, consistent with the static equilibrium conditions, the corresponding variations in complementary virtual work and complementary virtual strain energy density are

$$\Delta \bar{W} = \delta \bar{W} = q_n \, \delta Q_n \tag{6.25}$$

and

$$\Delta \bar{U}^* = \delta \bar{U}^* = \epsilon \, \delta \sigma \tag{6.26}$$

where

$$\delta \bar{U} = \int_V \delta \bar{U}^* \, dV \qquad \text{(volume integral)} \tag{6.27}$$

If we consider the variations in $\delta \bar{W}$ and $\delta \bar{U}$ for the variation of all forces Q_i ($i = 1, 2, \ldots, n$), Eqs. (6.25) and (6.26) become

$$\delta \bar{W} = \lfloor \delta Q \rfloor \{q\} \tag{6.28}$$

and

$$\delta \bar{U} = \int_V \lfloor \delta \Sigma \rfloor \{E\} \, dV \tag{6.29}$$

The principle of virtual forces[24] (principle of virtual work) dictates that an elastic structural system is in a compatible state of deformation if for every arbitrary virtual force δQ, the complementary virtual work is equal to the complementary strain energy consistent with static conditions of equilibrium. Mathematically, the principle may be expressed as

$$\delta \bar{W} = \delta \bar{U} \tag{6.30}$$

or

$$\lfloor \delta Q \rfloor \{q\} = \int_V \lfloor \delta \Sigma \rfloor \{E\} \, dV \tag{6.31}$$

where the displacements $\{q\}$ and the corresponding strains $\{E\}$ are compatible and the virtual forces $\{\delta Q\}$ and corresponding stresses $\{\delta \Sigma\}$ satisfy the equilibrium condition

$$\{\delta \Sigma\} = [\Phi]\{\delta Q\} \tag{6.32}$$

Substituting Eq. (6.32) into Eq. (6.31) yields

$$\lfloor \delta Q \rfloor \{q\} = \int_V \lfloor \delta Q \rfloor [\Phi]^T \{E\} \, dV \tag{6.33}$$

In Eq. (6.33), if every virtual force δQ_r $[(r = 1, 2, \ldots, n), r \neq i]$ is set equal to zero with the ith force δQ_i given a unit value, then the equation becomes

$$q_i = \int_V \lfloor \Phi_i \rfloor \{E\} \, dV \tag{6.34}$$

where the row matrix $[\Phi]$ is defined by Eq. (6.32), which under the condition

$$\delta Q_r = \begin{cases} 0 & \text{for } r \neq i \\ 1 & \text{for } r = i \end{cases}$$

becomes

$$\lfloor \Phi_i \rfloor = \lfloor \delta \Sigma_i \rfloor \tag{6.35}$$

Equation (6.34) is referred to by Argyris and Kelsey as the *unit-load-method*.

The principle of stationary total complementary potential may be deduced from Eq. (6.18) and is given by

$$\delta(\bar{P} + \bar{U}) = \delta \bar{V} = 0 \tag{6.36}$$

where

$$\delta \bar{W} = -\delta \bar{P} = \sum_{i=1}^{n} q_i \, \delta Q_i \tag{6.37}$$

Equation (6.36) may be shown to be equivalent to

$$\frac{\partial \bar{V}}{\partial Q_i} = 0 \qquad i = 1, 2, \ldots, n \tag{6.38}$$

which represents the compatibility conditions at all coordinates i.

If we consider Eq. (6.30), from Eq. (6.28) the complementary strain energy can be expressed as

$$\delta \bar{U} = \lfloor \delta Q \rfloor \{q\} \tag{6.39}$$

Expanding Eq. (6.39) in a Taylor series yields

$$\delta \bar{U} = \sum_{i=1}^{n} \frac{\partial \bar{U}}{\partial Q_i} \delta Q_i + \frac{1}{2} \sum_{i=1}^{n} \sum_{j=1}^{n} \frac{\partial 2 \bar{U}}{\partial Q_i \, \partial Q_j} \delta Q_i \, \delta Q_j + \cdots$$

or, in matrix form,

$$\delta \bar{U} = \lfloor \delta Q \rfloor \left\{ \frac{\partial \bar{U}}{\partial Q} \right\} + \tfrac{1}{2} [\delta Q][\alpha]\{\delta Q\} + \cdots \tag{6.40}$$

where the matrix $[\alpha]$ is the flexibility matrix whose coefficients are defined by

$$\alpha_{ij} = \frac{\partial^2 \bar{U}}{\partial Q_i \, \partial Q_j} \qquad (i, j = 1, 2, \ldots, n) \tag{6.41}$$

If only first-order variation in δQ is retained in Eq. (6.40), then from Eq. (6.39)

$$\lfloor \delta Q \rfloor \{q\} = \lfloor \delta Q \rfloor \left\{ \frac{\partial \bar{U}}{\partial Q} \right\}$$

or

$$\lfloor \delta Q \rfloor \left(\{q\} - \left\{ \frac{\partial \bar{U}}{\partial Q} \right\} \right) = 0 \tag{6.42}$$

Since $\lfloor \delta Q \rfloor$ are independent arbitrary forces, for Eq. (6.42) to be satisfied, the following must hold true:

$$\{q\} - \left\{ \frac{\partial \bar{U}}{\partial Q} \right\} = 0$$

or

$$\{q\} = \left\{ \frac{\partial \bar{U}}{\partial Q} \right\} \tag{6.43}$$

For $i = r$, Eq. (6.43) is equivalent to

$$q_r = \frac{\partial \bar{U}}{\partial Q_r} \tag{6.44}$$

which is the second theorem of Castigliano.

6.5 LINEAR ELASTIC STRUCTURAL SYSTEMS[10]

For structural systems whose behavior is linearly elastic, as shown in Fig. 6.2, the work done on the system is equal to the complementary work; likewise, the strain energy stored is equal to the complementary strain energy. As can be seen easily from Fig. 6.2, the expressions for the work and strain energy and their counterparts are

$$W = \bar{W} = \frac{1}{2} \lfloor Q_i \quad Q_2 \quad \cdots \quad Q_n \rfloor \begin{bmatrix} q_1 \\ q_2 \\ \vdots \\ q_n \end{bmatrix} \qquad (6.45)$$

or, in compact matrix form,

$$W = \bar{W} = \tfrac{1}{2} \lfloor Q \rfloor \{q\} \qquad (6.45a)$$

and

$$U = \bar{U} = \frac{1}{2} \int_V \lfloor \epsilon_{xx} \quad \epsilon_{yy} \quad \epsilon_{zz} \quad \epsilon_{xy} \quad \epsilon_{xz} \quad \epsilon_{yz} \rfloor \begin{bmatrix} \sigma_{xx} \\ \sigma_{yy} \\ \sigma_{zz} \\ \sigma_{xy} \\ \sigma_{xz} \\ \sigma_{yz} \end{bmatrix} dV \qquad (6.46)$$

or, in compact matrix form,

$$U = \bar{U} = \tfrac{1}{2} \int_V \lfloor E \rfloor \{\Sigma\} \, dV \qquad (6.46a)$$

where Q_i ($i = 1, 2, \ldots, n$) are the external applied loads and q_i ($i = 1, 2, \ldots, n$) are

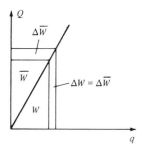 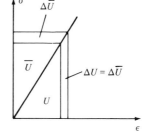

Figure 6.2 Linear elastic behavior.

the corresponding nodal coordinate displacements; the matrices $\lfloor E \rfloor$ and $\{\Sigma\}$ are the strain and stress fields, respectively.

By utilization of Hooke's law, Eq. (6.46) may be expressed in terms of stresses or strains alone as follows:

$$U = \bar{U} = \tfrac{1}{2} \int_V \lfloor \Sigma \rfloor [N]\{\Sigma\} \, dV \tag{6.47}$$

or

$$U = \bar{U} = \tfrac{1}{2} \int [\epsilon][\Theta]\{\epsilon\} \, dV \tag{6.48}$$

The matrices $[N]$ and $[\Theta]$ are defined by

$$[N] = \frac{1}{E}
\begin{bmatrix}
1 & & & & & \\
-v & 1 & & & \text{symmetric} & \\
-v & -v & 1 & & & \\
0 & 0 & 0 & 2(1+v) & & \\
0 & 0 & 0 & 0 & 2(1+v) & \\
0 & 0 & 0 & 0 & 0 & 2(1+v)
\end{bmatrix} \tag{6.49}$$

and

$$[\Theta] = \eta
\begin{bmatrix}
1-v & & & & & \\
v & 1-v & & & \text{symmetry} & \\
v & v & 1-v & & & \\
0 & 0 & 0 & (1-2v)/2 & & \\
0 & 0 & 0 & 0 & (1-2v)/2 & \\
0 & 0 & 0 & 0 & 0 & (1-2v)/2
\end{bmatrix} \tag{6.50}$$

where $\eta = E/(1+v)(1-2v)$
E = elastic modulus of elasticity
v = Poisson's ratio

In structural systems which are constructed from an assemblage of bar elements, it is more convenient to express the strain energy in terms of the element internal loads and then sum the contributions of each element to the total strain energy of the system.

To derive the strain energy expression, consider a linear elastic bar for which, at any point ζ along its length, the internal loads are given by $M_z(\zeta)$, $T(\zeta)$, $V_y(\zeta)$, and $S(\zeta)$, where M_z = bending moment about ζ axis, T = torque, V_y = shear force in y direction, and S = axial force. The normal stresses induced in the bar by the internal force system may be obtained from Chap. 5 and are given as

$$\sigma_{xxa} = \text{normal stress due to axial force} = \frac{S(\zeta)}{A(\zeta)} \tag{a}$$

$$\sigma_{xxb} = \text{normal stress due to bending} = \frac{M_z(\zeta)y}{I_z(\zeta)} \tag{b}$$

If we assume that σ_{xx} is the only nonzero stress, then Eq. (6.47) reduces to

$$U_\sigma = \frac{1}{2} \int_V \frac{\sigma_{xx}^2}{E} \, dV \qquad (c)$$

where U_σ = strain energy due to normal stresses only.

Substituting Eqs. (a) and (b) into Eq. (c) yields

$$U_\sigma = U_{\sigma_a} + U_{\sigma_b} = \frac{1}{2} \int_V \frac{S^2(\zeta)}{EA^2(\zeta)} \, dV + \frac{1}{2E} \int_V \frac{M_z^2(\zeta)}{EI_z^2(\zeta)} y^2 \, dV$$

$$= \frac{1}{2} \int_\zeta \int_A \frac{S^2(\zeta)}{EA^2(\zeta)} \, dA \, d\zeta + \frac{1}{2E} \int_\zeta \int_A \frac{M_z^2(\zeta)}{EI_z^2(\zeta)} y^2 \, dA \, d\zeta$$

but

$$\int_A dA = A(\zeta) = \text{cross-sectional area}$$

and

$$\int_A y^2 \, dA = I_z(\zeta) = \text{moment of inertia}$$

Therefore,

$$U_\sigma = \frac{1}{2} \int_\zeta \frac{S^2(\zeta)}{EA(\zeta)} \, d\zeta + \frac{1}{2} \int_\zeta \frac{M_z^2(\zeta)}{EI_z(\zeta)} \, d\zeta \qquad (d)$$

The strain energy due to transverse shear may be obtained by considering Fig. 6.3. The average shear stress acting on the right face of the bar can be expressed as

$$\sigma_{\zeta y} = \frac{V_y(\zeta)}{A_z(\zeta)} = \frac{V_y(\zeta)}{kA(\zeta)} \qquad (e)$$

where $A_s(\zeta)$ is the effective shear area, $A(\zeta)$ is the actual bar cross-sectional area, and k is the shear-form factor which accounts for the shear stress distribution across the bar depth.

From Hooke's law,

$$\epsilon_{\zeta y} = \frac{\sigma_{\zeta y}}{G} \qquad (f)$$

Utilizing (e), Eq. (f) becomes

$$\epsilon_{\zeta y} = \frac{V_y(\zeta)}{GkA(\zeta)} \qquad (g)$$

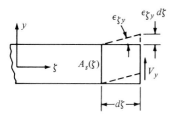

Figure 6.3

The work done on the differential element $d\zeta$ by the force $V_y(\zeta)$ is equal to the strain energy stored, or,

$$dW = dU_s = \tfrac{1}{2} V_y \epsilon_{\zeta y} \, d\zeta$$

$$= \frac{1}{2} \frac{V_y^2(\zeta)}{GkA(\zeta)} \, d\zeta$$

Therefore,

$$U_s = \frac{1}{2} \int_\zeta \frac{V_y^2(\zeta)}{GkA(\zeta)} \, d\zeta \tag{h}$$

The strain energy due to a torque T can be shown to be given by

$$U_t = \frac{1}{2} \int_\zeta \frac{T^2(\zeta)}{GJ} \, d\zeta \tag{i}$$

where J = torsional constant.

Thus, for a structural system which is made of m structural bar elements, the strain energy expression may be written as

$$U = \frac{1}{2} \sum_{i=1}^{m} \int_\zeta \left[\frac{S^2(\zeta)}{EA(\zeta)} + \frac{M^2(\zeta)}{EI(\zeta)} + \frac{V^2(\zeta)}{kGA(\zeta)} + \frac{T^2(\zeta)}{GJ(\zeta)} \right]_i d\zeta \tag{6.51}$$

where m = total number of elements in structure

S = axial internal load

M = internal bending moment

V = internal transverse shear

T = internal torque

AE, EI, AG, GJ = extensional, bending, shear and torsional rigidity, respectively

A, I, J = cross-sectional area, moment of inertia, and torsional constant, respectively

G = shear modulus

k = shear-form factor which accounts for distribution of shearing stresses across bar depth (for wide flange sections, $k \approx 1$, while for rectangular sections, $k \approx 0.833$)

ζ = generalized coordinate axis

6.6 CASTIGLIANO'S SECOND THEOREM IN DEFLECTION ANALYSIS OF STRUCTURES

For linear elastic structural systems, the second theorem of Castigliano may be written as

$$q_r = \frac{\partial U}{\partial Q_r} \qquad (U = \bar{U}) \tag{6.52}$$

This equation states that the displacement q at any point r on the structure and in any direction is equal to the first partial derivative of the total strain energy with respect to a corresponding force Q at r and in the same direction as q. The following examples illustrate applications of this widely used theorem in the deflection analysis of structures.

Example 6.1 Find the vertical deflection of point 3 on the structure shown in the figure. Assume all members to be the same material and to have the same cross-sectional properties. Points 2, 3, and 4 are pinned.

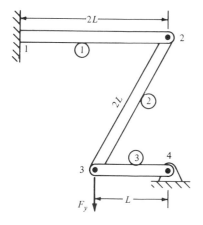

SOLUTION The total strain energy stored in the structure is

$$U = U_1 + U_2 + U_3$$

where U_1, U_2, and U_3 are the strain energies in members 1, 2, and 3, respectively. By noting that members 2 and 3 are axial rod elements, their strain energies may be expressed as

$$U_2 = \frac{1}{2} \int_0^{2L} \frac{S_2^2}{AE} \, dx_{32}$$

$$U_3 = \frac{1}{2} \int_0^{L} \frac{S_3^2}{AE} \, dx_{34}$$

where the coordinate x_{ij} denotes the axis being taken along the direction from i to j. For element 1, there exist internal loads in the form of bending, shear, and axial loads. Therefore the strain energy expression for this element is

$$U_1 = \frac{1}{2} \int_0^{2L} \frac{S_1^2}{AE} \, dx_{21} + \frac{1}{2} \int_0^{2L} \frac{M_1^2}{EI} \, dx_{21} + \frac{1}{2} \int_0^{2L} \frac{V_1^2}{kAG} \, dx_{21}$$

The internal loads in each element may be found from statics and are given by

$$S_3 = \frac{F_y}{\sqrt{3}}$$

$$S_2 = \frac{2F_y}{\sqrt{3}}$$

$$S_1 = \frac{F_y}{\sqrt{3}}$$

$$M_1 = F_y x_{21}$$

$$V_1 = F_y$$

Making the appropriate substitutions yields

$$U = \frac{1}{2AE} \left(\frac{1}{3} \int_0^{2L} F_y^2 \, dx_{21} + \frac{4}{3} \int_0^{2L} F_y^2 \, dx_{32} + \frac{1}{3} \int_0^L F_y^2 \, dx_{34} \right)$$

$$+ \frac{1}{2EI} \int_0^{2L} F_y^2 x_{21}^2 \, dx_{21} + \frac{1}{2kAG} \int_0^{2L} F_y^2 \, dx_{21}$$

or

$$U = \left(\frac{11L}{6AE} + \frac{4L^3}{3EI} + \frac{L}{6AG} \right) F_y^2$$

By using Castigliano's theorem, Eq. (6.52) yields

$$\delta_y = \frac{\partial U}{\partial F_y} = \left(\frac{11L}{3AE} + \frac{8L^3}{3EI} + \frac{2L}{kAG} \right) F_y$$

If shear deformation is neglected, then the deflection becomes

$$\delta y = \left(\frac{11L}{3AE} + \frac{8L^3}{3EI} \right) F_y$$

Example 6.2 Find the slope and vertical deflection at the tip of the cantilevered beam shown in the figure. Neglect shear deformation.

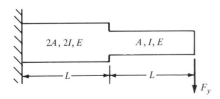

SOLUTION Recall that in order to find a particular coordinate displacement by using Castigliano's second theorem, there must exist a corresponding force at the point where the displacement is sought and in the direction of the displacement. Hence, in order to find the slope at the tip of the cantilevered beam shown in the figure, a fictitious moment must be applied at that point,

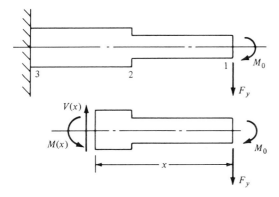

as shown. After the slope is found, this moment is set equal to zero, which implies its nonexistence on the original structure.

The strain energy expression for this case is

$$U = \frac{1}{2} \int_x \frac{M^2(x)}{EI} \, dx = \frac{1}{2} \int_0^L \frac{M_{12}^2}{EI_{12}} \, dx + \frac{1}{2} \int_L^{2L} \frac{M_{23}^2}{EI_{23}} \, dx$$

From the sketch above,

$$M_{12} = M_{23} = M_0 + F_y x$$

Therefore

$$U = \frac{1}{2EI} \int_0^L (M_0 + F_y x^2)^2 \, dx + \frac{1}{4EI} \int_0^{2L} (M_0 + F_y x)^2 \, dx$$

$$= \frac{3}{4EI} \left(M_0^2 L + \frac{5 M_0 F_y L^2}{3} + F_y^2 L^3 \right)$$

Using Eq. (6.52) yields

$$\delta_y = \frac{\partial U}{\partial F_y} \bigg|_{M_0 = 0} = \frac{3 F_y L^3}{2EI}$$

$$\text{slope} = \theta = \frac{\partial U}{\partial M_0} \bigg|_{M_0 = 0} = \frac{5 F_y L^2}{4EI}$$

Example 6.3 Find the horizontal motion of the right support of the semi-circular arch shown in the figure (p. 154). Assume bending deformations only, and assume EI to be constant.

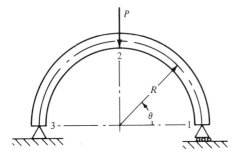

SOLUTION In order to find the horizontal motion at 1, a fictitious horizontal force must be applied at that point, as shown in the figure. The finding

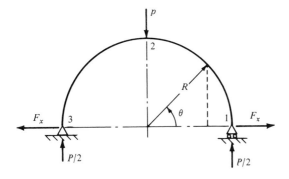

moment can be easily calculated and is given by

$$M_{12} = R(1 - \cos \theta)\frac{P}{2} + (R \sin \theta)F_x \qquad 0 \le \theta \le \frac{\pi}{2}$$

$$M_{32} = M_{12}$$

Therefore the strain energy is

$$U = 2\left(\frac{1}{2} \int_s \frac{M_{12}^2}{EI} \, ds\right) = 2\left(\frac{1}{2EI} \int_0^{\pi/2} M_{12}^2 R \, d\theta\right)$$

$$= \frac{1}{EI} \int_0^{\pi/2} \left[R(1 - \cos \theta)\frac{P}{2} + F_x R \sin \theta\right]^2 R \, d\theta$$

Using Castigliano's second theorem yields

$$\delta_x = \frac{\partial U}{\partial F_x}\bigg|_{F_x = 0} = \frac{2}{EI} \int_0^{\pi/2} \left[R(1 - \cos \theta)\frac{P}{2}\right]R^2 \sin \theta \, d\theta$$

or

$$\delta_x = \frac{PR^3}{2EI}$$

Example 6.4 Find the vertical deflection at 1 and the angular twist at 2 of the structure shown in the figure. Neglect shear deformation.

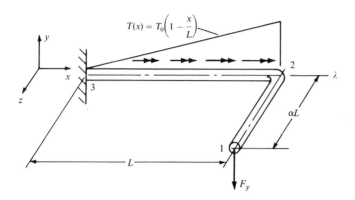

SOLUTION The strain energy stored in the structure is

$$U = \frac{1}{2} \int_0^{\alpha L} \frac{M_{12}^2}{EI} \, dz + \frac{1}{2} \int_0^L \frac{M_{23}^2}{EI} \, dx + \frac{1}{2} \int_0^L \frac{T_{23}^2}{GJ} \, dx$$

where $M_{12} = F_y z$

$M_{23} = F_y x$

$$T_{23} = T_F + T_0 \left(1 - \frac{x}{L}\right) + \alpha L F_y$$

T_F = fictitious torque applied at 2

$$\delta_y^{(1)} = \frac{\partial U}{\partial F_y} = \int_0^{\alpha L} \frac{M_{12}}{EI} \frac{\partial M_{12}}{\partial F_y} \, dz$$

$$+ \int_0^L \frac{M_{23}}{EI} \frac{\partial M_{23}}{\partial F_y} \, dx + \int_0^L \frac{T_{23}}{GJ} \frac{\partial T_{23}}{\partial F_y} \, dx$$

and

$$\theta_x^{(2)} = \frac{\partial U}{\partial T_F} = \int_0^{\alpha L} \frac{M_{12}}{EI} \frac{\partial M_{12}}{\partial T_F} \, dz$$

$$+ \int_0^L \frac{M_{23}}{EI} \frac{\partial M_{23}}{\partial T_F} \, dx + \int_0^L \frac{T_{23}}{GJ} \frac{\partial T_{23}}{\partial T_F} \, dx$$

The partials in the above equations may be easily calculated and are given by

$$\frac{\partial M_{12}}{\partial F_y} = z \qquad \frac{\partial M_{23}}{\partial F_y} = x \qquad \frac{\partial T_{23}}{\partial F_y} = \alpha L$$

$$\frac{\partial M_{12}}{\partial T_F} = 0 \qquad \frac{\partial M_{23}}{\partial T_F} = 0 \qquad \frac{\partial T_{23}}{\partial T_F} = 1$$

Substituting in the integrals and carrying out the integrations yield

$$\delta_y^{(1)} = \left(\frac{\alpha^3 + 1}{3EI} + \frac{\alpha^2}{GJ}\right) F_y L^3 + \frac{T_0 \alpha L^2}{2GJ}$$

$$\theta_x^{(2)} = \left(\frac{T_0 L}{2} + \alpha L^2 F_y\right) \Big/ GJ$$

6.7 RAYLEIGH-RITZ METHOD IN DEFLECTION ANALYSIS OF STRUCTURES

The Rayleigh-Ritz method of obtaining displacements of conservative elastic structural systems is based on the principle of conservation of energy, as shown previously. The method assumes that a deflection shape of the structure is known and may be taken as

$$q = \sum_{n=1}^{N} \beta_n \Phi_n \qquad (6.53)$$

where q is the generalized coordinate displacement, β_n are undetermined constant parameters and are chosen such that the total potential in the system becomes minimum, and Φ_n are a chosen set of functions which must satisfy only the displacement boundary conditions of the structure at hand.

The undetermined constants β_n are obtained by solution of a set of simultaneous linear algebraic equations which result when the principle of minimum potential is applied in conjunction with the assumed deflection shape given by Eq. (6.53). The method is best explained by considering the following examples.

Example 6.5 Using the Rayleigh-Ritz method, find the deflection of the simple beam shown below. Consider bending deformation only.

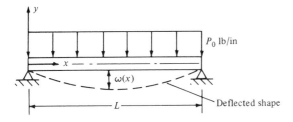

SOLUTION The boundary conditions of the beam are

$$\text{Deflection} = \omega(0) = \omega(L) = 0$$

$$\text{Bending moment} = M(0) = M(L) = 0 = EI \frac{d^2\omega}{dx^2} \tag{6.54}$$

Therefore, if Φ_n in Eq. (6.53) is taken as $\sin(n\pi x/L)$, where n is an integer, then it is obvious that both sets of boundary conditions are satisfied. Hence, the appropriate deflected shape of the beam may be taken as

$$\omega(x) = \sum_{n=1}^{N} \beta_n \sin \frac{n\pi x}{L} \tag{a}$$

The principle of minimum total potential states that

$$\frac{\partial}{\partial \beta_n} (U + P) = 0 \tag{b}$$

where

$$U = \frac{1}{2} \int_0^L \frac{M^2}{EI} \, dx \tag{c}$$

and

$$P = -W = -\int_0^L P_0(x)\omega(x) \, dx \tag{d}$$

The bending moment may be expressed in terms of ω as follows:

$$M = EI \frac{d^2\omega}{dx^2} = -EI \sum_{n=1}^{N} \left(\frac{n\pi}{L}\right)^2 \beta_n \sin \frac{n\pi x}{L}$$

By making the appropriate substitutions, Eqs. (c) and (d) become, respectively,

$$U = \frac{EI}{2} \int_0^L \left[\sum_{n=1}^{N} \left(\frac{n\pi}{L}\right)^2 \beta_n \sin \frac{n\pi x}{L}\right]^2 dx \tag{e}$$

and

$$P = -\int_0^L P_0 \sum_{n=1}^{N} \beta_n \sin \frac{n\pi x}{L} \, dx \tag{f}$$

Performing the integrations yields

$$U = \frac{L}{4EI} \sum_{n=1}^{N} \left(\frac{n\pi}{L}\right)^4 \beta_n^2 \tag{g}$$

and

$$P = -\frac{2L}{\pi} \sum \frac{\beta_n}{n} \qquad (n = 1, 3, 5, \ldots) \tag{h}$$

From Eq. (b)

$$\frac{\partial}{\partial \beta_n}(U + P) = 0 = \frac{L\alpha_n^2}{2EI}\beta_n - \frac{2L}{\pi n}$$

or
$$\beta_n = \frac{4P_0 L^4}{\pi^5 n^5 EI} \tag{i}$$

The deflection from Eq. (a) becomes

$$\omega = \frac{4P_0 L^4}{\pi^5 EI} \sum_{n=1,3,5,\ldots}^{N} \frac{1}{n^5} \sin \frac{n\pi x}{L} \tag{j}$$

The convergence of the above series is very rapid, and in most cases few terms are needed to obtain sufficient accuracy for the deflection. For example, if only the first term in the series is considered, the maximum deflection will be 0.0131 $P_0 L^4/(EI)$, as compared to the exact value of $0.0130 P_0 L^4/(EI)$.

6.8 FINITE DIFFERENCE METHOD IN DEFLECTION ANALYSIS OF STRUCTURES

Closed-form solutions of differential equations representing the deformations of structural systems are not always possible. Therefore, the application of some approximate numerical methods is not only permissible but, in many instances, desirable. The most commonly used is the method of finite differences. The method approximates differentials by finite differences; as a result, it transforms a set of differential equations to a set of simultaneous algebraic equations, which then can be solved easily.

The fundamental relations of finite differences may be established by considering the existence of a real, continuous simple function

$$\Phi = f(x) \tag{6.55}$$

Figure 6.4 shows a graphical representation of the function at equidistant values of Δx.

The first derivative of Φ for small values of Δx at some point i may be written

$$\frac{d\Phi}{dx}\bigg|_i \approx \frac{\Delta \Phi_i}{\Delta x} \tag{6.56}$$

where $\Delta \Phi_i$ can be found by considering either points i and $i + 1$ or i and $i - 1$:

$$\Delta \Phi_i^{(f)} \approx \Phi_{i+1} - \Phi_i \tag{6.57}$$

or
$$\Delta \Phi_i^{(b)} \approx \Phi_i - \Phi_{i-1} \tag{6.58}$$

The difference $\Delta \Phi_i^{(f)}$ represents the first forward difference approximation, while $\Delta \Phi_i^{(b)}$ represents the first backward difference approximation. The mean value of

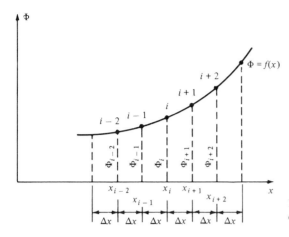

Figure 6.4 One-dimensional finite-difference approximation.

these two extremes may be expressed as

$$\Delta\Phi_i^{(c)} \approx \frac{\Delta\Phi_i^{(f)} + \Delta\Phi_i^{(b)}}{2}$$

or

$$\Delta\Phi_i^{(c)} \approx \frac{\Phi_{i+1} - \Phi_{i-1}}{2} \tag{6.59}$$

where $\Delta\Phi_i^{(c)}$ is referred to as the *first central difference approximation*.

By utilizing Eq. (6.59), Eq. (6.56) becomes

$$\left.\frac{d\Phi}{dx}\right|_i \approx \frac{\Phi_{i+1} - \Phi_{i-1}}{2\,\Delta x} \tag{6.60}$$

Equation (6.60) gives the approximate first derivative of Φ at point i in terms of central differences. Higher derivatives are obtained similarly,

$$\left.\frac{d^2\Phi}{dx^2}\right|_i \approx \frac{\Phi_{i+1} - 2\Phi_i + \Phi_{i-1}}{(\Delta x)^2}$$

$$\left.\frac{d^3\Phi}{dx^3}\right|_i \approx \frac{\Phi_{i+2} - 2\Phi_{i+1} + 2\Phi_{i-1} - \Phi_{i-2}}{2\,(\Delta x)^3} \tag{6.61}$$

$$\left.\frac{d^4\Phi}{dx^4}\right|_i \approx \frac{\Phi_{i+2} - 4\Phi_{i+1} + 6\Phi_i - 4\Phi_{i-1} + \Phi_{i-2}}{(\Delta x)^4}$$

etc.

For a two-dimensional function such as shown in Fig. 6.5, the partial derivatives may be obtained as in the case of one-dimensional function:

$$\left.\frac{\partial\Phi}{\partial x}\right|_{i,j} \approx \frac{\Phi_{i+1,j} - \Phi_{i-1,j}}{2\,\Delta x} \qquad \left.\frac{\partial\Phi}{\partial z}\right|_{i,j} \approx \frac{\Phi_{i,j+1} - \Phi_{i,j-1}}{2\,\Delta z}$$

$$\frac{\partial^2 \Phi}{\partial x^2}\bigg|_{i,j} \approx \frac{\Phi_{i+1,j} - 2\Phi_{i,j} + \Phi_{i-1,j}}{(\Delta x)^2} \qquad \frac{\partial^2 \Phi}{\partial z^2}\bigg|_{i,j} \approx \frac{\Phi_{i,j+1} - 2\Phi_{i,j} + \Phi_{i,j-1}}{(\Delta z)^2}$$

$$\frac{\partial^3 \Phi}{\partial x^3}\bigg|_{i,j} \approx \frac{\Phi_{i+2,j} - 2\Phi_{i+1,j} + 2\Phi_{i-1,j} - \Phi_{i-2,j}}{2(\Delta x)^3}$$

$$\frac{\partial^3 \Phi}{\partial z^3}\bigg|_{i,j} \approx \frac{\Phi_{i,j+2} - 2\Phi_{i,j+1} + 2\Phi_{i,j-1} - \Phi_{i,j-2}}{2(\Delta z)^3} \tag{6.62}$$

$$\frac{\partial^4 \Phi}{\partial x^4}\bigg|_{i,j} \approx \frac{\Phi_{i+2,j} - 4\Phi_{i+1,j} + \Phi_{i-2,j} - 4\Phi_{i-1,j} + 6\Phi_{i,j}}{(\Delta x)^4}$$

$$\frac{\partial^4 \Phi}{\partial z^4}\bigg|_{i,j} \approx \frac{\Phi_{i,j+2} - 4\Phi_{i,j+1} + \Phi_{i,j+1} - 4\Phi_{i-1,j} + 6\Phi i,j}{(\Delta z)^4}$$

$$\frac{\partial^2 \Phi}{\partial x \, \partial z}\bigg|_{i,j} \approx \frac{\Phi_{i+1,j+1} - \Phi_{i-1,j+1} - \Phi_{i+1,j-1} + \Phi_{i-1,j-1}}{4 \, \Delta x \, \Delta z}$$

$$\frac{\partial^3 \Phi}{\partial z^2 \, \partial y}\bigg|_{i,j} \approx \frac{\Phi_{i+1,j+1} - \Phi_{i+1,j-1} - 2\Phi_{i,j+1} + 2\Phi_{i,j-1} - \Phi_{i-1,j-1} + \Phi_{i-1,j+1}}{2(\Delta x)^2 \, \Delta z}$$

$$\frac{\partial^3 \Phi}{\partial x \, \partial y^2}\bigg|_{i,j} \approx \frac{\Phi_{i+1,j+1} - \Phi_{i-1,j-1} - 2\Phi_{i+1,j} + 2\Phi_{i-1,j} - \Phi_{i-1,j+1} + \Phi_{i+1,j-1}}{2 \, \Delta x \, (\Delta z)^2}$$

Example 6.6 Using the finite difference method, find the vertical deflection at

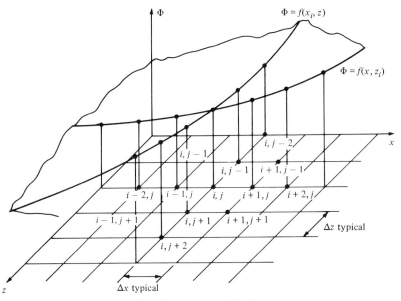

Figure 6.5 Two-dimensional finite-difference approximation.

midspan and the slope at the left support (see the figure). Assume $EI = $ constant.

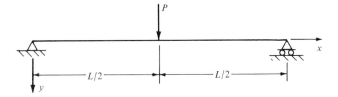

SOLUTION In order to use the finite difference method in structural analysis, the following steps must be taken:

1. Formulate all the governing differential equations which describe the behavior of the structure.
2. Express the governing differential equations and the equations which describe the boundary conditions in terms of finite differences.
3. Obtain a set of simultaneous algebraic equations by satisfying the resulting finite difference equations in step 2 at finite discrete points on the structure. This is equivalent to idealizing the structure to a finite number of elements connected at discrete points.
4. Solve for the finite difference values.

In this example, the governing differential equation of deformation for a simple beam structure may be obtained from the strength of material and is given by

$$EI \frac{d^4w}{dx^4} = p \qquad (a)$$

where $w = $ beam vertical deflection
$\quad EI = $ beam bending rigidity;
$\quad p = $ distributed normal applied load

The slope at any point along the beam is

$$\theta = \frac{dw}{dx} \qquad (b)$$

Expressing Eqs. (a) and (b) in terms of finite differences yields

$$w_{i+2} - 4w_{i+1} + 6w_i - 4w_{i-1} + w_{i-2} = \frac{p_i(\Delta x)^4}{(EI)_i} \qquad (a')$$

and

$$\theta_i = \frac{w_{i+1} - w_{i-1}}{2\,\Delta x} \qquad (b')$$

where $i = 1, 2, 3, \ldots, N$ are points along the beam span.

By examining Eq. (a'), it can be concluded that only a finite number of points may be satisfied if a feasible solution has to be reached. Thus, for simplicity, divide (idealize) the beam into four equal elements only, as shown in the figure below.

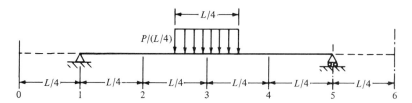

Note that the accuracy increases with an increase in the number of elements taken. The fictitious extension of the beam beyond its outer supports is necessary as can be seen from Eq. (a'). The displacements of points 0 and 6 are assumed to be equal to the negative displacements of points 2 and 4, respectively. For fixed supports, these displacements are equal in sign and magnitude. See sketches (a) and (b) below. Also note that the concentrated load P is assumed to be distributed uniformly over half the length of each element on both sides of the point where the load is concentrated. This is necessary because the load function in Eq. (a') is a distributed load.

Writing Eq. (a') for points 2, 3, and 4 yields

$$w_4 - 4w_3 + 6w_2 - 4w_1 + w_0 = 0$$

$$w_5 - 4w_4 + 6w_3 - 4w_2 + w_1 = \frac{PL^3}{64EI} \qquad (c)$$

$$w_6 - 4w_5 + 6w_4 - 4w_3 + w_2 = 0$$

From boundary conditions and symmetry, the following deflections are known:

$$w_1 = w_5 = 0 \qquad w_2 = w_4$$

Also, for the deformation to be continuous over the supports,

$$w_0 = -w_2 \qquad \text{and} \qquad w_6 = -w_4$$

Note that the first and the last equations in (c) are identical because of symmetry; therefore, by deleting one and utilizing the above condition, Eq. (c) becomes

$$6w_2 - 4w_3 = 0$$

$$-4w_2 + 3w_3 = \frac{PL^3}{128EI}$$

or, in matrix form,

$$\begin{bmatrix} 6 & -4 \\ -4 & 3 \end{bmatrix} \begin{bmatrix} w_2 \\ w_3 \end{bmatrix} = \begin{bmatrix} 0 \\ \dfrac{PL^3}{128EI} \end{bmatrix} \tag{d}$$

Solving Eq. (d) for w_2 and w_3 yields

$$w_2 = \frac{0.0156PL^3}{EI} \qquad \left(\text{exact:} \quad \frac{0.0143PL^3}{EI} \right)$$

$$w_3 = \frac{PL^3}{42.7EI} \qquad \left(\text{exact:} \quad \frac{PL^3}{48EI} \right)$$

The slope from Eq. (b') is

$$\theta_1 = \frac{w_2 - w_0}{2\,\Delta x} = \frac{w_2}{\Delta x} = \frac{0.0624PL^2}{EI} \qquad \left(\text{exact:} \quad \frac{0.0625PL^2}{EI} \right)$$

Example 6.7 Use the finite difference approach to find the maximum bending moments and vertical deflection on the plate structure shown. Assume that the plate is simply supported and acted on by a uniform pressure p. Also assume that the plate is of uniform thickness t.

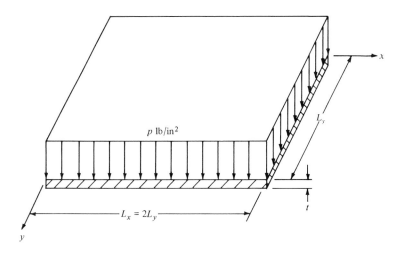

The governing differential equation of a flat plate is given by[25]

$$\frac{\partial^4 w}{\partial x^4} + 2\frac{\partial^4 w}{\partial x^2\,\partial y^2} + \frac{\partial^4 w}{\partial y^4} = \frac{p}{D} \qquad (a)$$

where $D = Et^3/[12(1 - v^2)]$ and $v =$ Poisson's ratio. The bending moments are given by[25]

$$M_x = -D\left(\frac{\partial^2 w}{\partial w^2} + v\,\frac{\partial^2 w}{\partial y^2}\right)$$

$$M^y = -D\left(\frac{\partial^2 w}{\partial y^2} + v\,\frac{\partial^2 w}{\partial x^2}\right) \qquad (b)$$

From Eq. (6.62), Eq. (a) may be easily expressed in terms of finite differences:

$$\begin{array}{c}
+ Cw_{i,\,j-2} \\
+ Ew_{i-1,\,j-1} + Fw_{i,\,j-1} + Ew_{i+1,\,j-1} \\
Aw_{i-2,\,j} + Bw_{i-1,\,j} \quad + w_{i,\,j} \quad + Bw_{i+1,\,j} + Aw_{i+2,\,j} = \bar{P} \\
+ Ew_{i-1,\,j+1} + Fw_{i,\,j+1} + Ew_{i+1,\,j+1} \\
+ Cw_{i,\,j+2}
\end{array} \qquad (c)$$

where $A = \dfrac{1}{H}$

$B = -\dfrac{4(1 + \alpha^2)}{H}$

$C = \dfrac{\alpha^4}{H}$

$E = \dfrac{2\alpha^2}{H}$

$F = \dfrac{4\alpha^2(1 + \alpha^2)}{H}$

$\bar{P} = \dfrac{P(\Delta x\,\Delta y\,\alpha)^2}{DH}$

$\alpha = \dfrac{\Delta x}{\Delta y}$

$H = 6 + 8\alpha^2 + 6\alpha^4$

As in the beam example, for simplicity, divide the plate into four equal parts along the x and y axes, as shown in the figure. From symmetry, only points 1, 2, 4, and 5 need to be considered. Thus, writing Eq. (c) for these points yields

$$
\begin{bmatrix}
1.0 & -0.149 & -0.597 & 0.0597 \\
-0.149 & 0.5 & 0.0597 & -0.299 \\
-0.597 & 0.0597 & 0.5 & -0.075 \\
0.0597 & -0.299 & -0.075 & 0.25
\end{bmatrix}
\begin{bmatrix}
w_1 \\
w_2 \\
w_4 \\
w_5
\end{bmatrix}
= \frac{10^{-4}PL_y^4}{D}
\begin{bmatrix}
4.664 \\
2.332 \\
2.332 \\
1.166
\end{bmatrix} \quad (d)
$$

Solving Eq. (d) yields

$$
w_1 = 0.00559 \frac{PL_y^4}{D}
$$

$$
w_2 = 0.00723 \frac{PL_y^4}{D}
$$

$$
w_4 = 0.00778 \frac{PL_y^4}{D}
$$

$$
w_5 = w_{max} = 0.01008 \frac{PL_y^4}{D} \quad \left(\text{exact:} \quad 0.01013 \frac{PL_y^4}{D} \right)
$$

Since we have the displacements, the moments are calculated as follows:

$$
M_{x,\,max} = M_{x,\,5} = -D\left(\frac{w_{i+1,\,j} - 2w_{i,\,j} + w_{i-\cdot,\,j}}{\Delta x^2} + v\, \frac{w_{i,\,j+1} - 2w_{i,\,j} + w_{i,\,j-1}}{\Delta y^2} \right)
$$

$$
= -\frac{16D}{L_y^2}\left[\frac{1}{\alpha^2}(w_6 - 2w_5 + w_4) + 0.3(w_8 - 2w_5 + w_2) \right]
$$

$$
= 0.0454 PL_y^2 \quad (\text{exact:} \quad 0.0479 PL_y^2)
$$

Likewise,

$$
M_{y,\,max} = M_{y,\,5} = 0.966 PL_y^2 \quad (\text{exact:} \quad 0.0948 PL_y^2)
$$

6.9 REDUNDANT STRUCTURES AND THE UNIT-LOAD METHOD

A redundant (statically indeterminate) structural system is one for which the external reactions or internal loads cannot be completely determined from the conditions for static equilibrium. A stable and statically determinate structural system contains only enough external support reactions or structural members for stability, and the equations of static equilibrium are sufficient to determine completely the reactions or the member internal loads. If one member or reaction is removed, the structure becomes unstable and hence incapable of resisting applied loads. If one member or reaction is added, the structure becomes singly redundant, and the reactions and member internal loads must be obtained by considering the deformation of the structure in addition to the conditions of static equilibrium.

Normally, a rigid coplanar structure requires three external reactions for stability, and they may be calculated from the three equations of statics. The number of reactions, however, is not the only criterion for stability, and it is necessary to examine each particular structure in order to determine whether it is stable, unstable, or statically indeterminate. For example, a horizontal simple beam normally requires three reaction components. However, if the beam is supported on rollers at three points along the span, the beam will be unstable for resisting horizontal forces and statically indeterminate for vertical forces. Similarly, if the three reactions of a simple beam act through any common point in the plane, the moments about that point will be zero regardless of the magnitudes of the forces, and the moment equation cannot be used to obtain the reactions. Such a structure, shown in Fig. 6.6, is a mechanism that is free to rotate through a small angle about point O as an instantaneous center, and it is unstable in resisting any load which does not act through point O. If a load acts through point O, the structure is statically indeterminate.

In most structures, the number of equations of statics may be compared with the number of redundants to determine the conditions of stability. For such special structures as that shown in Fig. 6.6, an attempt to find the three unknown reactions from the three equations of statics will yield equations which are not independent (one of the equations can be derived from the others). When an attempt to analyze a structure by the equations of statics results in such a condition, the structure must be examined for instability or redundancy. For simple redundant structures, the unit-load method is often used, and it is demonstrated here.

If a structure has one more member or reaction than is required for stability, the structure has single redundancy. In many cases, any one of several members or reactions may be removed without causing instability. Then one deflection equation must be used in addition to the equations of statics in order to analyze the structure. If a structure has several more members or reactions than are required for stability, it has multiple redundancy. The degree of redundancy is equal to the number of redundant members, and it is equal to the number of deflection conditions which must be used in the analysis.

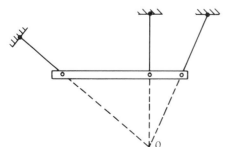

Figure 6.6

6.10 STRUCTURES WITH SINGLE REDUNDANCY

A truss which is composed of elastic members and which has single redundancy is considered first. A typical truss of this type is shown in Fig. 6.7a. The supports are assumed to be rigid, and the members are assumed to be unstressed before load P is applied. There are four external reactions, and only three would be required for stability. The horizontal reaction component X_1 is considered as the redundant, and the deflection equation will be obtained from the condition that the horizontal support deflection δ is zero. From Eq. (6.34) it can be shown that

$$\delta = \sum \frac{SsL}{AE} \tag{6.63}$$

where S represents the force in any member of the structure of Fig. 6.7a and s represents the force in any member of the structure due to a unit load applied in the direction of the desired deflection, as shown in Fig. 6.7c.

The force S in any member is found by superimposing the loading conditions shown in Fig. 6.7b and d. If the redundant force is removed, the resulting statically determinate structure is shown in Fig. 6.7b, and the applied loads produce a force S_0 in any member. If the redundant force X_1 is acting alone, it produces a force $X_1 s$ in any member, as shown in Fig. 6.7d, since the s forces result from a unit value of X_1. The total force S is obtained as the sum of the forces for the two conditions:

$$S = S_0 + X_1 s \tag{6.64}$$

Substituting from Eq. (6.64) into Eq. (6.63) yields

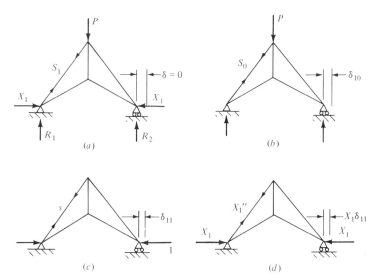

Figure 6.7

$$\delta = \sum \frac{S_0 \, sL}{AE} + X_1 \sum \frac{s^2 L}{AE} \tag{6.65}$$

or, for $\delta = 0$,

$$X_1 = -\frac{\sum S_0 \, sL/(AE)}{\sum s^2 L/(AE)} \tag{6.66}$$

where all the terms on the right side of the equation may be obtained from the loading and geometry of the structure.

Equation (6.66) is applicable to any elastic truss with single redundancy in which the deflection in the direction of the redundant is zero. If the deflection in the direction of the redundant δ is a known value other than zero, this deflection may be substituted into Eq. (6.65) and the value of X_0 determined for this condition.

The physical significance of the terms in Eqs. (6.65) and (6.66) is discussed in order to visualize the action of a redundant structure. If the redundant force is removed, the structure is statically determinate, and the deflection in the direction of the redundant due to the applied loads has the following value:

$$\delta_{10} = \sum \frac{S_0 \, sL}{AE} \tag{6.67}$$

This deflection is assumed positive in the direction of the redundant. A unit value of the redundant deflects the structure a distance

$$\delta_{11} = \sum \frac{s^2 L}{AE} \tag{6.68}$$

which is also positive in the direction of the redundant. The value of X_1 required to give a zero deflection is obtained by dividing the deflection resulting from the applied loads by the deflection resulting from the unit load, and it will be negative with the assumed sign conventions:

$$X_1 = -\frac{\delta_{10}}{\delta_{11}} \tag{6.69}$$

It is easier to visualize the deflection terms of Eq. (6.69) than the summation terms of Eq. (6.66). The form of Eq. (6.69) will apply identically for rigid frame structures with single redundant, except δ_{11} and δ_{10} will have different definitions.

Example 6.8 Find the reactions and the stresses in the members of the structure of Fig. 6.8. The areas of members AB, BC, and BD are 4 in², and the areas of members AD and DC are 3.6 in². Assume rigid supports and $E = 10,000$ kips/in².

SOLUTION The numerical solution of Eq. (6.66) is displayed in Table 6.1, with the notation as shown in Fig. 6.7. The horizontal reaction component X_1 is

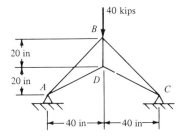

Figure 6.8

considered the redundant. The forces S_0 in the truss with the redundant removed are tabulated in column 1. The lengths L of the various members are calculated from the dimensions shown in Fig. 6.8 and are divided by the areas and modulus of elasticity. The values of $L/(AE)$ are tabulated in column 2. The forces s, resulting from a unit value of X_1 applied as shown in Fig. 6.7c, are tabulated in column 3. Positive signs indicate tension, and negative signs indicate compression. The summation terms of Eq. (6.66) are evaluated as the sum of the values for the individual members tabulated in columns 4 and 5.

$$X_1 = -\frac{\Sigma S_0 sL/(AE)}{\Sigma s^2 L/(AE)} = \frac{0.5150}{0.02008} = 25.6 \text{ kips}$$

This value is multiplied by the terms in column 3 to obtain values of $X_1 s$ for all members. The final forces S in the members are obtained as the algebraic sum of terms in columns 1 and 6 and are listed in column 7.

Example 6.9 Find the forces in the members of the structure of Example 6.8 if the support at point C is deflected 0.25 in to the right and the temperature is decreased $40°F$. Assume a thermal coefficient of expansion $\alpha = 10^{-5}$ in/(in · °F).

SOLUTION A temperature decrease would cause the right end of the statically determinate truss of Fig. 6.7b to move to the left a distance $(\alpha L)\, \Delta T$, where L

Table 6.1

Members	S_0, kips	$\dfrac{L}{AE}$, in/kip	s	$\dfrac{S_0 sL}{AE}$, in	$\dfrac{s^2 L}{AE}$, in/kip	$X_1 s$	S
	(1)	(2)	(3)	(4)	(5)	(6)	(7)
AB	−56.6	0.001414	1.414	−0.1130	0.00283	36.3	−20.3
BC	−56.6	0.001414	1.414	−0.1130	0.00283	36.3	−20.3
AD	44.8	0.001245	−2.233	−0.1245	0.00621	−57.3	−12.5
DC	44.8	0.001245	−2.233	−0.1245	0.00621	−57.3	−12.5
BD	40.0	0.000500	−2.0	−0.0400	0.00200	−51.3	−11.3
Total				−0.5150	0.02008		

Table 6.2

Member	S_0	$X_1 s$	S
AB	−56.6	16.4	−40.2
BC	−56.6	16.4	−40.2
AD	44.8	−25.9	18.9
DC	44.8	−25.9	18.9
BD	40.0	−23.2	16.8

is the distance between supports (80 in):

$$\alpha L\,(\Delta T) = 10^{-5} \times 80 \times 40 = 0.032 \text{ in}$$

The support of point C is displaced an additional 0.25 in to the right; therefore, the total displacement δ, which must be given to the right support of the truss by the strains in the members, is

$$\delta = -00.032 - 0.25 = -0.282 \text{ in}$$

where the negative sign indicates that the deflection due to the stresses is opposite to X_1. From Eq. (6.65),

$$\delta = \sum \frac{S_0\,sL}{AE} + X_1 \sum \frac{s^2 L}{AE} = -0.282$$

where the summation terms are obtained in Table 6.1 and

$$-0.515 + 0.02008 X_1 = -0.282$$

$$X_1 = 11.6 \text{ kips}$$

The values of the resulting forces in the members are found in Table 6.2 as $S = S_0 + X_1 s$, where values of S_0 and s are the same as for Example 6.8.

Example 6.10 Find the bending moment at any point of the semicircular arch of Fig. 6.9 if the supports do not move. The value of EI is constant for all cross sections. Neglect axial deformation.

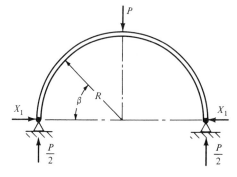

Figure 6.9

SOLUTION The structure is symmetrical about a vertical centerline, and all integrals are evaluated for the left half of the structure and multiplied by 2. The horizontal reaction X_1 is considered the redundant, and the value of M_0 is calculated for the statically determinate structure formed by supporting one end of the arch on frictionless rollers:

$$M_0 = \frac{PR}{2}(1 - \cos \beta)$$

The value of m is calculated for a unit load acting in the direction of X_1:

$$m = -R \sin \beta$$

Utilizing Eq. (6.34) and noting that $M = M_0 + X_1 m$ and $d\zeta = R\, d\beta$, yield

$$X_1 = -\frac{\int (M_0 m/EI)\, ds}{\int (m^2/EI)\, ds} = -\frac{2\int_0^{\pi/2} [PR/(2EI)](1 - \cos \beta)(-R \sin \beta)R\, d\beta}{2\int_0^{\pi/2} [(-R \sin \beta)^2 R/(EI)]\, d\beta} = \frac{P}{\pi}$$

The final bending moment is obtained by superimposing the values of M_0 for the applied loads and $X_1 m$ for the redundant:

$$M = M_0 + X_1 m$$

$$M = \frac{PR}{2}(1 - \cos \beta) - \frac{PR}{\pi} \sin \beta$$

This equation applies for $0 < \beta < \pi/2$, and the bending-moment diagram is symmetrical about a vertical centerline.

Example 6.11 The structure shown in Fig. 6.10a consists of a round tube in a horizontal plane, bent at an angle of 90°. The free end supports a load of 2 kips, and it is also supported by a vertical wire. Find the tension in the wire and make the bending-moment and torsion diagrams for the tube.

SOLUTION The wire is considered the redundant member. The vertical deflection of the free end of the tube is calculated by assuming the wire to be removed and the statically determinate tube to support the load of 2 kips. The bending moment and torsion in the tube under the loading are designated as M_0 and T_0, respectively, and are plotted in Fig. 6.10b. The deflection resulting from bending and torsional deformation of the tube is, from Eq. (6.34),

$$\delta_{10} = \int \frac{M_0 m}{EI}\, ds + \int \frac{T_0 m_t}{JG}\, ds$$

The values of m and m_t, the bending moment and torsional moment, respectively, in the statically determined structure for $X_1 = 1$, are plotted in Fig. 6.10c. The integrals of the above equation become

$$\delta_{10} = -\frac{36 \times 4 + 81 \times 6}{1000} - \frac{12 \times 6 \times 9}{800} = -1.44 \text{ in}$$

The negative sign indicates a deflection in the opposite direction to the unit load, or a downward deflection.

The deflection δ_{11}, resulting from a unit force X_1 in the separate wire, consists of parts due to torsion and bending of the tube and tension in the wire:

$$\delta_{11} = \int \frac{m^2}{EI}\,ds + \int \frac{m_t^2}{JG}\,ds + \frac{s^2 L}{AE}$$

The tension s in the wire is 1, and the values of m and m_t are shown in Fig.

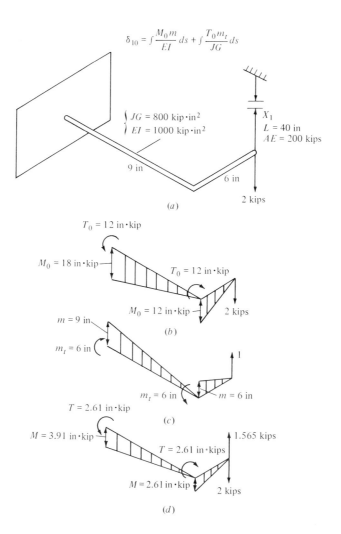

$$\delta_{10} = \int \frac{M_0 m}{EI}\,ds + \int \frac{T_0 m_t}{JG}\,ds$$

$\begin{cases} JG = 800 \text{ kip·in}^2 \\ EI = 1000 \text{ kip·in}^2 \end{cases}$

X_1
$L = 40$ in
$AE = 200$ kips

9 in

6 in

2 kips

(a)

$T_0 = 12$ in·kip

$M_0 = 18$ in·kip

$T_0 = 12$ in·kip

$M_0 = 12$ in·kip   2 kips

(b)

$m = 9$ in

$m_t = 6$ in

$m_t = 6$ in $m = 6$ in

1

$T = 2.61$ in·kip

(c)

$M = 3.91$ in·kip

1.565 kips

$T = 2.61$ in·kips

$M = 2.61$ in·kip   2 kips

(d)

Figure 6.10

6.10c:

$$\delta_{11} = \frac{18 \times 4 + 40.5 \times 6}{1000} + \frac{6 \times 6 \times 9}{800} + \frac{1^2 \times 40}{200} = 0.92 \text{ in/kip}$$

This is how far a 1-kip tension force in the wire would move the cut ends together. The force X_1 required to move the cut ends of the wire the distance $-\delta_{10}$ is found from Eq. (6.69):

$$X_1 = -\frac{\delta_{10}}{\delta_{11}} = \frac{1.44}{0.92} = 1.565 \text{ kips}$$

Now the bending-moment and torsion diagrams for the tube calculated from the equations of statics and are shown in Fig. 6.10d.

6.11 Structures with Multiple Redundancy

The procedure for analysis of structures with two or more redundants is similar to that used for a structure with one redundant member or reaction. The first step is to remove the redundant members or reactions in order to obtain a statically determinate base structure. Then the deflections of the statically determinate base structure in the directions of the redundants are calculated in terms of the redundant forces and are equated to the known deflections, which are usually zero. The number of known deflection conditions must be equal to the number of redundants. For a structure with n redundants, the deflection conditions yield n equations which must be solved simultaneously for the values of the redundants.

The truss shown in Fig. 6.11a has only three reactions and is statically determinate externally, but there are two more members than are required for stability; therefore, it is statically indeterminate internally. The deflection conditions which will be specified are that there are no stresses in the structure when it is not loaded or if two members of the unloaded structure are cut, the relative deflections δ_1 and δ_2 of the cut ends will be zero. The deflections are now expressed in terms of the forces X_1 and X_2 in the redundant members. All deflections are assumed to be elastic.

The statically determinate base structure, shown in Fig. 6.11b, is formed by cutting or removing the redundant members. The applied loads produce forces S_0 in the members of the base structure. The forces s_1 in the members are produced by a unit value of X_1 applied to base structure, as shown in Fig. 6.11c. Similarly, a force $X_2 = 1$ produces forces s_2 in the members when applied to the base structure, as shown in Fig. 6.11d. The final force S in any member may be obtained by superimposing the forces due to the applied loads and the redundant forces:

$$S = S_0 + X_1 s_1 + X_2 s_2 \tag{6.70}$$

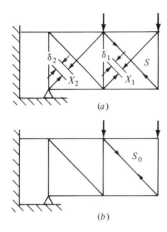

(a)

(b)

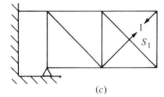

(c)

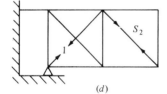

(d)   **Figure 6.11**

The deflections δ_1 and δ_2 of the cut ends of the members are now equated to zero:

$$\delta_1 = \sum \frac{S s_1 L}{AE} = 0$$

$$\delta_2 = \sum \frac{S s_2 L}{AE} = 0$$

$$\delta_1 = \sum \frac{S_0 s_1 L}{AE} + X_1 \sum \frac{s_1^2 L}{AE} + X_2 \sum \frac{s_1 s_2 L}{AE} = 0$$

$$\delta_2 = \sum \frac{S_0 s_2 L}{AE} + X_1 \sum \frac{s_1 s_2 L}{AE} + X_2 \sum \frac{s_2^2 L}{AE} = 0$$

These equations may be solved simultaneously for X_1 and X_2. The final values of the forces S may be obtained from Eq. (6.70).

The preceding equations have been derived from the superposition of the stress conditions, as stated in Eq. (6.70). They may also be obtained from a superposition of the deflection conditions. The applied loads are assumed to produce deflections δ_{10} and δ_{20} of the redundants. A unit value of X_1 produces deflections δ_{11} at X_1 and δ_{21} at X_2. A unit value of X_2 produces deflections δ_{22} at X_2 and δ_{12} at X_1. The total deflections in the directions of the redundants can be obtained by superimposing the effects of the various loads:

$$\delta_1 = \delta_{10} + X_1\delta_{11} + X_2\delta_{12}$$

$$\delta_2 = \delta_{z0} + X_1\delta_{21} + X_2\delta_{22}$$

In most problems, the deflections δ_1 and δ_2 are zero, but in some cases, known values of support deflections of similar deformations may be substituted.

For structure with n redundants, a number of deflections conditions n must be used. These may be written as

$$\delta_1 = \delta_{10} + X_1\delta_{11} + X_2\delta_{12} + \cdots + X_n\delta_{1n}$$

$$\delta_2 = \delta_{20} + X_1\delta_{21} + X_2\delta_{22} + \cdots + X_n\delta_{2n} \qquad (6.71)$$

$$\cdots\cdots\cdots\cdots\cdots\cdots\cdots$$

$$\delta_n = \delta_{n0} + X_1\delta_{n1} + X_2\delta_{n2} + \cdots + X_n\delta_{nn}$$

or, in matrix form,

$$\begin{bmatrix} \delta_1 \\ \delta_2 \\ \vdots \\ \delta_n \end{bmatrix} = \begin{bmatrix} \delta_{10} \\ \delta_{20} \\ \vdots \\ \delta_{n0} \end{bmatrix} + \begin{bmatrix} \delta_{11} & \delta_{12} \cdots \delta_{1n} \\ \delta_{21} & \delta_{22} \cdots \delta_{2n} \\ \vdots & \vdots \\ \delta_{n1} & \delta_{n2} \cdots \delta_{nn} \end{bmatrix} \begin{bmatrix} X_1 \\ X_2 \\ \vdots \\ X_n \end{bmatrix} \qquad (6.71a)$$

The terms in Eq. (6.71) may be defined in the following manner for truss structures:

$$\delta_{n0} = \sum \frac{S_0 s_n L}{AE} \qquad \delta_{mn} = \sum \frac{s_m s_n L}{AE} \qquad (6.72)$$

For rigid-frame structures where only bending deformation is considered, the terms in Eq. (6.71) are defined as

$$\delta_{n0} = \sum \int \frac{M_0 m_n}{EI} \, d\zeta \qquad \delta_{mn} = \sum \int \frac{m_n m_m}{EI} \, d\zeta \qquad (6.73)$$

From Maxwell's reciprocal theorem, $\delta_{ij} = \delta_{ji}$.

The internal loads in each member j may be obtained from the equations

$$S_j = S_0 + X_1 s_1 + X_2 s_2 + \cdots + X_n s_n$$

or

$$M_j = M_0 + X_1 m_1 + X_2 m_2 + \cdots + X_n m_n \qquad (6.74)$$

The applications of the preceding equations are illustrated by numerical examples.

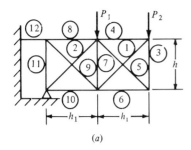

(a)

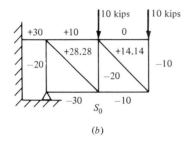

(b)

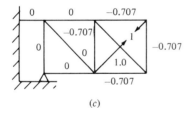

(c)

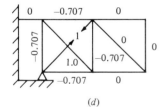

(d)

Figure 6.12

Example 6.12 Find the forces in the members of the truss shown in Fig. 6.12a if $P_1 = P_2 = 10$ kips, $h = h_1$, and $L/(AE)$ is the same for each member of the structure. The members are unstressed when $P_1 = P_2 = 0$, and stresses do not exceed the elastic limit.

SOLUTION: The numerical values of $L/(AE)$ are not required for the members, since only relative values are important. For $\delta_1 = \delta_2 = \cdots = \delta_n = 0$, in Eqs. (6.71) the summation terms may be multiplied by any constant value. If any

δ_n is not zero, it is necessary to know the numerical values. Thus it is assumed that $L/(AE)$ is unity for all members. The calculations of the summation terms are made in Table 6.3. The forces S_0 in the members of the statically determinate base structure are shown in Fig. 6.12b and tabulated in column 1. The values of s_1 and s_2 due to unit values of the redundant are shown in Fig. 6.12c and d and are tabulated in columns 2 and 3. The terms for the summations of Eqs. (6.72) are obtained as the totals of columns 4, 5, 6, 7, and 8. Substituting these totals into Eq. (6.71) yields

$$\begin{bmatrix} 4.0 & 0.5 \\ 0.5 & 4.0 \end{bmatrix} \begin{bmatrix} X_1 \\ X_2 \end{bmatrix} = \begin{bmatrix} -42.42 \\ -70.70 \end{bmatrix}$$

By carrying out the matrix inversion, the unknowns are obtained:

$$X_1 = -8.53 \quad \text{and} \quad X_2 = -16.60$$

Hence the internal load in each member is given by

$$S_j = (S_0 - 8.53 s_1 - 16.6 s_2)j$$

where j is the member number.

The values of S_j are shown in column 9 of Table 6.3.

Example 6.13 Find the forces in the truss of Fig. 6.13a. Assume that $L/(AE)$ is 0.01 in/kip for all members. The right-hand support deflects 0.5 in from the unstressed position of the truss. Assume $h = h_1$.

SOLUTION The redundants X_1 and X_2 are chosen as shown in Fig. 6.13a,

Table 6.3

	S_0 (1)	s_1 (2)	s_2 (3)	$\dfrac{S_0 s_1 L}{AE}$ (4)	$\dfrac{S_0 s_2 L}{AE}$ (5)	$\dfrac{s_1^2 L}{AE}$ (6)	$\dfrac{s_2^2 L}{AE}$ (7)	$\dfrac{s_1 s_2 L}{AE}$ (8)	S, kips (9)
1	0	1.000	0	0	0	1.0	0	0	−8.53
2	0	0	1.000	0	0	0	1.0	0	−16.60
3	−10	−0.707	0	7.07	0	0.5	0	0	−3.97
4	0	−0.707	0	0	0	0.5	0	0	+6.03
5	14.14	1.00	0	14.14	0	1.0	0	0	+5.61
6	−10	−0.707	0	7.07	0	0.5	0	0	−3.97
7	−20	−0.707	−0.707	14.14	14.14	0.5	0.5	0.5	−2.24
8	10	0	−0.707	0	−7.07	0	0.5	0	+21.73
9	28.28	0	1.000	0	28.28	0	1.0	0	+11.68
10	−30	0	−0.707	0	21.21	0	0.5	0	−18.27
11	−20	0	−0.707	0	14.14	0	0.5	0	−8.27
12	30	0	0	0	0	0	0	0	+30.00
Total				42.42	70.70	4.0	4.0	0.5	

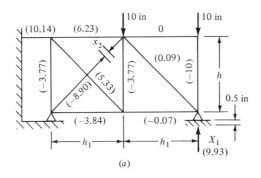

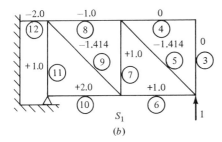

Figure 6.13

leaving the same statically determinate base structure as for Example 6.12. The values for S_0 and s_2 are the same as for Example 6.12, but values of s_1 must be calculated as shown in Fig. 6.13b. The terms involving S_0 and s_2 are therefore the same as those calculated in Table 6.3, except that the value of $L/(AE)$ is now 0.01 for each member, whereas a value of 1.0 was used in Table 6.3. For constant values of $L/(AE)$ for all members, Eq. (6.71) may be written as

$$\begin{bmatrix} 16.0 & -3.535 \\ -3.535 & 4.0 \end{bmatrix} \begin{bmatrix} X_1 \\ X_2 \end{bmatrix} = \begin{bmatrix} 190.0 \\ -70.70 \end{bmatrix}$$

Table 6.4

	s_1 (1)	$S_0 s_1$ (2)	$s_1 s_2$ (3)	s_1^2 (4)	S, kips (5)
2	0	0	0	0	−8.90
3	0	0	0	0	−10.00
4	0	0	0	0	0
5	−1.414	−20	0	2.0	0.09
6	1.0	−10	0	1.0	−0.07
7	1.0	−20	−0.707	1.0	−3.77
8	−1.0	−10	0.707	1.0	6.23
9	−1.414	−40	−1.414	2.0	5.33
10	2.0	−60	−1.414	4.0	−3.84
11	1.0	−20	−0.707	1.0	−3.77
12	−2.0	−60	0	4.0	10.14
Total		−240	−3.535	16.0	

Solving for X_1 and X_2 yields

$$X_1 = 9.93 \quad \text{and} \quad X_2 = -8.90$$

where $S_j = (S_0 + 9.93s_1 - 8.90s_2)j$

Table 6.4 shows the calculations for the quantities used in Eq. (6.71). It also gives the final results of the internal loads in each member.

Example 6.14 Find the bending-moment diagram for the frame shown in Fig. 6.14a. The value of EI is constant, and the members are fixed against rotation at the supports.

SOLUTION The structure has three redundant reactions. If the frame is cut at

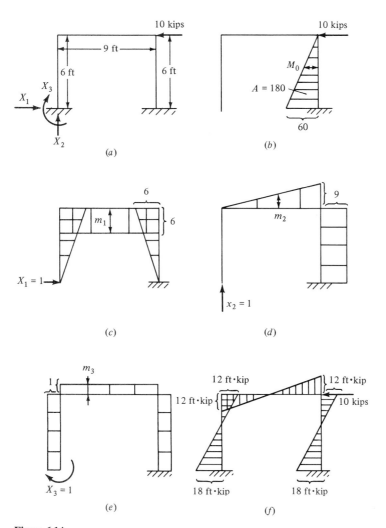

Figure 6.14

the left support, the remaining structure is stable and statically determinate. The two force components and the couple, X_1, X_2, and X_3, are assumed to be redundant reactions. The bending-moment diagram for the base structure under the action of the applied load is shown in Fig. 6.14b. All bending moments are plotted on the compression side of the members and are not designated as positive or negative. The product of two bending moments is positive if they are both plotted on the same side of the member. The bending moments m_1, m_2, and m_3 due to unit values of the redundants are plotted in Figs. 6.14c, d, and e. The various deflection terms are evaluated semi-graphically, by reference to the moment diagrams.

$$EI\delta_{10} = \int M_0 m_1 \, dx = 180 \times 2 = 360$$

$$EI\delta_{20} = \int M_0 m_2 \, dx = -180 \times 9 = -1620$$

$$EI\delta_{30} = \int M_0 m_3 \, dx = -180 \times 1 = -180$$

$$EI\delta_{11} = \int m_1^2 \, dx = 2 \times 18 \times 4 + 6 \times 9 \times 6 = 468$$

$$EI\delta_{22} = \int m_2^2 \, dx = 40.5 \times 6 + 54 \times 9 = 729$$

$$EI\delta_{33} = \int m_3^2 \, dx = 1 \times 1 \times 21 = 21$$

$$EI\delta_{12} = \int m_1 m_2 \, dx = -6 \times 40.5 - 18 \times 9 = -405$$

$$EI\delta_{13} = \int m_1 m_3 \, dx = -1 \times 18 \times 2 - 1 \times 54 = -90$$

$$EI\delta_{23} = \int m_2 m_3 \, dx = 1 \times 40.5 + 1 \times 54 = 94.5$$

Upon substituting in Eq. (6.71), the following is obtained:

$$\begin{bmatrix} 468.0 & -405.0 & -90.0 \\ -405.0 & 729.0 & 94.5 \\ -90.0 & 94.5 & 21.0 \end{bmatrix} \begin{bmatrix} X_1 \\ X_2 \\ X_3 \end{bmatrix} = \begin{bmatrix} -360.0 \\ 1620.0 \\ 180.0 \end{bmatrix}$$

Inverting the matrix and solving for the unknowns yield:

$$X_1 = 5.0 \qquad X_2 = 2.667 \qquad X_3 = 18.0$$

Now, the final bending moments in each member can be obtained from Eq. (6.74) and are shown in Fig. 6.14 f.

6.12 Shear Lag

It was pointed out in Chap. 5 that many of the assumptions made in deriving the simple beam flexure theory are somewhat in error. The assumptions that plane sections remain plane after bending and that bending stresses are proportional to the distance from the neutral axis are less accurate for semimonocoque structures than they are for heavy structures, because the shearing deformations in thin webs are not always negligible.

The effect of shearing deformations in redistributing the bending stresses in a box beam is commonly known as *shear lag*. The effect may be illustrated by considering the cantilever box beam shown in Fig. 6.15. For simplicity, it is assumed that the beam cross section is symmetrical about a vertical centerline and that the load is applied along this centerline, so that there is no torsional deformation. An analysis using the simple beam theory shows that all the stringers on the upper surface are the same for all cross sections. As a result of these shearing stresses, an originally plane cross section will deform to the position indicated by line a' b' c'.

At the support, however, the cross section is restrained from warping out of its original plane, and line abc of Fig. 6.15 remains straight. Since the distance cc' is greater than the distance aa', the stringer at c resists a smaller compressive stress than the stringer at a. Thus, the bending stress at a must be greater than that calculated by the simple flexure theory, and the bending stress at c must be less than indicated by the simple theory. In this case, all the cross sections at some distance from the support warp the same amount, and thus all the stringers have approximately the same bending stress and strain. The shear-lag effect is greatest at the support and is something of a local effect.

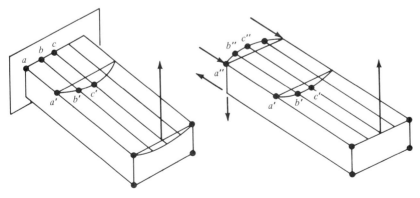

Figure 6.15   Figure 6.16

Many wing structures are spliced only at the spars, so that the stringers resist no bending stress at the splice. The box beam shown in Fig. 6.16 is spliced in this manner, so that only the corner flanges resist axial loads at the left-hand support. In this case, the cross section at the support deforms as indicated by line a'' b'' c'', in an opposite direction to the deformation a' b' c' of cross sections some distance from the support. The middle stringer has a final length c' c'' which is considerably greater than the final length a' a'' of the corner stringer. The shear-lag effect is greater in this beam than in the beam with the entire cross section restrained. The effect is also localized near the support, as the stringers at some distance from the support resist bending stresses, which are approximately as calculated by the simple flexure theory.

The effect of shear lag may be desirable, since it permits a structure to resist higher ultimate bonding moments than are calculated from the simple flexure theory. The allowable bending stresses for the stringers between the spars are smaller than the allowable stresses for the corner flanges, or spar caps. The stringers tend to fail as columns with lengths equal to the rib spacing. The spar caps are supported vertically by the spar web and horizontally by the skin, and they usually resist high compressive stresses.

When a rectangular box beam is subjected to torsion, a cross section tends to warp from its original plane. When one end is restrained against warping, axial loads are induced in the flanges, and the shear flows are redistributed near the fixed end. This also is an effect of the shear deformation and is sometimes referred to as a shear-lag effect.

The extent of the shear-lag effect can be studied by considering the stresses resulting from a few simple conditions of loading and then superimposing them with other stress conditions. The structure with two webs and three stringers, shown in Fig. 6.17a, is assumed to extend for an indefinite length in the x direction and to be loaded as shown. The distribution of the loads and deformations of the x direction are investigated.

The force P in the center stringer is a function of the distance x. In a length dx, the force changes an amount dP, and the web shearing deformation changes an amount $d\gamma$, as shown in Fig. 6.17b. From the spanwise equilibrium of a stringer, the load increase dP results from the shearing stress f_s in the web:

$$dP = -2f_s t\, dx \tag{6.75}$$

The deformation γ results from the web shearing stress:

$$f_s = G\gamma \tag{6.76}$$

The change in the angle γ results from the axial elongation of the stringers:

$$b\, d\gamma = -\left(\frac{P}{AE} + \frac{P}{2A_1 E}\right) dx \tag{6.77}$$

The variables f_s and γ may be eliminated from these three equations in order to obtain a differential equation for P as a function of x. Differentiating Eq. (6.75) and substituting from Eq. (6.76), we have

$$\frac{d^2 P}{dx^2} = -2tG \frac{d\gamma}{dx} \tag{6.78}$$

The value of $d\gamma/dx$ may be substituted from Eq. (6.77):

$$\frac{d^2 P}{dx^2} = k^2 P \tag{6.79}$$

where

$$k^2 = \frac{2tG}{bE} \left(\frac{1}{A} + \frac{1}{2A_1} \right) \tag{6.80}$$

Equation (6.79) may be integrated as follows:

$$P = C_1 e^{kx} + C_2 e^{-kx} \tag{6.81}$$

where C_1 and C_2 are constants of integration. The load P approaches zero at a large value of x; thus for $x = \infty$, $P = 0$ and $C_1 = 0$. At the loaded end, $x = 0$ and $P = P_0$ or $C_2 = P_0$. Therefore, Eq. (6.81) has the following value:

$$P = P_0 e^{-kx} \tag{6.82}$$

Equation (6.82) may be differentiated and equated to Eq. (6.75) in order to obtain an expression for f_s:

$$f_s = \frac{P_0 k}{2t} e^{-kx} \tag{6.83}$$

The displacement δ of the force P_0 is equal to γb, or $f_s b/G$ for $x = 0$:

$$\delta = \frac{P_0 kb}{2tG} \tag{6.84}$$

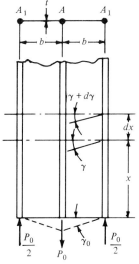

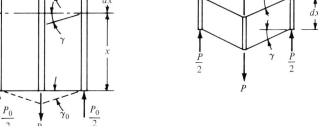

Figure 6.17

The structure shown in Fig. 6.18 may be analyzed in a similar manner to the structure of Fig. 6.17. The flange forces P at any cross section are defined by

$$P = P_0 e^{-kx} \tag{6.85}$$

where
$$k^2 = \frac{4G(1/b + 1/c)}{AE(1/t + 1/t_1)} \tag{6.86}$$

It is assumed that the ribs are closely spaced for Eq. (6.85) to be valid. The shear flows q may be similarly defined by

$$q = \frac{P_0 k}{2} e^{-kx} = q_0 e^{-kx} \tag{6.87}$$

where q_0 is the shear flow when $x = 0$. In the shear-lag portion only, the shear flow must be the same in all four webs to satisfy the equilibrium condition for torsional moments. The warping displacement δ of each of the forces P_0 from the plane of the original cross section is measured as shown in Fig. 6.18b:

$$\delta = \frac{q_0(1/t + 1/t_1)}{2G(1/b + 1/c)} = \frac{P_0 k(1/t + 1/t_1)}{4G(1/b + 1/c)} \tag{6.88}$$

To illustrate shear-lag calculations, consider the box beam of Fig. 6.19a. All the webs have thickness $t = 0.02$ in, and the material has the properties $E = 10^7$ lb/in^2 and $G = 0.4E$. The simple beam theory yields shear flows which are constant for the length of the span, with the values shown, and values of the axial stringer loads as shown. For this theory to apply, however, the cross section at the support must warp so that the middle stringer is displaced the distance δ_0 from the original plane:

$$\delta_0 = \frac{f_s}{G} b = \frac{200 \times 10}{0.020 \times 4,000,000} = 0.025 \text{ in}$$

If the cross section at the support is restrained from warping, the center stringer resists a compression force smaller than 40,000 lb and the corner stringers resist compression forces larger than 20,000 lb. The force P_0 acting as shown

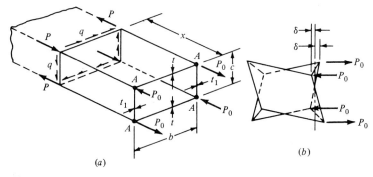

(a)

(b)

Figure 6.18

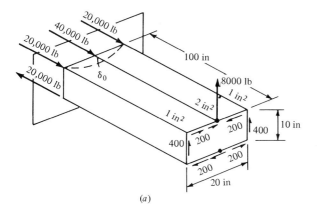

(a)

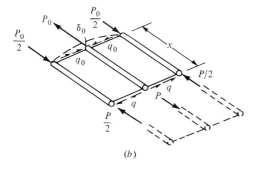

(b)

Figure 6.19

in Fig. 6.19b, which is required to displace the structure a distance δ_0, is calcu-lated from Eq. (6.84). Then the system of forces shown in Fig. 6.19b is superim-posed on those obtained by the simple flexure theory in Fig. 6.19a.

The structure of Fig. 6.19b is equivalent to that of Fig. 6.17. From Eq. (6.80)

$$k^2 = \frac{2 \times 0.020 \times 0.4}{10} \left(\frac{1}{2} + \frac{1}{2} \right) = 0.0016$$

or $\qquad\qquad k = 0.04$

Substituting $\delta = 0.025$ in into Eq. (6.84) and solving for P_0, we have

$$P_0 = \frac{2tG\delta}{kb} = \frac{2 \times 0.020 \times 4 \times 10^6 \times 0.025}{0.04 \times 10} = 10,000 \text{ lb}$$

From Eq. (6.82)

$$P = 10,000e^{-0.04x}$$

and from Eq. (6.83)

$$f_s = 10,000e^{-0.04x}$$

Table 6.5

x	$e^{-0.04x}$	$P = 10{,}000e^{-0.04x}$	$q = 200e^{-0.04x}$
0	1	10,000	200
5	0.817	8,170	163
10	0.670	6,720	134
20	0.450	4,500	90
40	0.202	2,020	40
100	0.019	190	4

or $q - f_s t = 200e^{-0.04x}$. Thus, at the support, the corner stringers each resist compression forces of 25,000 lb, and the center stringer resists a compression force of 30,000 lb. The shear flow is zero at this cross section, which is obviously necessary for the assumed condition of no shearing deformation.

The values of P and q at various distances x from the fixed support are calculated in Table 6.5. These tabulated values must be superimposed on the values shown in Fig. 6.19a. It is observed that the correction forces at a cross section 20 in from the support are less than one-half the values at the support.

The loading condition shown in Fig. 6.19 requires larger corrections for the effects of shear lag than are required in the normal airplane wing. The shear loads in an airplane wing usually are resisted at the side of the fuselage, but the cross section at the center of the fuselage is prevented from warping. Thus the cross section at the side of the fuselage, which has the maximum shear flows, is permitted to warp and distribute the shear flows in almost the same manner as predicted by the simple flexure theory.

PROBLEMS

6.1 Find the displacements of point 3 of the truss structure shown in Fig. P6.1. Assume 1 in^2 for the area of each member and $E = 10^7$ lb/in^2.

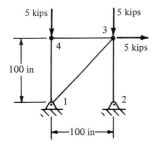

Figure P6.1

6.2 Find the rotation of member 4-5 of the truss structure shown in Fig. P6.2. Area = 1 in^2 and $E = 10^7$ lb/in^2 for all members.

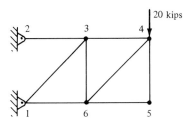

Figure P6.2

6.3 Find the displacements (translation and rotation) of points 2 and 3 of the beam structure shown in Fig. P6.3. Assume that the beam has a uniform area.

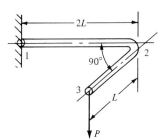

Figure P6.3

6.4 You are given a simply supported beam under the action of a uniformly distributed load, as shown in Fig. P6.4. Find the midspan deflection, using Castigliano's theorem.

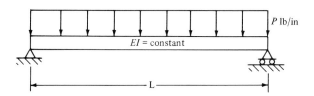

Figure P6.4

6.5 Solve Prob. 6.4 by using the Rayleigh-Ritz method.

6.6 Solve Prob. 6.4 by using the finite difference method.

6.7 Solve Prob. 6.4 by solving the classical beam differential equation $EI\, d^4w/dx^4 = p(x)$.

6.8 Using the finite difference method, solve for the vertical displacements at points a, b, and c and the slopes at the left and right supports of the multispan beam shown in Fig. P6.8. $EI = 10^6$.

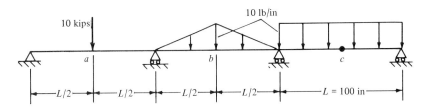

Figure P6.8

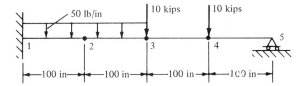

Figure P6.9

6.9 Use the finite difference method to find the moments and shears at the locations indicated on the beam structure shown in Fig. P6.9.

6.10 A square plate of uniform thickness t is fixed at two opposite edges and simply supported at the other two opposite edges. Using the finite difference method, find the maximum displacements and bending moments. Assume that the plate is under the action of a uniform normal pressure P. Use Poisson's ratio $v = 0.25$.

6.11 Analyze the truss of Fig. 6.8, assuming that $L/(AE)$ is constant for all members. Analyze the truss of Fig. 6-8 for a horizontal load of 20 kips at B, in addition to the vertical load shown. Assume AE is constant for all members.

6.12 Analyze the truss of Fig. 6.8 assuming there is an additional member AC with a cross-section area of 3 in^2. Take $B'D$ as the redundant.

6.13 Repeat Prob. 6.12, assuming BC is the redundant member.

6.14 Repeat Prob. 6.12, assuming that member BD is 0.1 in too long because of manufacturing tolerances. Assume no external loads on the structure.

6.15 Analyze the structure of Fig. 6.9 if load P is acting horizontally at the same point. Assume a constant value of EI.

6.16 Analyze the structure of Fig. 6.9 for the loading shown, assuming the supports to be spread horizontally a distance of 0.5 in. Use $R = 50$ in, $P = 2$ kips, $I = 1.0$ in^4, and $E = 10,000$ kips/in^2.

6.17 Analyze the structure of Fig. 6.10, assuming the 2-kip load to be applied at the point where the tube is bent.

6.18 Repeat Example 6.12, assuming $h = 40$ in, $h1 = 30$ in, and $AE = 10,000$ kips for each member.

6.19 Repeat Example 6.13, if $h = 40$ in, $h_1 = 30$ in, and $AE = 10,000$ kips for each member. Member 12 has a length of 10 in.

6.20 Repeat Prob. 6.19 assuming member 12 is a redundant in place of the reaction X_1.

6.21 Analyze the frame of Fig. 6.14a, assuming the left support to be pin-connected and the right support to be fixed.

6.22 Analyze the frame of Fig. 6.14a, assuming an additional load of 20 kips to act down at the center of the structure.

6.23 Obtain the bending-moment diagram for the frame shown in Fig. P6.23 to P6.25 if $P = 0$ and $W = 4.0$ kips/ft.

6.24 Obtain the bending-moment diagram for the frame shown in Fig. P6.23 to P6.25 if $W = 0$, $P = 10$ kips, and $a = b = 4.5$ ft.

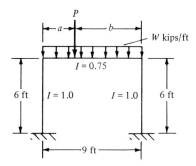

Figure P6.23 to P6.25

6.25 Obtain the bending-moment diagram for the frame shown in Fig. P623 to P6.25 if $W = 0$, $P = 10$ kips, $a = 3$ ft, and $b = 6$ ft.

6.26 The frame for a fuselage of rectangular cross section is shown in Fig. P6.26 to P6.28. Calculate the bending moments if $W = 3$ kips and $a = 1.5$ ft. Assume EI is constant.

6.27 Repeat Prob. 6.26 if EI for the bottom frame member is 4 times as large a₃ the value of EI for the other members.

6.28 Repeat Prob. 6.26 if $2W = 6$ kips and $a = 3.0$ ft.

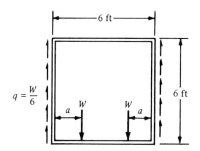

Figure P6.26 to P6.28

6.29 Use the principle of virtual displacements to establish the relationship between the applied forces and corresponding displacements on the beam structure shown in Fig. P6.29. This relationship is of the form $\{Q = [S]\{q\}$.

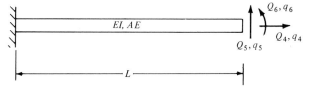

Figure P6.29

6.30 Repeat Prob. 6.29 except the beam is as shown in Fig. P6.30.

Figure P6.30

SEVEN

FINITE ELEMENT STIFFNESS METHOD IN STRUCTURAL ANALYSIS

7.1 INTRODUCTION

In real design cases, generally structural systems are composed of a large assemblage of various structural elements such as beams, plates, and shells or a combination of the three. Their overall geometry becomes extremely complex and cannot be represented by a single mathematical expression. In addition, these built-up structures are intrinsically characterized as having material and structural discontinuities such as cutout, thickness variations of members, etc., as well as discontinuities in loading and support conditions. Given these factors, relative to the structure geometry and discontinuities, it becomes apparent that the classical methods can no longer be used, particularly those whose prerequisites are the formulation and the solution of governing differential equations. Thus, for complex structures, the analyst has to resort to more general methods of analysis where the above factors place no difficulty in their application. These methods are the finite element stiffness method and the finite element flexibility method.

With the advent of high-speed, large-storage-capacity digital computers, the finite element matrix methods have become one of the most widely used tools in the analysis of complex structural systems. These methods are based on the formulations of a simultaneous set of linear algebraic equations relating forces to corresponding displacements (stiffness method) or displacements to corresponding forces (flexibility method) at discrete, preselected points on the structure.

These methods offer many advantages:

1. Capability for complete automation.
2. The structure geometry can be described easily.
3. The real structure can be represented easily by a mathematical model composed of various structural elements.
4. Ability to treat anisotropic material.
5. Ability to treat discontinuities.
6. Ability to implement residual stresses, prestress conditions, and thermal loadings.
7. Ability to treat nonlinear structural problems.
8. Ease of handling multiple load conditions.

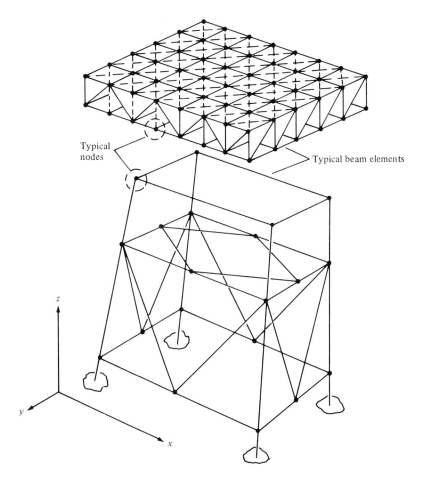

Figure 7.1

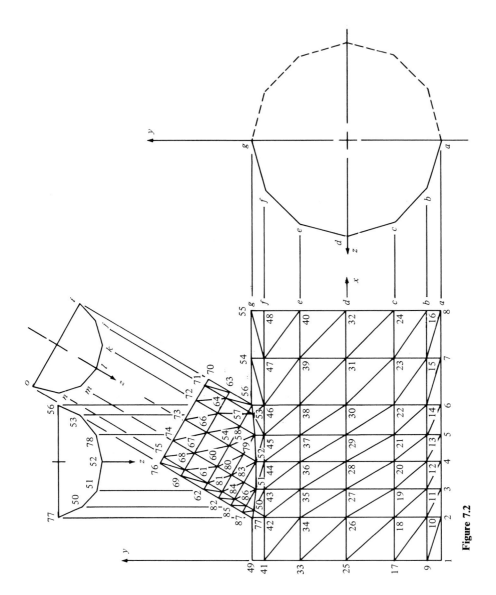

Figure 7.2

The finite element matrix methods have gained great prominence throughout the industries owing to their unlimited applications in the solution of practical design problems of high complexity. For further information on the finite element matrix methods, the reader is encouraged to consult Refs. 10, 24, and 27 to 35.

7.2 MATHEMATICAL MODEL OF THE STRUCTURE

The basic concept of the finite element matrix method in structural analysis is that the real structure can be represented by an equivalent mathematical model which consists of a discrete number of finite structural elements, as shown in Figs. 7.1 to 7.3. The structural behavior of each of these elements may be described by different sets of functions which normally are chosen such that continuity of stresses and/or strains throughout the structure is ensured .

The types of elements which are commonly used in structural idealization are the truss, beam, two-dimensional membrane, shell and plate bending, and three-dimensional solid elements. Figures 7.4 and 7.5 illustrate schematically each el-

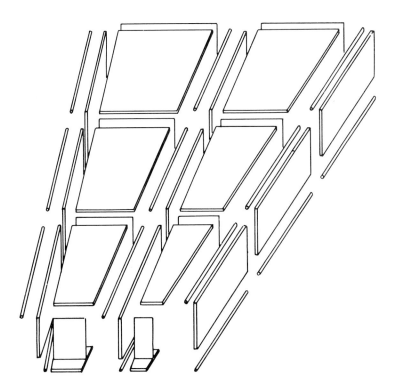

Figure 7.3

ement and the typical nodal forces and corresponding displacements associated with each type.

7.3 ELEMENT DISCRETIZATION

The mathematical relationships which govern the structural behavior of an element are derived on the basis of an idealized element model. For example, once the element shape is selected, it is discretized by placing a finite number of nodes at various locations on the element surface, as shown in Fig. 7.6. Generally

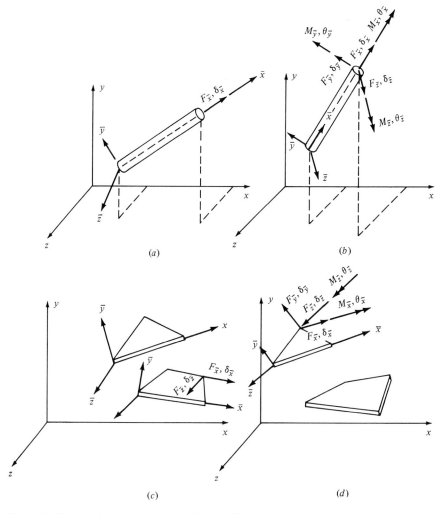

Figure 7.4 Structural elements. (a) Truss element; (b) beam element; (c) membrane plate element, (d) plate bending elements.

speaking, the accuracy increases with an increase in nodal points considered on the element. Likewise, the smaller the element size, the more accurate the analytical results become for a given structural system. Note that the core storage requirement increases rapidly with an increase in the number of element nodes and the number of elements considered in a structure.

7.4 APPLICATIONS OF FINITE ELEMENT MATRIX METHODS

The finite element matrix method has a wide range of applications in structural analysis. Its uses extend to every engineering field, from space structures to land-based and marine structures.

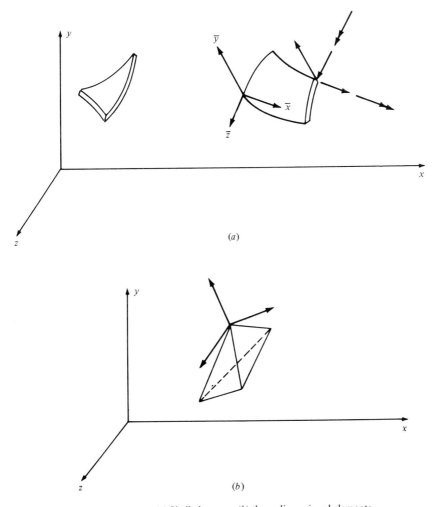

(a)

(b)

Figure 7.5 Structural elements. (a) Shell elements; (b) three-dimensional elements.

7.5 COORDINATE SYSTEM

The coordinate system used here in conjunction with the finite element matrix method is a set of orthogonal axes, x, y, and z, shown in Fig. 7.6a. These axes are referred to as either the global axes or the local axes, respectively, depending on whether the structure or the element is being treated.

In the formulation of element relationships, the element node geometry, forces, and displacements are referenced with respect to the local axes of the element. However, when relationships are being established for a structure, the node geometry, forces, and displacements are referenced with respect to the structure global axes. See Fig. 7.6b.

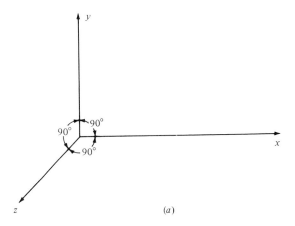

(a)

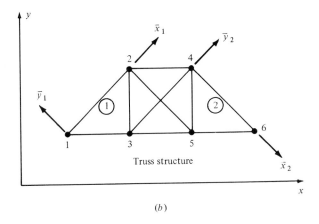

(b)

Figure 7.6 Global and local axes elements. (a) Coordinate system; (b) x, y = global axes for structure; x_1, y_1 = local axes for element 1; x_2, y_2 = local axes for element 2.

7.6 FORCES, DISPLACEMENTS, AND THEIR SIGN CONVENTION

The term *force* is used here to denote both forces and moments; likewise, *displacement* is used to mean both translational and rotational displacements. The forces and their corresponding displacements are assumed positive if they act in the positive direction of their respective axes, as shown in Fig. 7.7. The right-hand rule is used here to represent the vector notation of moments.

In formulating element relationships, it is convenient to adopt the concept of generalized forces and displacements as follows:

$$
\begin{bmatrix} f_1 \\ f_2 \\ f_3 \\ f_4 \\ f_5 \\ f_6 \end{bmatrix} = \begin{bmatrix} F_x \\ F_y \\ F_z \\ M_x \\ M_y \\ M_z \end{bmatrix}, \quad \begin{bmatrix} \delta_1 \\ \delta_2 \\ \delta_3 \\ \delta_4 \\ \delta_5 \\ \delta_6 \end{bmatrix} = \begin{bmatrix} \delta_x \\ \delta_y \\ \delta_z \\ \theta_x \\ \theta_y \\ \theta_z \end{bmatrix} \tag{7.1}
$$

For instance, consider the element shown in Fig. 7.8. The force and displace-

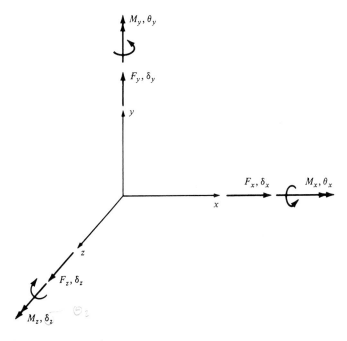

Figure 7.7 Positive forces and displacements.

ment vectors are represented as

$$
\begin{bmatrix} f_1 \\ f_2 \\ f_3 \\ f_4 \\ f_5 \\ f_6 \end{bmatrix}
=
\begin{bmatrix} F_{ix} \\ F_{iy} \\ F_{jx} \\ F_{jy} \\ F_{kx} \\ F_{ky} \end{bmatrix},
\qquad
\begin{bmatrix} \delta_1 \\ \delta_2 \\ \delta_3 \\ \delta_4 \\ \delta_5 \\ \delta_6 \end{bmatrix}
=
\begin{bmatrix} \delta_{ix} \\ \delta_{iy} \\ \delta_{jx} \\ \delta_{jy} \\ \delta_{kx} \\ \delta_{ky} \end{bmatrix}
$$

7.7 STIFFNESS METHOD CONCEPT

The stiffness method, or what is commonly referred to as the displacement method, is based on the principle of superposition of displacements. Consider the plane stress problem shown in Fig. 7.9. At every node of the solid such as i, j, and k, there exist two possible translational displacements in the x and y directions. Corresponding to each displacement there exists a set of induced forces at each node. The method of superposition for linear structures states that the total force induced at a given node in a given direction by all the possible nodal displacements that the solid element may experience can be obtained by simple algebraic summation:

$$f_1 = f_1^1 + f_1^2 + f_1^3 + f_1^4 + f_1^5 + f_1^6$$

$$f_2 = f_2^1 + f_2^2 + f_2^3 + f_2^4 + f_2^5 + f_2^6 \qquad (7.2)$$

$$\vdots$$

$$f_6 = f_6^2 + f_6^3 + f_6^4 + f_6^5 + f_6^6$$

or

$$f_m = \sum_{n=1}^{6} f_n^n = \sum_{n=1}^{6} f_{mn} \qquad (m = 1, 2, \ldots, 6) \qquad (7.3)$$

where $f_{mn} = f_m^n$ = force induced in the direction of coordinate force m as a result of the nodal coordinate displacement n and f_m is the total m coordinate force resulting from all possible coordinate displacements δ_n ($n = 1, 2, \ldots$), as shown in Fig. 7.9.

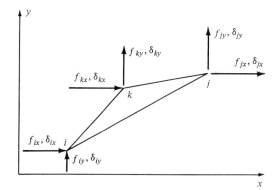

Figure 7.8 Element force and displacement notation.

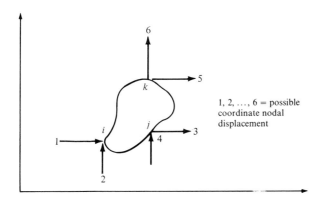

1, 2, ..., 6 = possible coordinate nodal displacement

Figure 7.9 Plane solid with three nodes showing all possible nodal coordinate displacements δ_n ($n = 1, 2, ..., 6$).

For linearly elastic structures, forces are directly proportional to displacements:

$$f = k\delta \qquad (7.4)$$

where f = force
δ = displacement
k = proportionality constant

By utilizing the basic relationship in Eq. (7.4), Eq. (7.3) can be written as

$$f_m = \sum_{n=1}^{6} k_{mn}\delta_n \qquad (m = 1, 2, ..., 6) \qquad (7.5)$$

where k_{mn} is referred to as a constant stiffness coefficient with dimensions of force per unit displacement and is defined as the force induced in the direction of coordinate m due to a unit displacement applied in the direction of nodal coordinate n.

In matrix form, Eq. (7.5) may be rewritten as

$$
\begin{bmatrix} f_1 \\ f_2 \\ f_3 \\ f_4 \\ f_5 \\ f_6 \end{bmatrix} =
\begin{bmatrix}
k_{11} & & & & & \\
k_{21} & k_{22} & & & & \\
k_{31} & k_{32} & k_{33} & & \text{symmetric} & \\
k_{41} & k_{42} & k_{43} & k_{44} & & \\
k_{51} & k_{52} & k_{53} & k_{54} & k_{55} & \\
k_{61} & k_{62} & k_{63} & k_{64} & k_{65} & k_{66}
\end{bmatrix}
\begin{bmatrix} \delta_1 \\ \delta_2 \\ \delta_3 \\ \delta_4 \\ \delta_5 \\ \delta_6 \end{bmatrix} \qquad (7.6)
$$

or, in compact matrix form,

$$\{f\} = [k]\{\delta\} \qquad (7.7)$$

where $\{f\}$ and $\{\delta\}$ are the columns of force and displacement vectors, respectively, and $\{k\}$ is a square matrix of the stiffness coefficients.

Equation (7.7) forms the basis of the stiffness method and establishes the conditions of static equilibrium.

For element relationships, Eq. (7.7) takes the form

$$\{f_m\} = [k_m]\{\delta_m\} \qquad (7.8)$$

where $\{f_m\}$ = internal nodal forces for member m
$\qquad \{\delta_m\}$ = structural nodal displacements associated with nodes of member m
$\qquad \{k_m\}$ = stiffness matrix for mth member

See Fig. 7.10a.

For a structural system such as shown in Fig. 7.10b, Eq. (7.7) becomes

$$\{F\} = [K]\{\Delta\} \qquad (7.9)$$

where $\{F\}$ = structure external applied nodal forces and/or reactions
$\qquad \{\Delta\}$ = structure nodal displacements
$\qquad [K]$ = structure aggregate stiffness matrix

Equation (7.9) set the sum of the internal element forces $[K]\{\Delta\}$ equal to generalized forces $\{F\}$ acting at the nodal points of the structure. The aggregate stiffness matrix $[K]$ can be formed by the direct addition of element stiffness matrices $[k_m]$;

$$[K] = \sum_{m=1}^{N} [k_m] \qquad (7.10)$$

where N = total number of elements in the structure under consideration.

Since Eq. (7.9) represents the relationship between all nodal point forces and corresponding displacements, it may be conveniently rewritten in a matrix partitioned form as

$$\begin{bmatrix} \{F_a\} \\ \{F_r\} \end{bmatrix} = \begin{bmatrix} [K_{aa}] & [K_{ar}] \\ [K_{ra}] & [K_{rr}] \end{bmatrix} \begin{bmatrix} \{\Delta_a\} \\ \{\Delta_r\} \end{bmatrix} \qquad (7.11)$$

where $\{F_a\}$ = specified external nodal forces
$\qquad \{F_r\}$ = unknown nodal reaction forces
$\qquad \{\Delta_a\}$ = unknown nodal displacements
$\qquad \{\Delta_r\}$ = specified nodal displacements

Expanding Eq. (7.11) yields

$$\{F_a\} = [K_{aa}]\{\Delta_a\} + [K_{ar}]\{\Delta_r\} \qquad (7.12)$$

and

$$\{F_r\} = [K_{ra}]\{\Delta_a\} + [K_{rr}]\{\Delta_r\} \qquad (7.13)$$

Equation (7.12) can be solved for the unknown displacement $\{\Delta_a\}$, and then the reactions may be determined by utilizing Eq. (7.13). Element internal forces are determined from element relationships.

To illustrate, consider the simple truss structure shown in Fig. 7.10b. Assume that elements 1 and 2 have cross-sectional areas of $4\sqrt{2}$ and 4 in^2, respectively, and a modulus of elasticity equal to 10^7 lb/in^2. As is shown later, the element relationships are given by

Element 1 :

$$
\begin{bmatrix} f_{1x} \\ f_{1y} \\ f_{2x} \\ f_{2y} \\ f_{3x} \\ f_{3y} \end{bmatrix} = 4 \times 10^5 \begin{bmatrix} 0.5 & 0.5 & 0 & 0 & -0.5 & -0.5 \\ 0.5 & 0.5 & 0 & 0 & -0.5 & -0.5 \\ 0 & 0 & 0 & 0 & 0 & 0 \\ 0 & 0 & 0 & 0 & 0 & 0 \\ -0.5 & -0.5 & 0 & 0 & 0.5 & 0.5 \\ -0.5 & -0.5 & 0 & 0 & 0.5 & 0.5 \end{bmatrix} \begin{bmatrix} \delta_{1x} \\ \delta_{1y} \\ \delta_{2x} \\ \delta_{2y} \\ \delta_{3x} \\ \delta_{3y} \end{bmatrix} \qquad (a)
$$

Element 2 :

$$
\begin{bmatrix} f_{1x} \\ f_{1y} \\ f_{2x} \\ f_{2y} \\ f_{3x} \\ f_{3y} \end{bmatrix} = 4 \times 10^5 \begin{bmatrix} 0 & 0 & 0 & 0 & 0 & 0 \\ 0 & 0 & 0 & 0 & 0 & 0 \\ 0 & 0 & 0 & 0 & 0 & 0 \\ 0 & 0 & 0 & 1 & 0 & -1 \\ 0 & 0 & 0 & 0 & 0 & 0 \\ 0 & 0 & 0 & -1 & 0 & 1 \end{bmatrix} \begin{bmatrix} \delta_{1x} \\ \delta_{1y} \\ \delta_{2x} \\ \delta_{2y} \\ \delta_{3x} \\ \delta_{3y} \end{bmatrix} \qquad (b)
$$

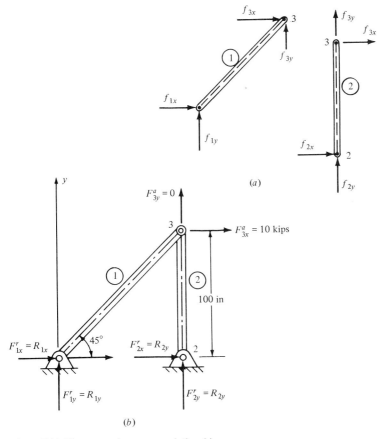

(a)

(b)

Figure 7.10 Element and structure relationships.

If we carry out the matrix addition of the element stiffness matrices in Eqs. (a) and (b), the stiffness matrix relationship for the entire structure becomes

$$
\begin{bmatrix} R_{1x} \\ R_{1y} \\ R_{2x} \\ R_{2y} \\ \hline 10 \\ 0 \end{bmatrix} = 4 \times 10^5 \begin{bmatrix} 0.5 & 0.5 & 0 & 0 & -0.5 & -0.5 \\ 0.5 & 0.5 & 0 & 0 & -0.5 & -0.5 \\ 0 & 0 & 0 & 0 & 0 & 0 \\ 0 & 0 & 0 & 1 & 0 & -1 \\ \hline -0.5 & -0.5 & 0 & 0 & 0.5 & 0.5 \\ -0.5 & -0.5 & 0 & .1 & 0.5 & 1.5 \end{bmatrix} \begin{bmatrix} 0 \\ 0 \\ 0 \\ 0 \\ \hline \Delta_{3x} \\ \Delta_{3y} \end{bmatrix}
\tag{c}
$$

Equation (c) may be expanded as

$$
\begin{bmatrix} 10 \\ 0 \end{bmatrix} = 4 \times 10^5 \begin{bmatrix} 0.5 & 0.5 \\ 0.5 & 1.5 \end{bmatrix} \begin{bmatrix} \Delta_{3x} \\ \Delta_{3y} \end{bmatrix}
\tag{d}
$$

$$
\begin{bmatrix} R_{1x} \\ R_{1y} \\ R_{2x} \\ R_{2y} \end{bmatrix} = 4 \times 10^5 \begin{bmatrix} -0.5 & -0.5 \\ -0.5 & -0.5 \\ 0 & 0 \\ 0 & -1 \end{bmatrix} \begin{bmatrix} \Delta_{3x} \\ \Delta_{3y} \end{bmatrix}
\tag{e}
$$

From Eq. (d), after the reduced aggregate stiffness matrix is inverted, the displacements are

$$\Delta_{3x} = 0.075 \text{ in}$$

$$\Delta_{3y} = 0.025 \text{ in}$$

Having determined the unknown displacements, we can calculate the reactions from Eq. (e):

$$R_{1x} = -10 \text{ kips} \qquad R_{2x} = 0$$

$$R_{1y} = -10 \text{ kips} \qquad R_{2y} = 10 \text{ kips}$$

From Eqs. (a) and (b), the internal loads acting on each element are

Element 1	Element 2
$f_{1x} = -10$ kips	$f_{2x} = 0$
$f_{1y} = -10$ kips	$f_{2y} = 10$ kips
$f_{3x} = 10$ kips	$f_{3x} = 0$
$f_{3y} = 10$ kips	$f_{3y} = -10$ kips

7.8 FORMULATION PROCEDURES FOR ELEMENT STRUCTURAL RELATIONSHIPS

The element stiffness matrix relationship can be derived by several different approaches, all of which will result in a set of algebraic simultaneous equations of the form

$$\{\bar{f}\} = [\bar{k}]\{\bar{\delta}\} \tag{7.14}$$

where $\{\bar{f}\}$ represents the element forces, $[\bar{k}]$ represents the element stiffness matrix, and $\{\bar{\delta}\}$ represents the element displacements.

Depending on the type of element being considered, one of three methods may be conveniently selected to establish the basic element relationship for finite-element analysis: direct method, method of weighted residuals, and energy methods in conjunction with variational principles. Although to cover theory development is not the intent of this text, a brief description of the first and third methods is presented.

Direct Method

The direct method of finite-element formulations consists of the following steps:

1. A set of functions q is chosen to define the element displacements in terms of its nodal displacements.
2. Element strains are expressed in terms of the chosen displacement functions in accordance with the basic strain-displacement relationships of the element.
3. Element stresses are expressed in terms of the displacement functions in accordance with the basic strain-stress relationships of the element.
4. Element node forces are expressed in terms of element stresses in accordance with the statical equivalence of element boundary stresses.

To illustrate, consider the simple element shown in Fig. 7.11.

The displacement of an axial rod element is described by two nodal displacements δ_{ix} and δ_{jx}, as shown in Fig. 7.11. The displacement function $q_x(x)$ may be chosen as a polynomial which must satisfy the nodal boundary conditions

$$q_x(0) = \delta_{ix} \quad \text{and} \quad q_y(L) = \delta_{jx} \tag{a}$$

Hence, the polynomial must be a linear function of x; that is,

$$q_x(x) = c_1 + c_2 x \tag{b}$$

or

$$q_x = \begin{bmatrix} 1 & x \end{bmatrix} \begin{bmatrix} c_1 \\ c_2 \end{bmatrix} \tag{c}$$

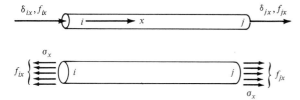

Figure 7.11 Axial rod element.

The displacement function q_x in Eq. (c) can be expressed in terms of the element nodal displacements by using Eq. (a):

$$q_x = \left[\left(1 - \frac{x}{L} \right) \ \frac{x}{L} \right] \begin{bmatrix} \delta_{ix} \\ \delta_{jx} \end{bmatrix} \qquad (d)$$

or

$$q_x = [N_i \ \ N_j] \begin{bmatrix} \delta_{ix} \\ \delta_{jx} \end{bmatrix} \qquad (e)$$

where $N_i = 1 - x/L$ and $N_j = x/L$ are referred to as shape functions, which play a major role in finite-element analysis.

A generalization of Eq. (e) may be written as follows:

$$\{q\} = [N_i \ \ N_j \ \cdots \ N_m] \begin{bmatrix} \delta_i \\ \delta_j \\ \delta_m \end{bmatrix} \qquad (7.15)$$

or

$$\{q\} = [N]\{\delta\} \qquad (7.15a)$$

where

$$\{\delta_i\} = \begin{bmatrix} \delta_{ix} \\ \delta_{iy} \\ \vdots \end{bmatrix} \qquad (7.16)$$

The linear strain-displacement relationship for an axial rod element is given by

$$\epsilon_x = \frac{\partial q_x}{\partial x} \qquad (f)$$

By utilizing Eq. (d), Eq. (f) becomes

$$\epsilon_x = \left[-\frac{1}{L} \ \ \frac{1}{L} \right] \begin{bmatrix} \delta_{ix} \\ \delta_{jx} \end{bmatrix} \qquad (g)$$

A generalization of Eq. (g) may be written as

$$\{\epsilon\} = [G]\{\delta\} \qquad (7.17)$$

The matrix $[G]$ can be calculated easily once the shape function matrix $[N]$ is determined.

For isotropic material, the strain-stress relationship for uniaxial stress is given by

$$\epsilon_x = \frac{\sigma_x}{E}$$

or

$$\sigma_x = E\epsilon_x \qquad (h)$$

For a two-dimensional state of stress, Eq. (h) becomes

$$\{\Sigma\} = [D]\{\epsilon\} \qquad (7.18)$$

where

$$\{\Sigma\} = 0 < \sigma_x \ \ \sigma_y \ \ \sigma_{xy}]^T$$

$$\{\epsilon\} = [\epsilon_x \ \ \epsilon_y \ \ \epsilon_{xy}]^T \qquad (7.19)$$

and $[D]$ is the elasticity matrix. The superscript T in Eq. (7.19) denotes the transpose.

In order to express the stress in terms of the nodal coordinate displacements, Eq. (g) is substituted into Eq. (h):

$$\sigma_x = E\left[-\frac{1}{L} \quad \frac{1}{L} \right]\begin{bmatrix} \delta_{ix} \\ \delta_{jx} \end{bmatrix} \tag{i}$$

The general form of Eq. (i) may be written as

$$\{\Sigma\} = [D][G]\{\delta\} = [S]\{\delta\} \tag{7.20}$$

where $[S]$ is the stress matrix.

As a final step, the stresses are transformed to equivalent nodal forces. This is done by simply multiplying the stress by the cross-sectional area A of the rod:

$$\begin{bmatrix} f_{ix} \\ f_{jx} \end{bmatrix} = \begin{bmatrix} -A \\ \quad A \end{bmatrix}[\sigma_x] \tag{j}$$

or, in general,

$$\{f\} = [H]\{\Sigma\} \tag{7.21}$$

where $[H]$ is a matrix which relates boundary stresses to equivalent element nodal forces.

Utilizing Eq. (i) in Eq. (j) yields the final result:

$$\begin{bmatrix} f_{ix} \\ f_{jx} \end{bmatrix} = \frac{AE}{L}\begin{bmatrix} 1 & -1 \\ -1 & 1 \end{bmatrix}\begin{bmatrix} \delta_{ix} \\ \delta_{jx} \end{bmatrix} \tag{k}$$

In general, Eq. (k) may be written as

$$\{f\} = [H][D][G]\{\delta\}$$

or $$\{f\} = [k]\{\delta\} \tag{7.22}$$

and $$\{k\} = [H][D][G] \tag{7.23}$$

where $\{k\}$ = element stiffness matrix
$[H]$ = matrix relating nodal forces to element boundary stresses
$[D]$ = matrix relating stresses to strains
$[G]$ = matrix relating strains to nodal displacements

Energy Methods

The energy methods in finite-element formulations are based on (1) work and strain energy and on (2) complementary work and complementary strain energy in conjunction with calculus of variation. In the first case, the methods yield the element stiffness matrix, while in the second the methods yield the element flexibility matrix. In this section, only element stiffness matrix formulation is presented.

From the principle of virtual work and virtual strain energy, the finite-element formulation proceeds as follows. The element basic relationships are

$$\{q\} = [N]\{\delta\} \qquad (7.15a)$$

$$\{\Sigma\} = [D]\{E\} \qquad (7.18)$$

$$\{E\} = [G]\{\delta\} \qquad (7.17)$$

The strain energy and work expressions are

$$U = \frac{1}{2} \int_V [E]\{\Sigma\} \, dV \qquad (7.24)$$

and

$$W = \frac{1}{2} [\delta]\{f\} \qquad (7.25)$$

From the principle of virtual displacements

$$\partial U = \frac{1}{2} \int_V [\partial E]\{\Sigma\} \, dV \qquad (7.26)$$

$$\partial W = \frac{1}{2} [\partial \delta]\{f\} \qquad (7.27)$$

$$\{\partial E\} = [G]\{\partial \delta\} \qquad (7.28)$$

where ∂ denotes the first variation.

By utilizing Eqs. (7.17), (7.18), and (7.28), Eq. (7.26) becomes

$$\partial U = \frac{1}{2} \int_V \lfloor \partial \delta \rfloor [G]^T [D][G]\{\delta\} \, dV \qquad (7.29)$$

Equating the virtual work to the virtual strain energy yields

$$\lfloor \partial \delta \rfloor \left(\int_V [G]^T [D][G]\{\delta\} \, dV - \{f\} \right) = 0 \qquad (7.30)$$

or

$$[k]\{\delta\} = \{f\} \qquad (7.31)$$

where $[k]$ is the stiffness matrix and is given by

$$[k] = \int_V [G]^T [D][G] \, dV \qquad (7.32)$$

As an illustration, consider the axial rod element in Fig. 7.11, where

$$[G] = \begin{bmatrix} -\dfrac{1}{L} & \dfrac{1}{L} \end{bmatrix} \qquad (a)$$

$$[G]^T = \begin{bmatrix} -\dfrac{1}{L} \\ \\ \dfrac{1}{L} \end{bmatrix} \qquad (b)$$

$$[D] = [E] \qquad (c)$$

Therefore, the stiffness matrix from Eq. (7.32) is

$$[k] = \int_V \begin{bmatrix} -\dfrac{1}{L} \\[2mm] \dfrac{1}{L} \end{bmatrix} [E] \begin{bmatrix} -\dfrac{1}{L} \\[2mm] \dfrac{1}{L} \end{bmatrix} dV$$

$$= \int_0^L \int_A \begin{bmatrix} \dfrac{E}{L^2} & -\dfrac{E}{L^2} \\[3mm] -\dfrac{E}{L^2} & \dfrac{E}{L^2} \end{bmatrix} dA\ dx$$

$$= \frac{AE}{L} \begin{bmatrix} 1 & -1 \\ -1 & 1 \end{bmatrix}$$

which, again, is the same result as in previous treatments.

7.9 ELEMENT SHAPE FUNCTIONS

We showed that the displacement shape functions play an important role in the formulation of finite-element relationships. Although stress or strain shape functions may be selected, in this section we discuss only those functions that are based on the displacement field.

The criteria for selecting displacement shape functions are:

1. The continuity of the chosen set of functions must prevail throughout the element design.
2. Free-strain conditions throughout the element must exist under rigid-body motion.
3. Constant strain or stress components must be present under the conditions of the chosen set of functions.

There are two means of arriving at shape functions: polynomial series and direct formulation through the interpolation technique.

Polynomial Methods:

Any function Φ can be represented by a polynomial series as follows:

$$\Phi = \sum_{n=0}^{N} P_n = \lfloor P_n \rfloor \{c\} \tag{7.33}$$

where the polynomial P_n of any order n may be written in two dimensions as

$$P_n = \sum_{i=0}^{N} C_{ni} x^{n-i} y^i \tag{7.34}$$

The coefficients C_{ni} are generalized parameters and are chosen such that their total number is at least equal to the total number of assumed nodal coordinate displacements of the element. For example, in case of the axial rod element, only two nodal coordinate displacements are possible. Therefore, for this case, the displacement function must be taken as

$$q_x(x) = P_0 + P_1 = C_{00} + C_{10}x + C_{11}^0 y$$

$$= C_{00} + C_{10}x = [1 \quad x]\begin{bmatrix} C_{00} \\ C_{10} \end{bmatrix} \tag{a}$$

where C_{11} is set equal to zero because q_x cannot be a function of y.

Equation (7.33) may be rewritten in terms of general displacement functions:

$$\{q\} = [P_n]\{c\} \tag{7.35}$$

The column matrix $\{c\}$ must be determined so that Eq. (7.35) is satisfied for all nodal displacements $\{\delta\}$; that is

$$\{\delta\} = [\lambda]\{c\} \tag{7.36}$$

and
$$\{c\} = [\lambda]^{-1}\{\delta\} \tag{7.37}$$

where $\{\delta\}$ = nodal coordinate displacements

$[\lambda]$ = matrix whose coefficients are constant functions of node coordinates

$\{c\}$ = sought coefficients of assumed polynomial

Hence Eq. (7.35) becomes

$$\{q\} = [P_n][\lambda]^{-1}\{\delta\} = [N]\{\delta\} \tag{7.38}$$

where $[N]$ is the matrix of the shape function.

To further illustrate the selection of displacement function polynomials, consider a plane stress triangular element as shown in Fig. 7.12. Consider that the element has only three nodal points, in which case the total number of nodal coordinate displacements is 6. The displacements at any point on the element are

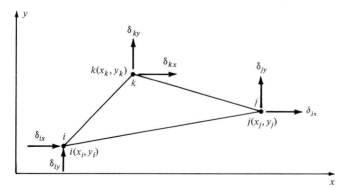

Figure 7.12 Plane stress triangular element.

described by $q_x(x, y)$ and $q_y(x, y)$, which are the displacements in the x and y directions, respectively. Since there are six nodal displacements, the polynomials for q_x and q_y must be chosen so that the total number of coefficient parameters C_{ij} does not exceed three constants in each case; i.e.,

$$\Phi = C_{00} + C_{10} x + C_{11} y$$

Then q_x and q_y will be of the form

$$q_x = C_1 + C_2 x + C_3 y$$

$$q_y = C_4 + C_5 x + C_6 y$$

The shape functions are determined in accordance with Eqs. (7.37) and (7.38).

7.10 ELEMENT STIFFNESS MATRICES

The stiffness matrix for the general beam element shown in Fig. 7.13 can be derived easily by any of the related energy methods discussed previously. One is Castigliano's second theorem:

$$\delta_i = \frac{\partial \bar{U}}{\partial f_i} \tag{7.39}$$

where $\bar{U}$ = complementary strain energy
= strain energy for linear elastic structures
δ_i = nodal coordinate displacements
f_i = corresponding nodal coordinate forces

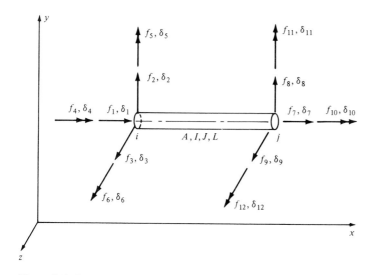

Figure 7.13 General beam element.

For the beam shown, the strain energy is given by

$$U = \frac{1}{2} \int_0^L \left(\frac{M_z^2}{EI_z} + \frac{M_y^2}{EI_y} + \frac{M_x^2}{GJ} + \frac{S^2}{AE} + \frac{V^2}{A_eG} \right) dx \qquad (7.40)$$

where M, S, and V are the internal moments, axial force, and transverse shear force, respectively. These internal loads may be expressed in terms of nodal forces at either end of the beam element:

$$M = g_1(f_i) \qquad S = g_2(f_i) \qquad V = g_3(f_i)$$

Thus

$$U = G(f_i)$$

The stiffness coefficients are obtained by systematically giving each nodal coordinate displacement, one at a time, a unit value, fixing all others, and then calculating the induced reactions. Thus, if δ_1 is set equal to 1 and all other δ's are set equal to 0, the resulting reactions are the stiffness coefficients $k_{i1}(i = 1, 2, \ldots, 12)$. The coefficient k_{i1} is defined as the force induced in the coordinate direction i due to a unit displacement applied in coordinate 1. By performing this operation for every possible displacement, the beam stiffness matrix can be obtained:

$$
\begin{bmatrix} f_{ix} \\ f_{iy} \\ f_{iz} \\ M_{ix} \\ M_{iy} \\ M_{iz} \\ f_{jx} \\ f_{jy} \\ f_{jz} \\ M_{jx} \\ M_{jy} \\ M_{jz} \end{bmatrix}
=
\begin{bmatrix}
k_1 & & & & & & & & & & & \\
0 & k_{4z} & & & & & & & & & & \\
0 & 0 & k_{4y} & & & & & & & & & \\
0 & 0 & 0 & k_5 & & & & & & & & \\
0 & 0 & -k_{2y} & 0 & k_{3y} & & & & & & & \\
0 & k_{2z} & 0 & 0 & 0 & k_{3z} & & & & & & \\
-k_1 & 0 & 0 & 0 & 0 & 0 & k_1 & & & & & \\
0 & -k_{4z} & 0 & 0 & 0 & -k_{2z} & 0 & k_{4z} & & & & \\
0 & 0 & -k_{4y} & 0 & k_{2y} & 0 & 0 & 0 & k_{4y} & & & \\
0 & 0 & 0 & -k_5 & 0 & 0 & 0 & 0 & 0 & k_5 & & \\
0 & 0 & -k_{2y} & 0 & k_{6y} & 0 & 0 & 0 & k_{2y} & 0 & k_{3y} & \\
0 & k_{2z} & 0 & 0 & 0 & k_{6z} & 0 & -k_{2z} & 0 & 0 & 0 & k_{3z}
\end{bmatrix}
\begin{bmatrix} \delta_{ix} \\ \delta_{iy} \\ \delta_{iz} \\ \theta_{ix} \\ \theta_{iy} \\ \theta_{iz} \\ \delta_{jx} \\ \delta_{jy} \\ \delta_{jz} \\ \theta_{jx} \\ \theta_{jy} \\ \theta_{jz} \end{bmatrix}
$$

$$(7.41)$$

Note that from Maxwell's reciprocal theorem[10] for linear elastic structures, $k_{ij} = k_{ji} (i \neq j)$. This can be easily proved by considering

$$\frac{\partial^2 U}{\partial q_i \, \partial q_j} = \frac{\partial^2 U}{\partial q_j \, \partial q_i}$$

where the order of differentiation with respect to q_i and q_j is immaterial. In compact form Eq. 7.41 becomes

$$\{\bar{f}\} = [\bar{k}]\{\bar{\delta}\} \qquad (7.41a)$$

where

$$k_1 = \frac{1}{a_1} \qquad\qquad k_{2z} = \frac{a_2}{b} \qquad k_{3z} = \frac{a_3}{b}$$

$$k_{4z} = \frac{a_4}{b} \qquad\qquad k_5 = \frac{1}{a_5} \qquad k_{6z} = \frac{a_2 L - a_3}{b}$$

$$b = a_3 a_4 - a_2^2 \qquad a_1 = \frac{L}{AE} \qquad a_2 = \frac{L^2}{2EI_z}$$

$$a_3 = \frac{L^3}{3EI_z} + \frac{L}{A_e G} \qquad a_4 = \frac{L}{EI_z} \qquad a_5 = \frac{L}{GJ}$$

The k_{iy} expressions can be obtained by replacing the subscripts z and y in the above definitions. The bars which appear in Eq. (7.41a) denote relationship in local axes of the element.

Equation (7.41) is valid only when nonlinearities, material and/or geometric, do not exist. When large displacements are encountered, the structure behavior is referred to as geometrically nonlinear behavior. In this case, the force equilibrium equations must be formulated while the structure is in its deformed position. The stiffness matrix for the geometrically nonlinear beam element behavior can be derived by any of the previously presented methods in conjunction with the nonlinear strain-displacement relationships.

7.11 FROM ELEMENT TO SYSTEM FORMULATIONS

Element formulations generally are carried out in reference to an element local set of coordinate axes $\bar{x}$, $\bar{y}$, and $\bar{z}$. The stiffness matrix relation for such elements is

$$\{\bar{f}\} = [\bar{k}]\{\bar{\delta}\} \tag{7.42}$$

where the overbar denotes reference to the local axes.

Structural systems are defined as an assemblage of structural elements which may be arbitrarily oriented in space. In order to formulate the structural system force-displacement equations, all element relationships such as Eq. (7.42) must be transformed to a common set of structure global axes x, y, and z.

From vector mathematics it can be shown that any set of three orthogonal axes $\bar{x}$, $\bar{y}$, and $\bar{z}$ can be expressed in terms of another set of orthogonal axes x, y, and z by

$$\begin{bmatrix} \bar{x} \\ \bar{y} \\ \bar{z} \end{bmatrix} = \begin{bmatrix} \lambda_{\bar{x}x} & \lambda_{\bar{x}y} & \lambda_{\bar{x}z} \\ \lambda_{\bar{y}x} & \lambda_{\bar{y}y} & \lambda_{\bar{y}z} \\ \lambda_{\bar{z}x} & \lambda_{\bar{z}y} & \lambda_{\bar{z}z} \end{bmatrix} \begin{bmatrix} x \\ y \\ z \end{bmatrix} \tag{7.43}$$

where $\lambda_{\bar{x}x}$ = cosine of angle between x and $\bar{x}$ axes, etc. See Fig. 7.14.

In general, Eq. (7.43) states that any set of three vectors $\{\bar{\Gamma}\}$ in local coordi-

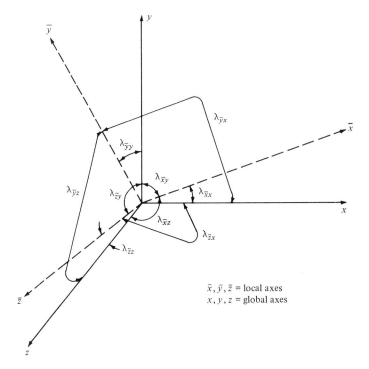

Figure 7.14 Local and global axes.

nates can be transformed to a vector $\{\Gamma\}$ in global coordinates by

$$\{\bar{\Gamma}\} = [\lambda]\{\Gamma\} \tag{7.44}$$

where $[\lambda]$ is the 3×3 transformation matrix given in Eq. (7.43).

In accordance with Eq. (7.44), the following transformations may be accomplished on any force and displacement vectors $\{\bar{f}\}$ and $\{\bar{\delta}\}$:

$$\{\bar{f}\} = [T]\{f\} \qquad \text{and} \qquad \{\bar{\delta}\} = [T]\{\delta\} \tag{7.45}$$

where $[T]$ is a matrix whose diagonal elements are the submatrices $[\lambda]$. The number of elements in the diagonal submatrices depends on the order of the matrix vectors $\{\bar{f}\}$ and $\{\bar{\delta}\}$.

By utilizing Eq. (7.45), Eq. (7.42) becomes

$$[T]\{f\} = [\bar{k}][T]\{\delta\}$$

or
$$\{f\} = [T]^{-1}[\bar{k}][T]\{\delta\} = [k]\{\delta\} \tag{7.46}$$

where $[k]$ is referred to as the element global stiffness matrix and is defined by

$$[k] = [T]^{-1}[\bar{k}][T] = [T]^{T}[\bar{k}][T] \tag{7.47}$$

where the inverse of $[T]$ is its transpose:

$$[T]^{-1} = [T]^T$$

This property holds true for any orthogonal square matrix.

Once the element global stiffness matrix has been found, the global structure stiffness matrix can be constructed easily by simple algebraic summation:

$$[K] = \sum_{i=1}^{m} [k_i] \qquad (7.48)$$

where m is the total number of elements which make up the structure. Note that in order to carry out the summation in Eq. (7.48), the order of $[k_i]$ must be made compatible with the order of the anticipated structure matrix $[K]$. This can be done by filling zero arrays corresponding to all nodal forces and corresponding displacements not connected with the nodes which appear on the element being considered.

To illustrate, consider the structure shown in Fig. 7.15. For each element, the stiffness matrix equation is formulated in accordance with Eq. (7.46). Each element is identified by its connectivity. For example, element 1 has connectivities 1, 3, and 4; element 2 has the connectivities 1 and 2; etc. Each element stiffness matrix $[k]$ is placed in the appropriate location of the global matrix in accord-

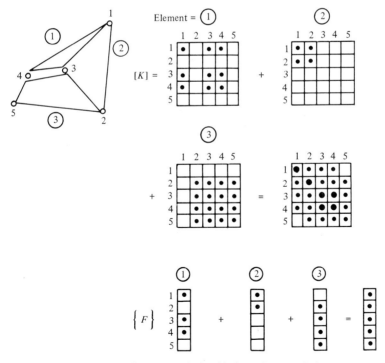

Figure 7.15 Construction of structure relationship from element relations.

ance with its connectivity. Each dot may represent a single coefficient or a sub-matrix of coefficients if more than one degree of freedom is being considered at the node. Once this is done for each element, a simple algebraic summation is performed which yields the structure stiffness matrix equation. Applying the boundary conditions will result in a reduced aggregate structure stiffness matrix equation which can be solved for the unknown displacements. Back substitution of these displacements into each element equation will yield element internal forces.

7.12 GLOBAL STIFFNESS MATRICES FOR SPECIAL BEAM ELEMENTS

The space truss element is shown in Fig. 7.16a. Its global stiffness matrix is

$$
[k]_{st} = \frac{AE}{L}
\begin{bmatrix}
R_{11}^2 & & & & \text{symmetric} & \\
R_{11}R_{12} & R_{12}^2 & & & & \\
R_{11}R_{13} & R_{12}R_{13} & R_{13}^2 & & & \\
-R_{11}^2 & -R_{11}R_{12} & R_{11}R_{13} & R_{11}^2 & & \\
-R_{11}R_{12} & -R_{12}^2 & -R_{12}R_{13} & R_{11}R_{12} & R_{12}^2 & \\
-R_{11}R_{13} & -R_{12}R_{13} & R_{13}^2 & R_{11}R_{13} & R_{12}R_{13} & R_{13}^2
\end{bmatrix}
\begin{matrix}
\delta_{ix}\\ \delta_{iy}\\ \delta_{iz}\\ \delta_{jx}\\ \delta_{jy}\\ \delta_{jz}
\end{matrix}
\tag{7.49}
$$

where $R_{11} = \dfrac{x_j - x_i}{L}$

$R_{12} = \dfrac{y_j - y_i}{L}$

$R_{13} = \dfrac{z_j - z_i}{L}$

$L = [(x_j - x_i)^2 + (y_j - y_i)^2 + (z_j - z_i)^2]^{1/2}$ (7.49a)

$A =$ cross-sectional area of member

$E =$ Young's modulus

The plane truss element is shown in Fig. 7.16b. Its global stiffness matrix is

$$
[k]_{pt} = \frac{AE}{L}
\begin{bmatrix}
R_{11}^2 & & \text{symmetric} & \\
R_{11}R_{12} & R_{12}^2 & & \\
-R_{11}^2 & -R_{11}R_{12} & R_{11}^2 & \\
-R_{11}R_{12} & -R_{12}^2 & R_{11}R_{12} & R_{12}^3
\end{bmatrix}
\begin{matrix}
\delta_{ix}\\ \delta_{iy}\\ \delta_{jx}\\ \delta_{jy}
\end{matrix}
\tag{7.50}
$$

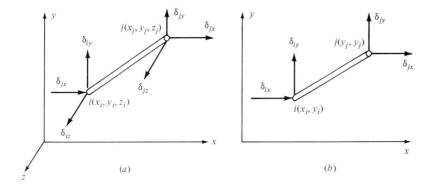

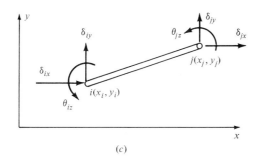

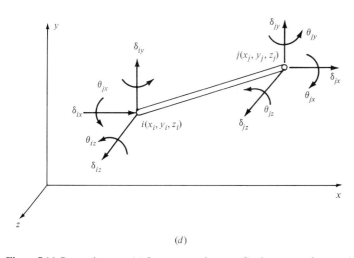

Figure 7.16 Beam elements. (*a*) Space truss element; (*b*) plane truss element; (*c*) plane frame element; (*d*) space frame element.

215

The plane frame element is shown in Fig. 7.16c. Its global stiffness matrix is

$$
[k]_{\text{pf}} =
\begin{array}{c}
\quad\delta_{ix}\quad \delta_{iy}\quad \theta_{iz}\quad \delta_{jx}\quad \delta_{jy}\quad \theta_{jz}\\
\begin{array}{|c|c|c|c|c|c|}
\hline
a & & & \multicolumn{3}{c}{\text{symmetric}}\\
\hline
b & d & & & &\\
\hline
c & e & f & & &\\
\hline
-a & -b & -c & a & &\\
\hline
-b & -d & -e & b & d &\\
\hline
c & e & g & -c & -e & f\\
\hline
\end{array}
\end{array}
\tag{7.51}
$$

where $a = R_{11}^2 \alpha_1 + R_{21}^2 \alpha_{4z}$ $\qquad b = R_{11}R_{12}\alpha_1 + R_{21}R_{22}\alpha_{4z}$

$\quad c = R_{21}\alpha_{2z}$ $\qquad\qquad\qquad d = R_{12}^2\alpha_1 + R_{22}^2\alpha_{4z}$

$\quad e = R_{22}\alpha_{2z}$ $\qquad\qquad\qquad f = \alpha_{3z} \qquad g = \alpha_{6z}$

$$R_{21} = -\frac{y_j - y_i}{L} \qquad\qquad R_{11} = \frac{x_j - x_i}{L}$$

$$R_{22} = \frac{x_j - x_i}{L} \qquad\qquad R_{12} = \frac{y_j - y_i}{L}$$

$$L = [(x_j - x_i)^2 + (y_j - y_i)^2]^{1/2}$$

$$\alpha_1 = \frac{AE}{L} \qquad \alpha_{2z} = \frac{6EI_z}{L^2} \qquad \alpha_{3z} = \frac{4EI_z}{L}$$

$$\alpha_{4z} = \frac{12EI_z}{L^3} \qquad \alpha_{6z} = \frac{2EI_z}{L}$$

I_z = cross-sectional moment of inertia about z axis

Example 7.1 Use the finite element stiffness method to analyze the structures shown in Fig. 7.17.

Case 1   Plane truss problem The joints and their geometries are:

Joint no.	x coordinate, in	y coordinate, in
1	0	0
2	100	0
3	100	100
4	0	100

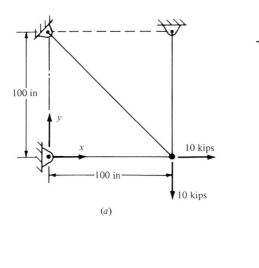

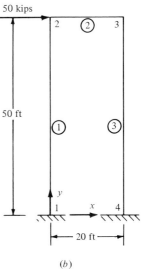

Figure 7.17 Plane truss and frame structures.

The following information is given about each member:

Member no. (and its nodes)	A, in^2	E, lb/in^2	L, in
1 (1–2)	2	10^7	100
2 (2–3)	2	10^7	100
3 (2–4)	$2\sqrt{2}$	10^7	$100\sqrt{2}$

The rotation matrix coefficients for member 1

$$R_{11} = \frac{x_2 - x_1}{L} = \frac{100}{100} = 1$$

and
$$R_{12} = \frac{y_2 - y_1}{L} = 0$$

For member 2, the vertical member with $y_3 - y_2 = 100 - 0 > 0$,

$$R_{11} = 0 \quad \text{and} \quad R_{12} = 1$$

For member 3,

$$R_{11} = \frac{x_4 - x_2}{L} = 0 - 100/100\sqrt{2} = -\sqrt{1/2}$$

$$R_{12} = \frac{y_4 - y_2}{L} = (100 - 0)/100\sqrt{2} = \sqrt{1/2}$$

The known boundary conditions are:

Displacements: $\qquad\qquad \delta_{4x} = \delta_{4y} = \delta_{1x} = \delta_{1y} = \delta_{3x} = \delta_{3y} = 0$

External forces: $\qquad\qquad F_{2x} = 10^k \qquad F_{2y} = -10^k$

The global stiffness matrices [Eq. (7.50)] are

Member 1: $\quad 2 \times 10^5 \begin{bmatrix} 1 & 0 & -1 & 0 \\ 0 & 0 & 0 & 0 \\ -1 & 0 & 1 & 0 \\ 0 & 0 & 0 & 0 \end{bmatrix}$ $\qquad\qquad$ (a)

Member 2: $\qquad 2 \times 10^5 \begin{bmatrix} 0 & 0 & 0 & 0 \\ 0 & 1 & 0 & -1 \\ 0 & 0 & 0 & 0 \\ 0 & -1 & 0 & 1 \end{bmatrix}$ $\qquad\qquad$ (b)

Member 3: $\quad 2 \times 10^5 \begin{bmatrix} 0.5 & -0.5 & -0.5 & 0.5 \\ -0.5 & 0.5 & 0.5 & -0.5 \\ -0.5 & 0.5 & 0.5 & -0.5 \\ 0.5 & -0.5 & -0.5 & 0.5 \end{bmatrix}$ $\qquad\qquad$ (c)

Since there are four joints on the structure with two degrees of freedom at each joint, the order of the structure stiffness matrix is 8 × 8. Expanding each member matrix to 8 × 8 by adding rows and corresponding columns of zeros associated with joints not appearing in the member itself, thus making all matrices compatible in size with the structure matrix, and then using matrix addition yield

$$[k]^{8 \times 8} = \sum_{i=1}^{3} [k_i]^{8 \times 8}$$

$$\textcircled{1} \begin{bmatrix} 1 & & & & & \text{symmetric} & & \\ 0 & 0 & & & & & & \\ -1 & 0 & 1 & & & & & \\ 0 & 0 & 0 & 0 & & & & \\ 0 & 0 & 0 & 0 & 0 & & & \\ 0 & 0 & 0 & 0 & 0 & 0 & & \\ 0 & 0 & 0 & 0 & 0 & 0 & 0 & \\ 0 & 0 & 0 & 0 & 0 & 0 & 0 & 0 \end{bmatrix} \quad +$$

②
$$\begin{bmatrix}
0 & & & & & \text{symmetric} \\
0 & 0 & & & & \\
0 & 0 & 0 & & & \\
0 & 0 & 0 & 1 & & \\
0 & 0 & 0 & 0 & 0 & \\
0 & 0 & 0 & -1 & 0 & 1 \\
0 & 0 & 0 & 0 & 0 & 0 & 0 \\
0 & 0 & 0 & 0 & 0 & 0 & 0 & 0
\end{bmatrix} +$$

③
$$\begin{bmatrix}
0 & & & & & \text{symmetric} \\
0 & 0 & & & & \\
0 & 0 & 0.5 & & & \\
0 & 0 & -0.5 & 0.5 & & \\
0 & 0 & 0 & 0 & 0 & \\
0 & 0 & 0 & 0 & 0 & 0 \\
0 & 0 & -0.5 & 0.5 & 0 & 0 & 0.5 \\
0 & 0 & 0.5 & -0.5 & 0 & 0 & -0.5 & 0.5
\end{bmatrix} =$$

$$[k] = \begin{bmatrix}
1 & & & & & \text{symmetric} \\
0 & 0 & & & & \\
-1 & 0 & 1.5 & & & \\
0 & 0 & -0.5 & 1.5 & & \\
0 & 0 & 0 & 0 & 0 & \\
0 & 0 & 0 & -1 & 0 & 1 \\
0 & 0 & -0.5 & 0.5 & 0 & 0 & 0.5 \\
0 & 0 & 0.5 & -0.5 & 0 & 0 & -0.5 & 0.5
\end{bmatrix} \qquad (d)$$

Hence the matrix equation for the structure becomes

$$\frac{1}{2\times10^5}\begin{bmatrix} R_{1x} \\ R_{1y} \\ \hline 10^k \\ -10^k \\ \hline R_{3x} \\ R_{3y} \\ R_{4x} \\ R_{4y} \end{bmatrix} = \begin{bmatrix}
1 & & & & & & & \text{symmetric} \\
0 & 0 & & & & & & \\ \hline
-1 & 0 & 1.5 & & & & & \\
0 & 0 & -0.5 & 1.5 & & & & \\ \hline
0 & 0 & 0 & 0 & 0 & & & \\
0 & 0 & 0 & -1 & 0 & 1 & & \\
0 & 0 & -0.5 & 0.5 & 0 & 0 & 0.5 & \\
0 & 0 & 0.5 & -0.5 & 0 & 0 & -0.5 & 0.5
\end{bmatrix}\begin{bmatrix} 0 \\ 0 \\ \hline \delta_{2x} \\ \delta_{2y} \\ \hline 0 \\ 0 \\ 0 \\ 0 \end{bmatrix}$$

$$(e)$$

The unknowns are δ_{2x}, δ_{2y}, and all the reactions R_{1x}, R_{1y}, etc. Thus, extracting the third and fourth equations from the above yields

$$\frac{1}{2 \times 10^5} \begin{bmatrix} 10^k \\ -10^k \end{bmatrix} = \begin{bmatrix} 1.5 & -0.5 \\ -0.5 & 1.5 \end{bmatrix} \begin{bmatrix} \delta_{2x} \\ \delta_{2y} \end{bmatrix}$$

Inverting the matrix gives

$$\begin{bmatrix} \delta_{2x} \\ \delta_{2y} \end{bmatrix} = \frac{1}{2 \times 10^5} \begin{bmatrix} 0.75 & 0.25 \\ 0.25 & 0.75 \end{bmatrix} \begin{bmatrix} 10^k \\ -10^k \end{bmatrix} = \begin{bmatrix} 0.025 \\ -0.025 \end{bmatrix} \text{ in}$$

The reactions from Eq. (e) are as follows (see the sketch below):

$$R_{1x} = -(2 \times 10^5)(\delta_{2x}) = -5^k \qquad R_{3x} = 0$$

$$R_{1y} = 0 \qquad\qquad\qquad R_{3y} = 5^k$$

$$R_{4x} = -5^k \qquad\qquad\qquad R_{4y} = 5^k$$

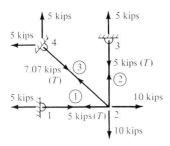

The internal loads are found from the element relationships:

$$\begin{bmatrix} f_{1x} \\ f_{1y} \\ f_{2x} \\ f_{2y} \end{bmatrix} = 2 \times 10^5 \begin{bmatrix} 1 & 0 & -1 & 0 \\ 0 & 0 & 0 & 0 \\ -1 & 0 & 1 & 0 \\ 0 & 0 & 0 & 0 \end{bmatrix} \begin{bmatrix} 0 \\ 0 \\ 0.025 \\ -0.025 \end{bmatrix} = \begin{bmatrix} -5^k \\ 0 \\ 5^k \\ 0 \end{bmatrix}$$

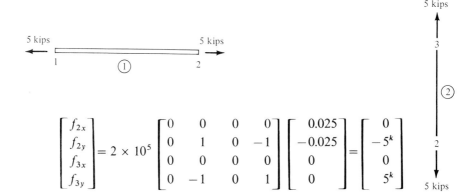

$$\begin{bmatrix} f_{2x} \\ f_{2y} \\ f_{3x} \\ f_{3y} \end{bmatrix} = 2 \times 10^5 \begin{bmatrix} 0 & 0 & 0 & 0 \\ 0 & 1 & 0 & -1 \\ 0 & 0 & 0 & 0 \\ 0 & -1 & 0 & 1 \end{bmatrix} \begin{bmatrix} 0.025 \\ -0.025 \\ 0 \\ 0 \end{bmatrix} = \begin{bmatrix} 0 \\ -5^k \\ 0 \\ 5^k \end{bmatrix}$$

$$
\begin{bmatrix} f_{2x} \\ f_{2y} \\ f_{4x} \\ f_{4y} \end{bmatrix} = 2 \times 10^5
\begin{bmatrix}
0.5 & -0.5 & -0.5 & 0.5 \\
-0.5 & 0.5 & 0.5 & -0.5 \\
-0.5 & 0.5 & 0.5 & -0.5 \\
0.5 & -0.5 & -0.5 & 0.5
\end{bmatrix}
\begin{bmatrix} 0.025 \\ -0.025 \\ 0 \\ 0 \end{bmatrix} =
\begin{bmatrix} 5^k \\ -5^k \\ -5^k \\ 5^k \end{bmatrix}
$$

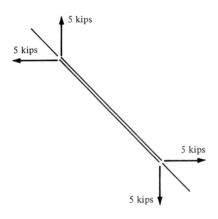

Case 2 Plane frame structure

The joints and their geometrics are as follows:

Joint no.	x coordinate, in	y coordinate, in
1	0	0
2	0	600
3	240	600
4	240	0

The following information is given for each member:

Member no. (and its nodes)	A, in^2	I_z, in^4	E, lb/in^2	L, in
1 (1–2)	20	360	30×10^6	600
2 (2–3)	20	360	30×10^6	240
3 (3–4)	20	360	30×10^6	600

The rotation matrix coefficients are as follows:
Member 1:

$$
R_{11} = \frac{x_2 - x_1}{L} = 0 \qquad R_{12} = \frac{y_2 - y_1}{L} = 1
$$

$$
R_{21} = -\frac{y_2 - y_1}{L} = -1 \qquad R_{22} = \frac{x_2 - x_1}{L} = 0
$$

Member 2: $R_{11} = \dfrac{x_3 - x_2}{L} = 1$ $R_{12} = \dfrac{y_3 - y_2}{L} = 0$

$R_{21} = 0$ $R_{22} = 1$

Member 3: $R_{11} = \dfrac{x_4 - x_3}{L} = 0$ $R_{12} = \dfrac{y_4 - y_2}{L} = -1$

$R_{21} = 1$ $R_{22} = 0$

The rest of the problem is left as an exercise for the student.

Example 7.2 Determine the displacements and internal loads for the truss structure shown in the figure. Assume $E = 10^7$ lb/in² for both rods; $A(\text{rod } 1) = 4\sqrt{2}$ in² and $A(\text{rod } 2) = 4$ in².

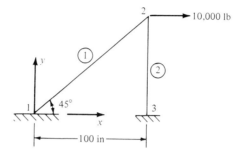

SOLUTION By choosing the system axes as shown, from Eq. (7.49a) the transformation coefficients R_{ij} for each element may be obtained:

Element 1: $R_{11} = \dfrac{x_2 - x_1}{[(x_2 - x_1)^2 + (y_2 - y_1)^2]^{1/2}} = \dfrac{1}{\sqrt{2}}$

$R_{12} = \dfrac{y_2 - y_1}{[(x_2 - x_1)^2 + (y_2 - y_1)^2]^{1/2}} = \dfrac{1}{\sqrt{2}}$

Element 2: $y_3 - y_2 < 0$

$\therefore R_{11} = 0 \qquad R_{12} = -1$

From Eq. (7.50) the element relationships are the following

Element 1: $\alpha_1 = \dfrac{A_1 E}{L} = 4\sqrt{2} \times 10^7 100\sqrt{2} = 4 \times 10^5$

$$
\begin{bmatrix} F_1^x \\ F_1^y \\ F_2^x \\ F_2^y \\ F_3^x \\ F_3^y \end{bmatrix} = 4 \times 10^5
\begin{bmatrix}
0.5 & & & & & \\
0.5 & 0.5 & & & & \\
-0.5 & -0.5 & 0.5 & & & \\
-0.5 & -0.5 & 0.5 & 0.5 & & \\
0 & 0 & 0 & 0 & 0 & \\
0 & 0 & 0 & 0 & 0 & 0
\end{bmatrix}
\begin{bmatrix} \delta_1^x \\ \delta_1^y \\ \delta_2^x \\ \delta_2^y \\ \delta_3^x \\ \delta_3^y \end{bmatrix}
\qquad (a)
$$

Element 2:
$$\alpha_1 = \frac{A_2 E}{L_2} = \frac{4 \times 10^7}{100} = 4 \times 10^5$$

$$
\begin{bmatrix} F_1^x \\ F_1^y \\ F_2^x \\ F_2^y \\ F_3^x \\ F_3^y \end{bmatrix} = 4 \times 10^5
\begin{bmatrix}
0 & & & & & \\
0 & 0 & & \text{symmetric} & & \\
0 & 0 & 0 & & & \\
0 & 0 & 0 & 1.0 & & \\
0 & 0 & 0 & 0 & 0 & \\
0 & 0 & 0 & -1.0 & 0 & 1.0
\end{bmatrix}
\begin{bmatrix} \delta_1^x \\ \delta_1^y \\ \delta_2^x \\ \delta_2^y \\ \delta_3^x \\ \delta_3^y \end{bmatrix}
\qquad (b)
$$

Carrying out the matrix addition of the element stiffness matrices in Eqs. (*a*) and (*b*), the stiffness matrix relationship for the entire structure becomes

$$
\begin{bmatrix} R_{1x} \\ R_{1y} \\ 10^k \\ 0 \\ R_{3x} \\ R_{3y} \end{bmatrix} = 4 \times 10^5
\begin{bmatrix}
0.5 & & & & & \\
0.5 & 0.5 & & \text{symmetric} & & \\
-0.5 & -0.5 & 0.5 & & & \\
-0.5 & -0.5 & 0.5 & 1.5 & & \\
0 & 0 & 0 & 0 & 0 & \\
0 & 0 & 0 & -1.0 & 0 & 1.0
\end{bmatrix}
\begin{bmatrix} 0 \\ 0 \\ \delta_2^x \\ \delta_2^y \\ 0 \\ 0 \end{bmatrix}
\qquad (c)
$$

Extracting the two middle equations from (*c*) yields

$$
\begin{bmatrix} 10^k \\ 0 \end{bmatrix} =
\begin{bmatrix} 2 \times 10^5 & 2 \times 10^5 \\ 2 \times 10^5 & 6 \times 10^5 \end{bmatrix}
\begin{bmatrix} \delta_2^x \\ \delta_2^y \end{bmatrix}
\qquad (d)
$$

Solving for the unknown displacements gives

$$
\begin{bmatrix} \delta_2^x \\ \delta_2^y \end{bmatrix} =
\begin{bmatrix} 2 \times 10^5 & 2 \times 10^5 \\ 2 \times 10^5 & 6 \times 10^5 \end{bmatrix}
\begin{bmatrix} 10^k \\ 0 \end{bmatrix}
\qquad (e)
$$

where the inverse of the matrix in (*e*) is

$$
10^{-5}\begin{bmatrix} 0.75 & -0.25 \\ -0.25 & 0.25 \end{bmatrix}
$$

Therefore the displacements are

$$\delta_2^x = 0.075 \text{ in}$$

$$\delta_2^y = -0.025 \text{ in}$$

Displacements known, the external reactions on the structure and the element internal loads may be found from Eqs. (*c*), (*a*), and (*b*), respectively:

$$
\begin{bmatrix} R_{1x} \\ R_{1y} \end{bmatrix} =
\begin{bmatrix} -2 \times 10^5 & -2 \times 10^5 \\ -2 \times 10^5 & -2 \times 10^5 \end{bmatrix}
\begin{bmatrix} 0.075 \\ -0.025 \end{bmatrix} =
\begin{bmatrix} -10^4 \\ -10^4 \end{bmatrix} \text{ lb}
$$

$$
\begin{bmatrix} R_{3x} \\ R_{3y} \end{bmatrix} =
\begin{bmatrix} 0 & 0 \\ 0 & -4 \times 10^5 \end{bmatrix}
\begin{bmatrix} 0.075 \\ -0.025 \end{bmatrix} =
\begin{bmatrix} 0 \\ 10^4 \end{bmatrix} \text{ lb}
$$

$$\begin{bmatrix} F_1^x \\ F_1^y \\ F_2^x \\ F_2^y \end{bmatrix} = 4 \times 10^5 \begin{bmatrix} 0.5 & & & 0 \\ 0.5 & 0.5 & \text{symmetric} & 0 \\ -0.5 & -0.5 & 0.5 & & 0.075 \\ -0.5 & -0.5 & 0.5 & 0.5 & -0.025 \end{bmatrix} = \begin{bmatrix} -10^4 \\ -10^4 \\ 10^4 \\ 10^4 \end{bmatrix} \text{lb}$$

$$\begin{bmatrix} F_2^x \\ F_2^y \\ F_3^x \\ F_3^y \end{bmatrix} = 4 \times 10^5 \begin{bmatrix} 0 & & & 0.075 \\ 0 & 1 & \text{symmetric} & -0.25 \\ 0 & 0 & 0 & & 0 \\ 0 & -1 & 0 & 1 & 0 \end{bmatrix} = \begin{bmatrix} 0 \\ -10^4 \\ 0 \\ 10^4 \end{bmatrix} \text{lb}$$

A free-body sketch of the internal loads is shown below.

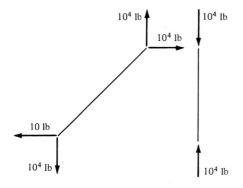

Example 7.3 Assume that the structure of Example 7.2 has rigid joints as shown in the figure below. Determine the displacements and internal loads. Neglect shear deformation. I(beam 1) $= 10\sqrt{2}$ in^4, I(beam 2) $= 10$ in^4.

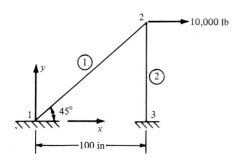

SOLUTION The transformation coefficients R_{ij} are obtained from Eq. (7.51):

Beam 1:

$$R_{11} = \frac{1}{\sqrt{2}} \qquad R_{12} = \frac{1}{\sqrt{2}}$$

$$R_{21} = -\frac{1}{\sqrt{2}} \qquad R_{22} = \frac{1}{\sqrt{2}}$$

Beam 2:

$$R_{11} = 0 \qquad R_{12} = -1$$
$$R_{21} = 1 \qquad R_{22} = 0$$

From Eq. (7.51) the element stiffness matrix relationships are as follows:
Beam 1:

$$
\begin{bmatrix} F_1^x \\ F_1^y \\ M_1^z \\ F_2^x \\ F_2^y \\ M_2^z \end{bmatrix} = 10^6
\begin{bmatrix}
0.2003 & & & & & \\
0.1997 & 0.2003 & & \text{symmetric} & & \\
-0.03 & 0.03 & 4 & & & \\
-0.2003 & -0.1997 & 0.03 & 0.2003 & & \\
-0.1997 & -0.2003 & -0.03 & 0.1997 & 0.2003 & \\
-0.03 & 0.03 & 2 & 0.03 & -0.03 & 4
\end{bmatrix}
\begin{bmatrix} \delta_1^x \\ \delta_1^y \\ \theta_1^z \\ \delta_2^x \\ \delta_2^y \\ \theta_2^z \end{bmatrix} \qquad (a)
$$

Beam 2:

$$
\begin{bmatrix} F_2^x \\ F_2^y \\ M_2^z \\ F_3^x \\ F_3^y \\ M_3^z \end{bmatrix} = 10^6
\begin{bmatrix}
0.0012 & & & & & \\
0 & 0.4 & & \text{symmetric} & & \\
0.06 & 0 & 4 & & & \\
-0.0012 & 0 & -0.06 & 0.0012 & & \\
0 & -0.4 & 0 & 0 & 0.4 & \\
0.06 & 0 & 2 & -0.06 & 0 & 4
\end{bmatrix}
\begin{bmatrix} \delta_2^x \\ \delta_2^y \\ \theta_2^x \\ \delta_3^x \\ \delta_3^y \\ \theta_3^z \end{bmatrix} \qquad (b)
$$

As in Example 7.1, the overall stiffness matrix relationship for the whole structure may be easily constructed from Eqs. (*a*) and (*b*):

$$
\begin{bmatrix} R_1^x \\ R_1^y \\ FM_1^z \\ 10_3 \\ 0 \\ 0 \\ R_3^x \\ R_3^y \\ FM_3^z \end{bmatrix} = 10^6
\begin{bmatrix}
0.2003 & & & & \text{symmetric} & & & & \\
0.1997 & 0.2003 & & & & & & & \\
0.03 & 0.03 & 4.0 & & & & & & \\
0.2003 & -0.2003 & -0.03 & 0.2015 & & & & & \\
0.1997 & -0.2003 & -0.036 & 0.1997 & 0.6003 & & & & \\
0.03 & 0.03 & 2.0 & 0.091 & -0.03 & 8.0 & & & \\
0 & 0 & 0 & -0.0012 & -0.0012 & -0.06 & 0.0012 & & \\
0 & 0 & 0 & & -0.4 & 0 & 0 & 0.4 & \\
0 & 0 & 0 & 0.06 & 0 & 2.0 & -0.06 & 0 & 4
\end{bmatrix}
\begin{bmatrix} 0 \\ 0 \\ 0 \\ \delta_2^x \\ \delta_2^y \\ \theta_2^z \\ 0 \\ 0 \\ 0 \end{bmatrix} \qquad (c)
$$

where *FM* denotes fixed end moment reactions.

Extracting from (*c*) the equations corresponding to the nonzero displacements yields

$$
\begin{bmatrix} 10^3 \\ 0 \\ 0 \end{bmatrix} = 10^6
\begin{bmatrix}
0.2015 & \text{symmetric} & \\
0.1997 & 0.6003 & \\
0.09 & -0.03 & 8.0
\end{bmatrix}
\begin{bmatrix} \delta_2^x \\ \delta_2^y \\ \theta_2^z \end{bmatrix} \qquad (d)
$$

Carrying out matrix inversion produces the unknown displacements:

$$
\begin{bmatrix} \delta_2^x \\ \delta_2^y \\ \theta_2^z \end{bmatrix} = 10^{-6}
\begin{bmatrix}
7.415 & -2.468 & -0.0371 \\
-2.468 & 2.488 & 0.01859 \\
-0.0371 & 0.01859 & 0.1249
\end{bmatrix}
\begin{bmatrix} 10^4 \\ 0 \\ 0 \end{bmatrix} =
\begin{bmatrix} 0.07415 \text{ in} \\ -0.02468 \text{ in} \\ -0.000371 \text{ rad} \end{bmatrix}
$$

Note that for large structural systems, the matrix order becomes quite large and the conventional matrix inversion approach is avoided. Other means of solving large sets of simultaneous algebraic equations are available.

The internal loads are calculated from Eqs. (a) and (b) and are shown in the sketch below.

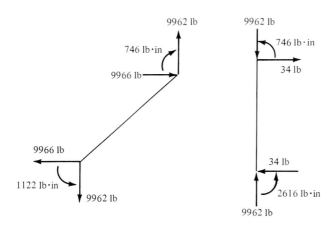

PROBLEMS

7.1 In the following set of algebraic equations, R_1, δ_2, and δ_3 are the unknowns. Put them in a matrix form, and solve for the unknowns. *Use only matrix notation and manipulation.*

$$R_1 = \delta_1 - \delta_2$$
$$P = -\delta_1 + 1.5\delta_2 - 0.5\delta_3$$
$$2P = \delta_1 - 0.5\delta_2 + 0.5\delta_3$$

7.2 A structure is acted on by the forces shown in Fig. P7.2. If the material behavior is given by $\sigma = K\varepsilon^2$, where K is a constant, find the strain energy stored in the system. Neglect the dead weight, and assume AE to be constant.

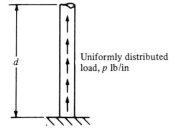

Uniformly distributed load, p lb/in

Figure P7.2

7.3 Find the matrix relationship $\{Q\} = [A]\{q\}$ for the truss structure shown in Fig. P7.3. *Note:* Assume AE to be constant and the same for both members.

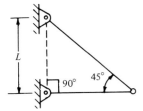

Figure P7.3

7.4 Given the truss structures shown in Fig. P7.4, determine the following information for each:
  (*a*) The external nodal applied loads to be used
  (*b*) The known displacement boundary conditions
  (*c*) The size of the element stiffness matrix
  (*d*) The size of the structure reduced stiffness matrix

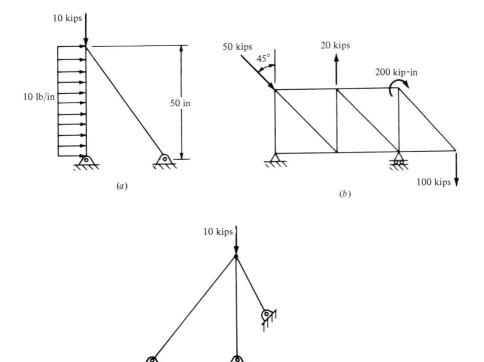

Figure P7.4 (*a*) Planar truss; (*b*) planar truss; (*c*) space truss.

7.5 Using the finite element method, determine the reactions, displacements and internal loads for the planar truss structure shown below. *Show all the details.*

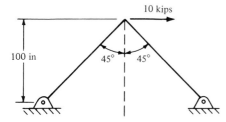

Figure P7.5 Cross-sectional area $= 2$ in^2 for each member. Material $E = 10^7$ lb/in^2.

7.6 Write mathematical expressions which will determine for any structure (a) the order of the aggregate stiffness matrix and (b) the order of the reduced aggregate stiffness matrix. Illustrate the use of the derived expressions on structures shown in Fig. P7.6.

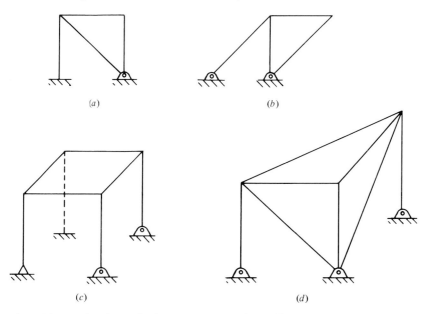

(a)

(b)

(c)

(d)

Figure P7.6 (a) Plane frame; (b) plane truss; (c) space frame; (d) space truss.

7.7 Find the joint loads which must be used in the finite-element analysis of the frame structure shown in Fig. P7.7.

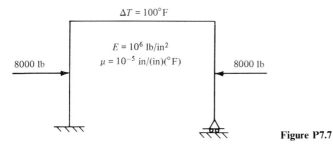

$\Delta T = 100°F$

$E = 10^6$ lb/in^2

$\mu = 10^{-5}$ in/(in)(°F)

8000 lb

8000 lb

Figure P7.7

7.8 Show how you would go about analyzing the beam on continuous elastic foundation shown in Fig. P7.8 by using the finite-element technique.

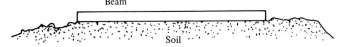

Beam

Soil

Figure P7.8

7.9 What displacement functions would you use to develop the stiffness matrix relationship for each of the elements shown in Fig. P7.9?

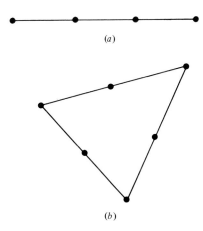

(a)

(b)

Figure P7.9 (a) Axial rod element idealized into four nodes; (b) plate element (in plane forces only); six nodes.

7.10 Find the strain energy expression per unit volume of a structural member whose stress-strain behavior is related as

$$\sigma = c\epsilon^n$$

where c and n are constants.

7.11 Find the total displacement of point 2 on the truss structure in Fig. P7.11. Assume $E = 10^7$ lb/in^2 and cross-sectional area of each member $= 2$ in^2.

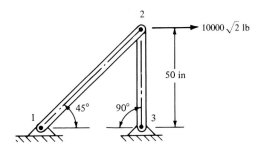

2

$10000\sqrt{2}$ lb

50 in

1   45°   90°   3

Figure P7.11

7.12 The displacements for a two-dimensional solid in plane stress are given by

$$q_x = a_1 + a_2 x + a_3 y \qquad q_y = b_1 + b_2 x + b_3 y$$

Find the strain energy per unit volume of the solid. Assume the solid is homogeneous and isotropic.

7.13 Show the structure global axes and each element's axes for the structure in Fig. P7.13.

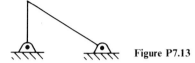

Figure P7.13

7.14 An axial rod element is idealized as shown in Fig. P7-14. The displacement function $q(x)$ is assumed to be

$$q(x) = c_1 + c_2 x + c_3 x^2$$

How would you derive the element stiffness matrix equation?

Figure P7.14

7.15 Derive the stiffness coefficients for the element shown in Fig. P7.15.

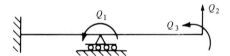

Figure P7.15

7.16 Illustrate by means of sketches the physical significance of the stiffness coefficients corresponding to the actions shown in Fig. P7.16.

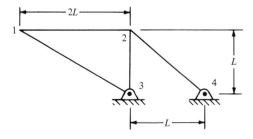

Figure P7.16

7.17 Find the coefficients of the transformation matrices for each of the truss structure elements shown in Fig. P7.17.

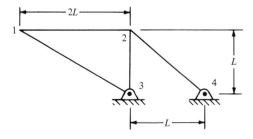

Figure P7.17

7.18 Find the equivalent nodal loads to be used in the finite-element analysis for each of the structures in Fig. P7.18. You are given that

$$\Delta T = 100°F, \qquad \alpha = 10^{-6} \text{ in}/(\text{in} \cdot °F)$$

$$E = 10^7 \text{lb/in}^2 \qquad A = 10 \text{ in}^2 \qquad I = 100 \text{ in}^4$$

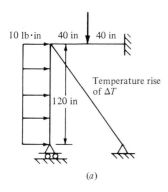

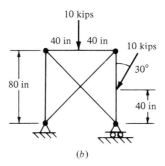

(a)   (b)

Figure P7.18 (a) Frame.

7.19 The radius of a tapered, solid, tubular shaft element varies as follows:

$$R = R_0 e^{\beta x} \qquad R_0, \beta = \text{const}$$

Assuming that only torsional loads can be transmitted, find the stiffness matrix of the element.

7.20 The nodal displacements of the truss structure shown in Fig. P7.20 were found by the finite element to be

$$\delta_{2x} = -0.02 \text{ in} \qquad \delta_{2y} = -0.01 \text{ in}$$

Find the internal loads on element 2. $AE = 10^7$ lb.

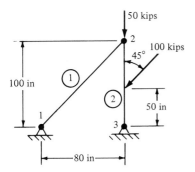

Figure P7.20

7.21 Determine the displacements and the internal loads for the truss structure shown in Fig. P7.21. Assume a cross-sectional area of 1 in² for each rod and a modulus of elasticity of 10^7 lb/in². Use the stiffness method.

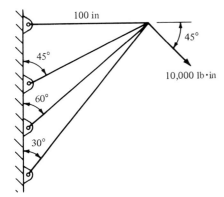

Figure P7.21

7.22 In each of the structures in Fig. P7.22, determine the internal loads and the deflections. Use the stiffness matrix method only.

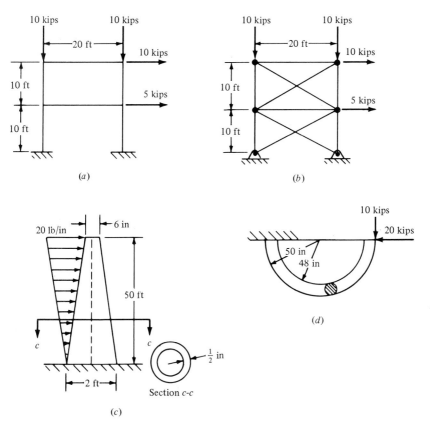

Figure P7.22 (a) Rigid frame; all members are made of 2-in tubing with 3/8-in wall thickness. (b) Truss: 2-in tubing; 3/8-in wall thickness.

EIGHT

ANALYSIS OF TYPICAL MEMBERS
OF SEMIMONOCOQUE STRUCTURES

8.1 INTRODUCTION

This chapter presents approximate methods for the analysis of typical members of semimonocoque structures. Inherently, these structures are highly redundant, and an accurate analysis would require the use of a computer in conjunction with the matrix methods discussed in Chap. 6. Cutouts, shear lag, warping restraint, discontinuity of loads, etc. are some of the factors which affect the accuracy of the analysis, and their inclusion is what makes computer solutions inevitable.

8.2 DISTRIBUTION OF CONCENTRATED LOADS TO THIN WEBS

Modern aircraft structures are constructed primarily from sheet metal. The metal is necessary for a covering and this is utilized for structure as well. The thin sheets or webs are very efficient in resisting shear or tension loads on the planes of the webs, but usually they must be stiffened by members more capable of resisting compression loads and loads normal to the web. When no stiffening members are used and the skin or shell is designed to resist all loads, the construction is called *monocoque*, or *full monocoque*, from the French word meaning "shell only." Usually it is not feasible to have the skin thick enough to resist compression loads, and stiffeners are provided to form semimonocoque structures. In such structures, the thin webs resist tension and shearing forces in the planes of the webs. The stiffeners resist either compression forces in the plane of the web or small, distributed loads normal to the plane of the web.

When semimonocoque structures must resist large, concentrated loads, it is necessary to transmit the loads to the planes of the webs. Since the concentrated loads may have components along three mutually perpendicular axes, it is necessary to provide webs in different planes, so that the loads may be applied at the intersection of two planes. A fuselage structure, for example, has closely spaced rings or bulkheads which resist loads in transverse planes, while the fuselage shell resists loads in the fore-and-aft direction. Concentrated loads must be applied at the intersection of the plane of the bulkhead and the shell, or else additional structure members must be provided to span between bulkheads and transfer the loads to two such intersecting planes.

When a concentrated load is applied to the plane of a web, a stiffening member is required to distribute this load to the web, as shown in Fig. 8.1a. This member should be in the direction of the load, or the load should be applied at the intersection of two stiffeners, so that each stiffener resists the load component in its direction. The load P shown in Fig. 8.1 is distributed to the web by the stiffener AB. The shear flows q_1 and q_2 in the adjacent webs are approximately constant for the length of the stiffener. The axial load in the stiffener, therefore, varies linearly from P at point B to 0 at point A, as shown in Fig. 8.1c. From the

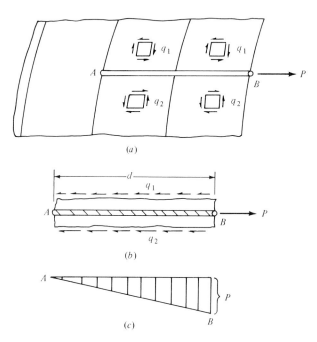

(a)

(b)

(c)

Figure 8.1

equilibrium of the forces shown in Fig. 8.1b, $P = (q_1 + q_2)d$. Thus the required length d of the stiffener depends on the ability of the webs to resist shear, since a longer stiffener reduces shear flows q_1 and q_2. The end of the stiffener, point A, should always be at a transverse stiffener. If a stiffener ends in the center of a web, it produces abrupt change in the shear flows at the end of the stiffener and undesirable concentration conditions.

In this chapter, thin webs are assumed to resist pure shear along their boundaries. In actual structures, the thin webs may wrinkle in shear, thus introducing tension field stresses in addition to those calculated. The effects of tension field stresses are calculated in later chapters. It is found at that time that the tension field stresses can be superimposed readily on those calculated by the methods used here, and the methods used in this chapter remain valid for obtaining the shear distribution in tension field webs. In some cases, the tension field stresses produced by wrinkling of the webs induce additional axial compression loads in stiffeners. These loads should be computed separately and added algebraically to the loads obtained in this chapter.

A study of a simple numerical example demonstrates the method by which loads are distributed to shear webs. The beam shown in Fig. 8.2a is similar to a wing rib which is supported by spars at the ends and which resists the load of 3000 lb, as shown. The stiffener AB transmits this load to the two webs in inverse proportion to the horizontal lengths of the webs, since the vertical shear at any cross section of the beam must be in equilibrium with the external reaction on the beam. The axial load in AB is shown in Fig. 8.2c; it varies from 3000 at B to 0 at A. The axial load in the upper flange of the beam can be obtained from either the

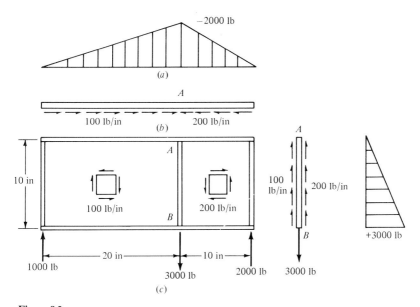

Figure 8.2

bending-moment diagram of the beam or a summation of the shear flows, as shown in Fig. 8.2*b*. The compression at point *A* of 2000 lb can be obtained from the shear flow of 100 lb/in for 20 in or from the shear flow of 200 lb/in for 10 in.

The cantilever beam shown in Fig. 8.3 resists a load *R* which has a horizontal component of 1500 lb and a vertical component of 3000 lb. The horizontal stiffener *AB* must be provided to resist the horizontal component of the load, and the vertical stiffener *CBD* must resist the vertical component. The intersection of these stiffeners, point *B*, should be on the line of action of *R*. The shear flows q_1 and q_2 can be obtained from the equilibrium of these stiffeners. For stiffener *AB* to be in equilibrium under the forces shown in Fig. 8.3*c*,

$$10q_1 - 10q_2 = 1500$$

Similarly, for member *CBD* to be in equilibrium under the forces shown in Fig. 8.3*b*,

$$5q_1 + 10q_2 = 3000$$

Solving the above two equations simultaneously yields $q_1 = 300$ and $q_2 = 150$ lb/in. These values also can be derived by analyzing the beam separately for each of the two load components and then superimposing the results. The vertical load alone would produce shear flows of 200 lb/in in each web while the horizontal

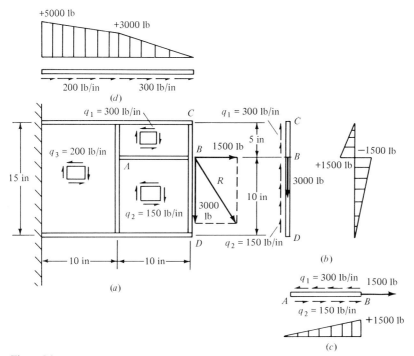

Figure 8.3

load would produce a shear flow of 100 lb/in in the upper web and − 50 lb/in in the lower web. The axial loads in the upper flange member, shown in Fig. 8.3*d*, could not be obtained readily from a bending moment diagram of the member.

The loads considered above were assumed to act in the plane of the web. When loads have components along all three references axes, the structure should be arranged so that the loads act as the intersection of two webs, as shown in Fig. 8.4*a*. Here each of the three components of the force *R* is distributed to the webs by a stiffener in the direction of the force component. In some cases, this is not practical and a load normal to a web, as shown in Fig. 8.4*b*, cannot be avoided. If the load is small, the stiffener may be designed to have enough bending strength to resist the load. In many cases, the loads are such that it is necessary to provide an additional member, such as web *ABCD* in Fig. 8.4*c*, to resist the load. This member spans between ribs or bulkheads and can resist any load in its plane by means of the three reactions F_1, F_2, and F_3 shown in Fig. 8.4*d*. Even small loads such as those from brackets supporting control pulleys should not be applied as normal loads to an unsupported web. Such brackets may be attached to stiffeners or may be located at the intersections of webs.

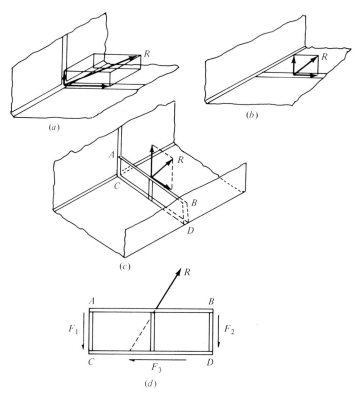

Figure 8.4

8.3 LOADS ON FUSELAGE BULKHEADS

The structural unit which transfers concentrated loads to the shell of an airplane fuselage or wing is commonly called a *bulkhead*. Bulkheads are attached to the wing or fuselage skin continuously around their perimeters. They may be solid webs with stiffeners or beads, webs with access holes, or truss structures. Fuselage bulkheads usually are open rings or frames, so that the fuselage interior is not obstructed. Normally the chordwise bulkheads in wings are called *ribs*, while fuselage bulkheads are called rings or frames. In addition to transferring loads to the skin, wing and fuselage bulkheads supply column support to stringers and redistribute shear flows in the skin. The first step in the design of a bulkhead is to obtain the loads which act on the bulkhead and thus hold it in static equilibrium. In the case of fuselage rings, this step is simpler than the next problem—to obtain the unit stresses from the loads. Unit stresses in fuselage rings and similar structures are analyzed in a later chapter on statically indeterminate structures.

Fuselage shells normally are symmetrical about a vertical centerline and often are loaded symmetrically with respect to the centerline. The fuselage bending stresses can be obtained by the simple flexure formula $f = My/I$, and the fuselage shear flows can be found from the related expression derived in Chap. 5,

$$q = \frac{V_w}{I} \int y \, dA \tag{5.31}$$

In applying Eq. (5.31) to a symmetrical box structure, often it is convenient to consider only half of the structure, since the shear flow must be zero at the top and bottom centerlines. Thus, each term of Eq. (5.31) applies to only half the fuselage shell. If stringers or longerons are located on the top or bottom centerlines, half of their area is considered to act with each side of the structure.

The fuselage ring shown in Fig. 8.5 is loaded by a vertical load P on the centerline of the airplane. This vertical load P must be in equilibrium with the running loads q which are applied to the perimeter of the ring, as shown in Fig. 8.5c. The present problem is to obtain the distribution of the forces q. The fuselage cross section just forward of the ring has an external shear V_a, and the cross section aft of the ring has a shear V_b, as shown in Fig. 8.5a and b. The load P on the ring must be equal to the difference of these shears:

$$V_a - V_b = P$$

If, for the moment, the shear resisted by the in-plane components of the stringer loads is neglected, the shear flows on the two cross sections adjacent to the ring are

$$q_a = \frac{V_a}{I} \int y \, dA \tag{8.2}$$

and

$$q_b = \frac{V_b}{I} \int y \, dA \tag{8.3}$$

The load q transmitted to the perimeter of the ring must equal the difference between q_a and q_b, or

$$q = q_a - q_b \tag{8.4}$$

From Eqs. (8.1) to (8.4),

$$q = \frac{V_a - V_b}{I} \int y\, dA$$

or

$$q = \frac{P}{I} \int y\, dA \tag{8.5}$$

When the areas resisting bending of the shell are concentrated as flange areas A_f, the integral is replaced by a summation, as in the following equation:

$$q = \frac{P}{I} \sum y A_f \tag{8.6}$$

Equations (8.5) and (8.6) are correct even when the relieving effects of the in-plane components of the stringer forces are considered, since this shear resisted by the stringers must be the same on both fuselage cross sections adjacent to the ring, if the stringers have no abrupt change in direction at the ring. Thus the difference

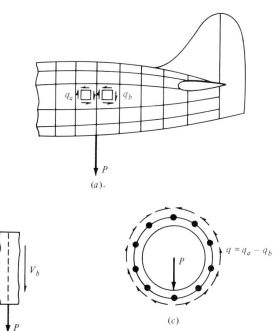

Figure 8.5

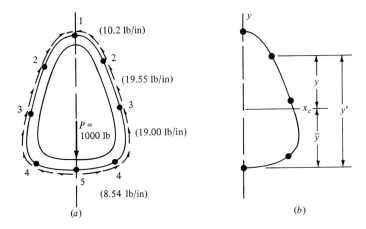

Figure 8.6

in total shear forces, $V_a - V_b$, must equal the difference in the shears resisted by the webs.

In many cases, a fuselage structure may be symmetrical but the loads may not be symmetrical. Any unsymmetrical vertical load may be resolved into a vertical load at the centerline and a couple. The couple applied to the ring will be resisted by a constant shear flow

$$q_T = \frac{T}{2A} \tag{8.7}$$

where T is the magnitude of the couple and A is the area enclosed by the fuselage skin in the plane of the bulkhead.

A fuselage ring also may resist loads which have horizontal components. In this case it is not possible to find a web with zero shear flow by inspection, as in the case for symmetrical vertical loads. It is necessary first to obtain all the shear flows in terms of one unknown and then to find this unknown from the equilibrium of moments, as was done in the analysis of box beams. The method will be obvious after you study Example 8.2.

Example 8.1 The fuselage bulkhead is shown in Fig. 8.6. Only one-half of the shell is considered, as shown in Fig. 8.6*b*. The value of *P* resisted on this half of the structure is 500 lb, and the moment of inertia is found for only one-half of the structure. Of course, the value of P/I in Eq. (8.8) will be the same if both values are obtained for the entire structure, since both will be doubled.

SOLUTION The solution is shown in Table 8.1. The areas A_f listed in column 2 are the total areas of stringers 2, 3, and 4 but only half the areas of stringers 1 and 5, since the structure shown in Fig. 8.6*b* is being considered. The

Table 8.1

Stringer no. (1)	A_f (2)	y' (3)	$A_f y'$ (4)	y (5)	yA_f (6)	$y^2 A_f$ (7)	$\Sigma\, yA_f$ (8)	q, lb/in (9)
1	0.05	34.0	1.7	18.5	0.925	17.12		
							0.925	10.20
2	0.10	24.0	2.4	8.5	0.85	7.23		
							1.775	19.55
3	0.10	15.0	1.5	−0.5	−0.05	0.02		
							1.725	19.00
4	0.10	6.0	0.6	−9.5	−0.95	9.02		
							0.775	8.54
5	0.05	0.0	0.0	−15.5	−0.775	12.01		
Σ	0.4		6.2			45.40		

centroid $\bar{y}$ is determined from the summations of columns 4 and 2:

$$\bar{y} = \frac{\Sigma\, A_f y'}{\Sigma\, A_f} = \frac{6.2}{0.4} = 15.5 \text{ in}$$

It is now necessary to obtain coordinates of the stringers with respect to the centroidal axis. These values, $y = y' - \bar{y}$, obtained by subtracting 15.5 from terms in column 3, are shown in column 5. The terms yA_f and $y^2 A_f$ are calculated in columns 6 and 7. The summation of column 7 yields the moment of inertia I. Equation (8.6) becomes

$$q = \frac{P}{I} \Sigma\, yA_f = \frac{500}{45.4} \Sigma\, yA_f$$

The values of q are calculated in column 9 and are shown in Fig. 8.6a.

Example 8.2 The fuselage bulkhead shown in Fig. 8.7 resists a horizontal load as shown. The stringer coordinates are given in column 3 of Table 8.2.

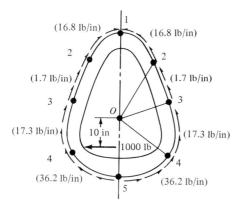

Figure 8.7

Table 8.2

Stringer no. (1)	A_f (2)	x (3)	xA_f (4)	x^2A_f (5)	$\Sigma\, xA_f$ (6)	q' (7)	$2A$ (8)	$2Aq'$ (9)	q (10)
1	0.05	0	0	0			140		+16.8
					0	0		0	
2	0.10	8	0.8	6.4			100		+1.7
					0.8	−15.1		−1,510	
3	0.10	10	1.0	10.0			100		−17.3
					1.8	−34.1		−3,410	
4	0.10	10	1.0	10.0			160		−36.2
					2.8	−53.0		−8,480	
5	0.05	0	0	0					
Σ	0.4			26.4			500	−13,400	

The areas enclosed by the skin segments and the lines to reference point O are indicated by the double areas listed in column 8 of Table 8.2. Find the reactions of the skin on the bulkhead.

SOLUTION First the shear flows are obtained with the assumption that web 1-2 resists a zero shear flow. The resulting shear flows q' are

$$q' = \frac{P}{I_y} \Sigma\, xA_f \tag{8.8}$$

The calculations are performed in Table 8.2. The value of I_y for half the structure is obtained in column 5 as 26.4 in⁴. If the shear flows are positive clockwise around the ring, the force P is considered as negative in Eq. (8.8), or

$$q' = \frac{-1000}{2 \times 26.4} \Sigma\, xA_f$$

The values of q' are calculated in column 7. These values are obviously the same for the left half of the structure because of symmetry. The shear flows q' produce moments of $2Aq'$ about point O. These moments are calculated in column 9. It is now necessary to superimpose a constant shear flow q_0 around the ring so that the external moments on the ring are in equilibrium. Taking moments about O and considering clockwise moments as positive, we have

$$q_0 \Sigma\, 2A + \Sigma\, 2Aq' + 1000 \times 10 = 0$$

$$1000q_0 - 2 \times 13,400 + 1000 \times 10 = 0$$

or

$$q_0 = 16.8 \text{ lb/in}$$

The resulting shear flows q are found by adding 16.8 to each value in column 7. These values are shown in column 10 and in Fig. 8.7.

8.4 ANALYSIS OF WING RIBS

In the simplest type of wing structure, that in which the bending stresses are resisted by only three concentrated flange members, the skin reactions on the ribs can be obtained from the equations of statics. There will be only three unknown shear flows, and these may be obtained readily from the equations for the equilibrium of forces in the vertical and drag directions and for the equilibrium of moments about a spanwise axis.

Then the internal stresses in the rib are obtained from the shears and bending moments at the various cross sections. Normally, there will be axial loads in the rib in addition to the shears and bending moments; thus it is necessary to calculate bending moments about a point with a vertical position corresponding to the neutral axis of the rib. For the rib analyzed in Example 8.3, we assume that all bending moments are resisted in the rib flange members and that all shears are resisted by the webs. With these assumptions, it is more convenient to calculate bending moments about the neutral axis.

In the more general case of a wing in which the bending moments are resisted by more than three flange members, it is necessary to determine the section properties of the wing cross section before the shear reactions on the ribs can be found. The problem is similar to that of calculating reactions on fuselage bulkheads, but it differs in the condition that the fuselage cross section usually is symmetrical, whereas the wing cross section seldom is symmetrical. Thus the skin shear flows, and consequently the skin reactions on the ribs, must be obtained by the more general methods which involve the product of inertia of the cross section.

Example 8.3 Find the shear flows acting on the rib of Fig. 8.8. The wing bending moments are resisted by the three flange areas shown in Fig. 8.8a to

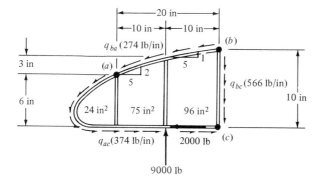

c. Calculate the loads in the rib flanges and the shear flows in the rib webs at a vertical cross section through flange *a* and at vertical cross sections a short distance to either side of the applied loads.

SOLUTION The reactions of the wing skin on the rib must be in equilibrium with the applied loads of 9000 and 2000 lb. From a summation of moments about point *c* and a summation of forces in the vertical and drag directions, the following equations are obtained:

$$\Sigma M_C = 9000 \times 10 - 168q_{ac} - 222q_{ba} = 0$$
$$\Sigma F_H = 20q_{ac} - 20q_{ba} - 2000 = 0$$
$$\Sigma F_V = 9000 - 10q_{bc} - 4q_{ba} - 6q_{ac} = 0$$

The solution of these equations yields $q_{ba} = 274$, $q_{ac} = 374$, and $q_{bc} = 566$ lb/in.

At a vertical cross section through flange *a*, the stresses are obtained by considering the free body shown in Fig. 8.9*a*. The total shear at the cross section is $V = 374 \times 6 = 2244$ lb, and the bending moment is $M = 2 \times 24 \times 374 = 17,952$ in · lb. The horizontal components of the axial loads in the flanges are found from the bending moment as $P_1 = P_2 = M/6 = 2990$ lb. The lower rib flange is horizontal at this point, but the upper member has a slope of 0.4. The shear carried by the flange is therefore

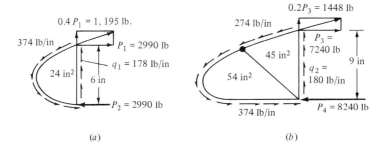

(*a*)   (*b*)

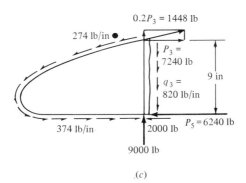

(*c*)

Figure 8.9

$V_f = 0.4 \times 2990 = 1195$ lb. The remaining shear, which is resisted by the web, is $V_w = V - V_f = 2244 - 1195 = 1049$ lb. The shear flow at this section is $q = 1049/b = 178$ lb/in. These values are shown in Fig. 8.9a.

The stresses at a vertical section to the left of the applied loads are obtained by considering the free body shown in Fig. 8.9b. The shear at the cross section is $V = 274 \times 3 + 374 \times 6 = 3066$ lb. The bending moment about the lower flange is $M = 2 \times 45 \times 275 + 2 \times 54 \times 374 = 65,100$ in · lb. The horizontal component of the upper flange load is $P_3 = 65,100/9 = 7240$ lb. The lower flange load is obtained from a summation of horizontal forces as $P_4 = 7240 - 10 \times 274 + 10 \times 374 = 8240$ lb. It is obvious that the bending moment in the rib depends on the vertical location of the center of moments, since there is a resultant horizontal load at the section. If the entire depth of a beam resists bending moment, the centroid of the section is used as the center of moments, and the stresses resulting from the axial load then are uniformly distributed over the area of the cross section. The vertical component of the upper flange load is $0.2 \times 7240 = 1448$ lb. The shear flow in the web is therefore $q = (3066 - 1448)/9 = 180$ lb/in. These values are shown in Fig. 8.9b.

The stresses at a vertical cross section just to the right of the applied loads are obtained in similar manner and are shown in Fig. 8.9c. Since the bending moment was computed about the intersection of the two applied loads, the value of P_3 is the same as for the previous case. The axial load in the lower flange and the web shear flow differ from those shown in Fig. 8.9b.

Example 8.4 The rib shown in Fig. 8.10 transfers the vertical load of 10,000 lb to the wing spars and to the wing skin. Find the reacting shear flows around the perimeter of the rib, the shear flows in the rib web, and the axial loads in the top and bottom rib flanges.

SOLUTION The distribution of shear flows depends on the spar-cap areas. The section properties are: $I_z = 200$ in^4, $I_y = 800$ in^4, $I_{yz} = +200$ in^4. The reacting shear flows on the rib are equal to the shear flows in the skin of a box which resists an external shear of 10,000 lb, but have opposite directions. The change in shear flow at each flange area is found from Eq. (5.25):

$$\Delta p_f = \left[\frac{V_z I_z + V_y I_{yz}}{I_z I_y - I_{yz}^2} z - \frac{V_y I_y + V_z I_{yz}}{I_z I_y - I_{yz}^2} y \right] A_f$$

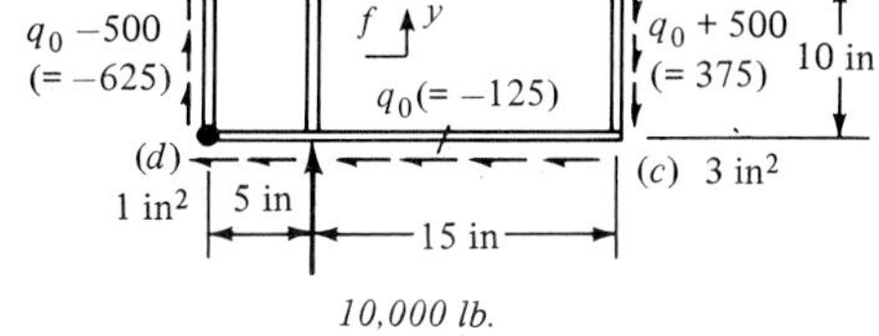

Figure 8.10

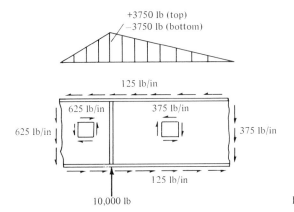

Figure 8.11

If we assume q positive clockwise, the following values are obtained:

$$\Delta q = -500 \text{ at flanges } a \text{ and } b$$

$$\Delta q = +500 \text{ at flanges } c \text{ and } d$$

If a shear flow q_0 is assumed in the top skin, the other shear flows are found in terms of q_0, as shown in Fig. 8.10. The final shear flows can be obtained now from the equilibrium of moments about some convenient point, say flange a:

$$200q_0 + 200(q_0 + 500) = 10,000 \times 5$$

or

$$q_0 = -125 \text{ lb/in}$$

The shear flows in the remaining webs can be found now from these values of q_0 and are shown in the proper directions in Fig. 8.11. The shear flows in the rib webs are obtained from the equilibrium of forces on the vertical cross sections and are shown in Fig. 8.11 as 625 lb/in. on the left-hand web and 375 lb/in on the right-hand web. The axial loads in the rib flanges are shown in Fig. 8.11 and are derived by a summation of the shear-flow forces acting on the rib flanges, including shear flows from both the wing and the rib web. A comparison of the rib web shears and flange loads shown in Fig. 8.11 with those for a simple beam of the same dimensions shows that the flange loads will be the same in both cases, but the web shears will be different.

8.5 SHEAR FLOW IN TAPERED WEBS

The shear flow in a tapered beam with two concentrated flanges is considered in Sec. 5.9, in the discussion of the unit method of shear-flow analysis for tapered box beams. The distribution of the shear flow in a tapered web is considered now in greater detail, since a large proportion of the shear webs in an airplane structure are tapered rather than rectangular. The shear flows in the web of the beam shown in Fig. 8.12a are obtained in Sec. 5.9. From Eq. (5.30) the shear V_w resisted by the web will be

$$V_w = V_y \frac{h_0}{h} = V \frac{x_0}{x}$$ (8.9)

where the notation corresponds to that shown in Fig. 8.12. The shear flow q may be expressed in terms of the shear flow $q_0 = V_y/h_0$ at the free end by the following equations:

$$q = \frac{V_w}{h} = \frac{V_y h_0}{h^2} = q_0 \left(\frac{h_0}{h}\right)^2 = q_0 \left(\frac{x_0}{x}\right)^2$$ (8.10)

The distribution of the shear flow q along the span of the beam is shown in Fig. 8.12b.

In many problems it is necessary to obtain the average shear flow in a tapered web. The average shear flow between the free end and the point x of the beam shown in Fig. 8.12a can be found from the spanwise equilibrium of the flange shown in Fig. 8.12c. The horizontal component of the flange load is found by dividing the bending moment $q_0 h_0 b$ by the beam depth h. The average shear flow in this length, q_{av}, is therefore obtained by dividing this force by the horizontal length b:

$$q_{av} = q_0 \frac{h_0}{h} \quad \text{or} \quad q_0 \frac{x_0}{x}$$ (8.11)

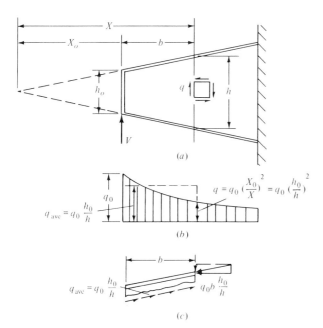

Figure 8.12

If the shear flow on one side of a tapered web is known, the shear flows on the other three sides may be obtained from Eqs. (8.10) and (8.11).

It is assumed in the derivation of Eqs. (8.10) and (8.11) that the stresses existing on all four boundaries of the tapered plate are pure shearing stresses. It has been shown that pure shearing stresses can exist on only two planes, which must be at right angles to each other. Since the corners of the tapered webs do not form right angles, it is necessary for some normal stresses to act at the boundary of the web. In order to estimate the magnitude of these normal stresses, a tapered web in which pure shearing stresses may exist at all the boundaries is considered.

It can be shown by the theory of elasticity that a sector such as shown in Fig. 8.13 may have pure shearing stresses on all the boundaries. Under these boundary conditions, any element such as that shown will have no normal stress in the radial direction σ_{rr} and no normal stress in the tangential direction $\sigma_{\theta\theta}$. The shearing stresses on these radial and tangential faces must satisfy the equation

$$\sigma_{r\theta} = \frac{K}{r^2} \tag{8.12}$$

where K is an undetermined constant. This equation is similar to Eq. (8.10) if the taper is small.

By comparing the sector of Fig. 8.13 with the tapered web of Fig. 8.12, it is seen that the assumption of pure shear on the top and bottom boundaries of the tapered web was correct. The left and right boundaries also must resist some normal stresses, however. The magnitude of these normal stresses may be determined for the Mohr circle of Fig. 8.14b. The element under pure shearing stresses has faces A and B which are inclined at an angle θ with the vertical and horizontal. The Mohr circle for the pure shear condition will have a center at the origin and a radius τ_s. Point A will be at the top of the circle, and point C, representing stresses on the vertical plane, will be clockwise at an angle 2θ from point A. The coordinates of point C represent a tensile stress of $\tau_s \sin 2\theta$ and a shearing stress of $\tau_s \cos 2\theta$ on the vertical plane. The normal stresses obviously are negligible for small values of the angle θ.

The equations for shear flow in tapered webs first were derived for the web of

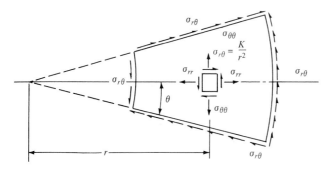

Figure 8.13

a beam with two concentrated flanges and then were shown to be approximately correct for any web that resists no normal loads at its boundaries. It can be shown by examples of other structures containing tapered webs that the shear flows may be applied to the webs by members other than beam flanges. Tapered webs often are used in torque boxes, such as shown in Fig. 8.15. For this box, all four sides are tapered in such a way that the corners of the box would intersect if extended. The enclosed area at any cross section varies with x according to

$$\frac{A}{A_0} = \left(\frac{h}{h_0}\right)^2 = \left(\frac{x}{x_0}\right)^2 \tag{8.13}$$

The shear flow at any cross section, for the pure torsion loading condition shown, is obtained from

$$q = \frac{T}{2A} \tag{8.14}$$

From Eqs. (8.13) and (8.14) and from the value of the shear flow at the left end, $q_0 = T/(2A_0)$, the following expression for q is obtained:

$$q = \frac{T}{2A_0}\left(\frac{x_0}{x}\right)^2 = q_0\left(\frac{x_0}{x}\right)^2 \tag{8.15}$$

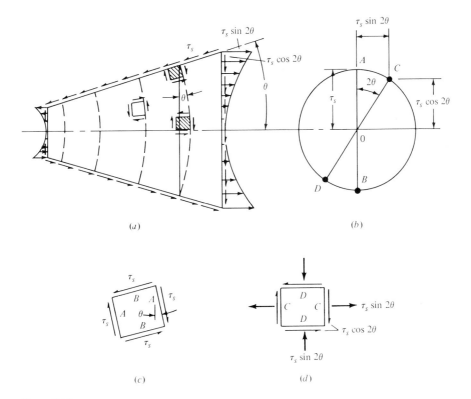

Figure 8.14

This corresponds to the value obtained in Eq. (8.10) for the two-flange beam. The shear flow q given by Eq. (8.15) applies for all four webs of the box. Consequently, there is no axial load in the flange members at the corners of the box, since the shear flow at the sides of the box will be transmitted directly to the top and bottom webs.

In the structure shown in Fig. 8.15, all four webs are tapered in the same ratio, so that the shear flow obtained from Eq. (8.15) will be the same for all four webs. When the taper ratio for the horizontal webs is not the same as the taper ratio for the vertical webs, the shear flows will not have the same distribution for all webs. If, for example, the top and bottom webs are rectangular and the side webs are tapered, as shown in Fig. 8.16, the shear in the rectangular web must remain constant for the entire length, while the shear in the tapered web must vary according to Eq. (8.10). The shear flow for this structure which has ribs only at the ends cannot be obtained from Eq. (8.14), although Eq. (8.14) is quite accurate for the common airplane wing structure with closely spaced ribs. The ribs divide the tapered web into several smaller webs and distribute shear flows so that they are approximately equal in the horizontal and vertical webs.

The shear flows in the tapered webs of Fig. 8.16 vary according to Eq. (8.10):

$$q_1 = q_0 \left(\frac{h_0}{h_1}\right)^2 \tag{8.16}$$

Since the flange members at the corners of the box must be in equilibrium for spanwise forces, the shear flows q_a in the top and bottom webs must equal the average shear flows for the tapered webs, as obtained from Eq. (8.11),

$$q_a = q_0 \frac{h_0}{h_1} \tag{8.17}$$

and from Eq. (8.16),

$$q_a = q_1 \frac{h_1}{h_0} \tag{8.18}$$

The difference in shear flows between two adjacent webs produces axial loads in the flange member between these webs. At any intermediate cross section of the box, the in-plane components of the flange loads must be considered in addition to the web shears, in order to check the equilibrium with the external torque on the box. At the end cross sections, the shear flows are in equilibrium with the

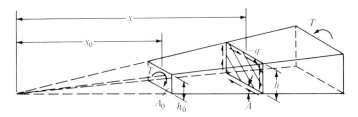

Figure 8.15

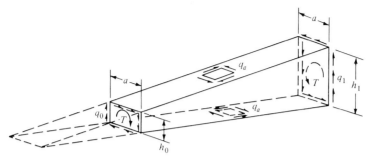

Figure 8.16

external torque. For the left end of the box,

$$(q_0 + q_a)ah_0 = T$$

and substituting values from Eq. (8.17) gives

$$q_a = \frac{T}{a(h_1 + h_0)} \tag{8.19}$$

This equation also can be derived from the equilibrium of forces at the right end of the box

$$T = ah_1(q_1 + q_a)$$

and substituting values from Eq. (8.18) yields

$$q_a = \frac{T}{a(h_1 + h_0)}$$

which checks the previous value. The denominator of Eq. (8.19) represents the average value of $2A$ for the box, as might be expected from Eq. (8.14). For most conventional wing or fuselage structures, the ribs and bulkheads are closely spaced, and it is seldom necessary to consider the taper of the structure when torsional shear flows are obtained. Equation (8.14) may be used, and the shear flows in all webs will be approximately equal at a cross section. For unconventional structures, however, where the ribs cannot distribute shear flows, it may be necessary to use methods similar to those employed for the structure of Fig. 8.16. If the top and bottom webs also are tapered, but have a different taper ratio than the side webs, then the flows may be obtained by applying Eq. (8.10) to each web, equating the average shear flows for all four webs, and then equating the torsional moments of the shear flows at one end to the external torque on the structure.

8.6 CUTOUTS IN SEMIMONOCOQUE STRUCTURES

Typical aircraft structures which consist of closed boxes with longitudinal stiffeners and transverse bulkheads are analyzed in preceding sections. In actual aircraft structures, however, it is necessary to provide many openings in the ideal continuous structure. Wing structures usually must be interrupted to provide wheel wells for retraction of the main landing gear. Other openings may be necessary for armament installations, fuel tanks, or engine nacelles. Fuselage structures often must be discontinuous for doors, windows, cockpit openings, bomb bays, gun turrets, or landing-gear doors. It is also necessary to provide holes and doors for access during manufacture and for inspection and maintenance in service. These "cutouts" are undesirable from a structural standpoint, but are always necessary. Often they occur in regions where high loads must be resisted, and frequently considerable structural weight is required for reinforcements around the cutouts.

A simplified example of a structure with a large cutout is shown in Fig. 8.17. This corresponds to a wing structure with four flange members in which the lower skin is completely removed. In previous sections we stated that a closed torque box was necessary to provide stability for resisting torsional loads. In order for the structure of Fig. 8.17 to be stable, one end must be built in, so that the torsion may be resisted by the two side webs acting independently as cantilever beams, as shown in Fig. 8.17b. The flange members resist axial loads, which have the values $P = TL/(bh)$ at the support. The shear flows q in the vertical webs are double the values obtained in a closed torque box with the same dimensions. The horizontal web resists no shear flow in the case of the pure torsion loading, but it is necessary for stability in resisting horizontal loads. The torque box, with webs on all six faces, is capable of resisting torsion with no axial loads in the flange members. Hence the torque box is much more rigid in torsion, since the shear deformations of the web are negligible in comparison with bending deformations of a cantilever beam.

In a full-cantilever airplane wing, it is not feasible to have an open structure for the entire span, for the wing tip would twist to an excessive angle of attack under some flight conditions. A closed torque box is necessary for most of the span, but may be omitted for a short length, such as the length of a wheel-well opening. When the lower skin is omitted for such a region, the torsion is resisted

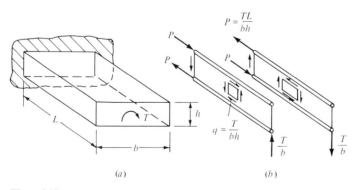

(a)   (b)

Figure 8.17

by "differential bending" of the spars, as indicated in Fig. 8.17*b*. The axial loads in the spar flanges usually are developed at both sides of the opening, since the closed torque boxes inboard and outboard of the opening both resist the warping deformation of the wing cross section. For the torsion loading shown, often it would be assumed that flange loads were zero at the midpoint of the opening and that loads of $P/2$ were developed at both sides of the opening.

The open box with three webs and four flange areas is stable for any loading if one or both ends are restrained. The shear flows in the three webs may be obtained from three equations of statics. The method of obtaining the shear flows is obvious from a numerical example such as that indicated by Fig. 8.18. From the equilibrium of moments about point C of Fig. 8.18*b*,

$$10q_1 \times 20 = 10,000 \times 10 + 2000 \times 5 + 40,000$$

or $\qquad q_1 = 750 \text{ lb/in}$

From the equilibrium of vertical forces,

$$10q_3 + 10 \times 750 = 10,000$$

or $\qquad q_3 = 250 \text{ lb/in}$

Similarly, from the equilibrium of horizontal forces,

$$20q_2 = 2000$$

or $\qquad q_2 = 100 \text{ lb/in}$

The axial loads in the flange members can be found from a summation of span-wise forces:

$$P_a = 40q_1 = 30,000 \text{ lb} \qquad \text{(tension)}$$

$$P_b = 40q_1 + 40q_2 = 34,000 \text{ lb} \qquad \text{(compression)}$$

$$P_c = 40q_3 - 40q_2 = 6000 \text{ lb} \qquad \text{(compression)}$$

$$P_d = 40q_3 = 10,000 \text{ lb} \qquad \text{(tension)}$$

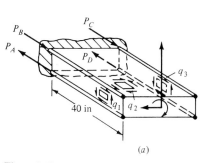

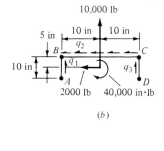

(a)

(b)

Figure 8.18

While these forces satisfy the conditions for static equilibrium, they cannot be obtained from the flexure formula. For an open beam containing only four flange members, the flange loads are independent of the flange areas. If the beam has more than four flanges, one must consider the flange areas in estimating the distribution of axial loads. This problem is statically indeterminate, and usually it is solved by approximate methods.

A cutout in a short length of the wing structure affects the shear flows in the adjacent sections of the wing which have closed torque boxes. First consider a case in which the wing resists pure torsion. The shear flow in a continuous closed box is

$$q_t = \frac{T}{2A} \tag{8.20}$$

This equation is derived from the assumption that the flange members resist no axial loads. At the edges of the cutout, however, the flange loads resulting from differential bending have their maximum values. These flange loads are distributed to the webs, and at some distance from the cutout, the flange loads become zero for the box in pure torsion. The distance along the span required for the distribution of the flange loads depends on the relative rigidities of the members, but it will be approximately equal to the width of the cutout. The shear flows in the torque box are affected considerably by this distribution of load.

The rectangular torque box shown in Fig. 8.19a resists pure torsion. The lower skin is cut out for the entire width of the box, for a length L. The effect of the cutout is assumed to extend a distance L along the span on either side of the cutout; therefore, it is necessary to consider only the length $3L$, which is shown. The shear flows at the section through the cutout are similar to those obtained in Fig. 8.17b, or they will be zero in the upper skin and $2q_t$ in the spar webs, where q_t is the shear flow in a continuous box, as obtained from Eq. (8.20). The axial loads P in the spar flanges are assumed to be equal on the inboard and outboard sides of the cutout and thus are half the value shown in Fig. 8.17b, or

$$P = q_t L \tag{8.21}$$

This axial load must be transferred to the webs adjacent to the flange in the assumed length L. From the equilibrium of the flange member shown in Fig. 8.19b,

$$q_1 L - q_2 L = P \tag{8.22}$$

or, from Eqs. (8.21) and (8.22),

$$q_1 - q_2 = q_t \tag{8.23}$$

The shear flows q_1 and q_2 must satisfy the conditions of equilibrium of the structure shown in Fig. 8.19c. For the vertical forces to be in equilibrium at a cross section, the shear flows in the spars must have equal and opposite values q_2. For horizontal forces to be in equilibrium, the shear flows in top and bottom skins must have equal and opposite values q_1. For the shear flows in all four webs to react the torque T, the following condition must be satisfied:

$$q_1 A + q_2 A = T$$

or, from Eq. (8.20),

$$q_1 + q_2 = 2q_t \tag{8.24}$$

Solving Eqs. (8.23) and (8.24) yields

$$q_1 = 1.5q_t \quad \text{and} \quad q_2 = 0.5q_t$$

The values of these shear flows are shown in parentheses in Fig. 8.19a.

From these resulting values of shear flows q_1 and q_2, the cutout is seen to have a serious effect on the shear flows in the closed torque boxes adjacent to the cutout. The top and bottom skins have shear flows of $1\frac{1}{2}$ times the magnitude of those for a continuous box, while the shear flows in the spars are only one-half as much. The ribs adjacent to the cutout also resist high shear flows. The rib just outboard of the cutout is shown in Fig. 8.19d. The rib receives the shear flow of $1.5q_t$ from the top and bottom skins of the torque box. The spars transfer shear flows of $2q_t$ from the cutout section and $0.5q_t$ from the torque box section. Since

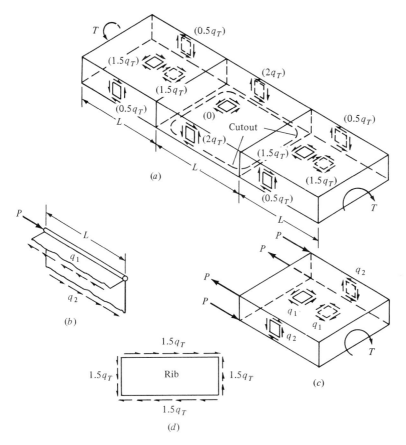

Figure 8.19

the shear flows are in opposite directions, the resultant shear flow applied to the rib is $1.5q_t$.

The problem of a box beam's resisting a more general condition of loading usually is analyzed by another method. The method employed for the structure resisting pure torsion becomes more difficult when the spar flanges resist axial loads resulting from wing bending in addition to those resulting from the differential bending. The common procedure for the general case is first to analyze the continuous wing structure as if there were no cutout. Then a system of correcting shear flows must be obtained and superimposed on the original shear flows found for the continuous structure. In finding the correcting shear flows, only a short length on either side of the cutout need be considered, since the loads applied to the wing in arriving at these shear flows are in equilibrium with themselves. One of the established principles of mechanics, formulated by Saint Venant, states that the stresses resulting from such a system of forces will be negligible at a distance from the forces. The distance is approximately equal to the width of the opening.

The method of obtaining correcting shear flows is illustrated for the wing structure shown in Fig. 8.20. The wing is assumed to have a constant shear of 30,000 lb in the vertical direction and -9000 lb in the chordwise direction for the entire length from station 30 (30 in from the airplane centerline) to station 120. The lower skin is removed for the entire width between the spars from station 60 to station 90. The wing bending moments affect the flange loads but not the shear flows; therefore, the bending moments are not considered. The dimensions of the cross section are shown in Fig. 8.20b. The shear flows in the continuous closed box with no cutout are shown in Fig. 8.20c. These are computed by the methods discussed in Chap. 7, and the computations are not discussed here. Since the external shear is constant, the shear flows in all webs between stations 30 and 120 would have the values shown in Fig. 8.20c if there were no cutout.

The correcting shear flows are now found by applying the loads of 660 lb/in in the cutout region, as shown in Fig. 8.21a, and finding the shear flows in the remaining webs. It is obvious that the loads of 660 lb/in around all four sides of the cutout are in equilibrium with one another. The shear flows at the cross section through the cutout are assumed to be q_1, q_2, and q_3, as shown in Fig. 8.21b, and must have a resultant equal to the applied load of 660 lb/in at the lower skin. From a summation of horizontal forces,

$$30q_2 = 30 \times 660$$

or

$$q_2 = 660 \text{ lb/in}$$

From a summation of moments about point O of Fig. 8.20b,

$$2 \times 90 \times 660 = 2 \times 75 \times q_1 - 2 \times 200 \times 660 + 2 \times 90 \times q_3$$

Solving these equations simultaneously yields $q_1 = 1340$ lb/in and $q_3 = 1010$ lb/in. These correcting shear flows are shown in Fig. 8.21a for the structure between stations 60 and 90. The final shear flows in the cutout region are now

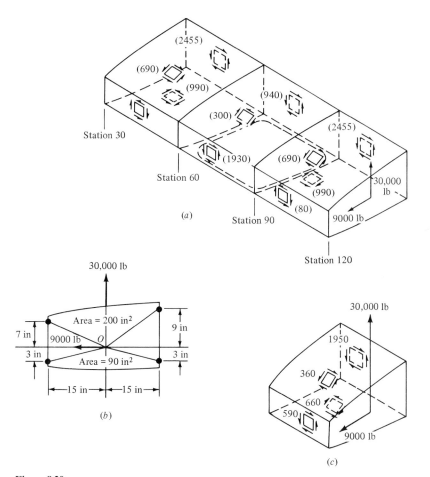

Figure 8.20

obtained by superimposing the values shown in Fig. 8.20c and 8.21a. This super-position yields a shear flow of 300 lb/in in the upper skin, 1930 lb/in in the front spar web, and 940 lb/in in the rear spar web, as shown in parentheses on Fig. 8.20a.

The correcting shear flows between stations 90 and 120 are found from the equilibrium of the forces on a cross section. The shear flows are shown in Fig. 8.21c, and the following equations are derived from the equilibrium of the shear flows on the cross section:

$$\Sigma F_z = 30q_5 - 30q_7 = 0$$

$$\Sigma F_y = 10q_4 + 2q_5 - 12q_6 = 0$$

$$\Sigma M_0 = 2 \times 75q_4 + 2 \times 200q_5 + 2 \times 90q_6 + 2 \times 90q_7 = 0$$

One additional equation may be derived from the spanwise equilibrium of forces on one of the flange areas. For the flange member shown in Fig. 8.21d, the axial load at station 90 is obtained by assuming no axial loads at the center of the cutout, station 75. From the shear flows shown in Fig. 8.21a, $P = 15(1340 + 660) = 30,000$ lb. From Fig. 8.21d,

$$30q_7 - 30q_4 = 30,000$$

Solving these four equations simultaneously yields $q_4 = -670$, $q_5 = 330$, $q_6 = -505$, and $q_7 = 330$ lb/in. These values of the correcting shears are shown in Fig. 8.21a. The final shear flows are obtained by superimposing the correcting shear

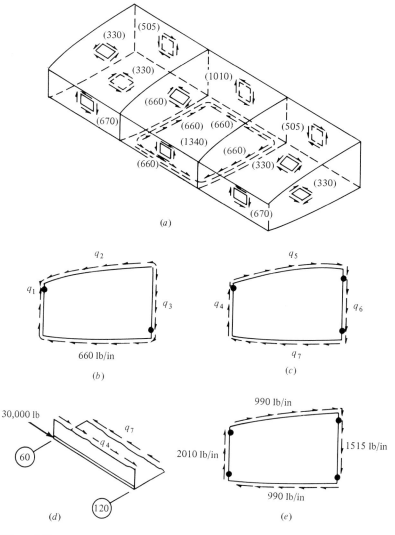

Figure 8.21

flows and those for the continuous structure, shown in Fig. 8.20c. The corrected values are shown in parentheses on Fig. 8.20a.

The loads acting on the rib at station 90 are found from the differences in shear flow on the two sides of the rib and shown in Fig. 8.21e. The shear flows transferred to the rib by the wing skin are seen to be greater than the shear flows in the skin, since the skin shears act in the same direction on the rib and must be added. The rib at station 60 will resist the same loads as the rib at station 90, but the directions of all loads will be reversed.

Cutouts in fuselage structures are treated in essentially the same manner as cutouts in wing structures. Fuselage structures usually have lighter stringers and skin and resist smaller loads, particularly torsional loads. The torsional rigidity of fuselages is not as important as the torsional rigidity of wings, although flutter problems may develop in high-speed aircraft if the fuselage is too flexible torsionally. Fuselage structures often are open for a large proportion of their length in order to permit long cockpit openings or long bomb bays. These structures are able to resist the torsional loads by differential bending of the sides of the fuselage.

Fuselages of large passenger airplanes often contain rows of windows, as shown in Fig. 8.22. If these windows are equally spaced and have equal sizes, the shear flows in webs adjacent to the windows can be obtained in terms of the average shear flow q_0 which would exist in a continuous structure with no windows. If the windows have a spacing w and the webs between them have a width w_1, the shear in these webs q_1 can be derived from a summation of forces on a horizontal section through the windows:

$$q_1 = \frac{w}{w_1} q_0 \qquad (8.25)$$

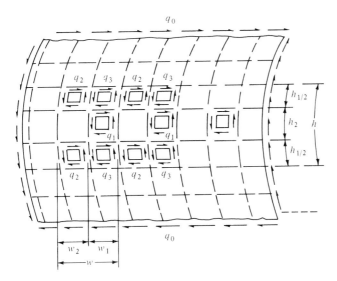

Figure 8.22

Similarly, if the effect of the cutouts is assumed to extend over a vertical distance h, as shown, the shear flows in webs above and below the windows may be obtained by considering a vertical cross section through a window:

$$q_2 = \frac{h}{h_1} q_0 \tag{8.26}$$

The shear flows q_3, shown in Fig. 8.22, can be found by considering either a horizontal section through the webs or a vertical cross section through the webs. The two equations are

$$q_2 w_2 + q_3 w_1 = q_0 w$$

and

$$q_1 h_2 + q_3 h_1 = q_0 h$$

Either of these two equations, when values from Eqs. (8.25) and (8.26) are substituted, reduces to

$$q_3 = q_0 \left(1 - \frac{h_2}{h_1} \frac{w_2}{w_1} \right) \tag{8.27}$$

The notation is shown in Fig. 8.22.

Openings for large fuselage doors may be analyzed in the same manner as wing cutouts. Sometimes it is difficult to provide rigid fuselage bulkheads on either side of an opening, because of interior space limitations. In such cases, a rigid doorframe can be provided so that the doorframe itself resists the shear loads in place of the cutout structure. If such a structure is provided, it is no longer necessary to have the heavy bulkheads adjacent to the opening. A fuselage doorframe usually must follow the curvature of the fuselage and hence does not lie in a plane. Thus, the structure of the doorframe must be capable of resisting torsion as well as bending, and the frame must be a closed-box structure.

8.7 SHEARING DEFORMATIONS

In the analysis of semimonocoque vehicle structures, the shear stress distribution is of great importance. Much of the classical theory of statically indeterminate structures has been developed for the analysis of heavy structures in which shearing deformations are of minor importance. Consequently, much of the published work on structural deflections and indeterminate structures does not treat shearing deformations. The deflections caused by shearing deformations can be determined by the method of virtual work, in the same manner as other types of deflections are analyzed.

The shearing deformation of an elastic rectangular plate with thickness t, width L_x, and length L_y is indicated in Fig. 8.23a. The shearing strain ϵ_{xy} is obtained from the relation

$$\epsilon_{xy} = \frac{\sigma_{xy}}{G} = \frac{q}{tG} \tag{8.28}$$

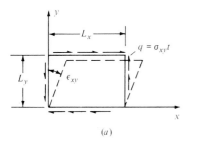
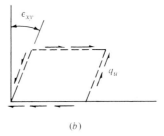

Figure 8.23 Plate deformation under shear action.

where G = shear modulus

$\quad \sigma_{xy}$ = shearing stress

$\quad q = \sigma_{xy} t$ = shear flow

In finding the deflections of a structure resulting from shear deformations of the webs, it is convenient to use a unit virtual load applied at the point of the desired deflection Δ. Thus from Eq. (6.31), replacing the generalized displacement q by Δ yields

$$\lfloor \delta Q \rfloor \{\Delta\} = \int_V \lfloor \delta \Sigma \rfloor \{E'\}\, dV$$

or

$$1 \cdot \Delta = \int_V \delta\sigma_{xy}\, \epsilon_{xy}\, dV \tag{8.29}$$

Utilizing Eq. (8.28) in Eq. (8.29) yields

$$1 \cdot \Delta = \int_V \frac{\delta q}{t} \frac{q}{tG}\, dV$$

or

$$\Delta = \int_V \frac{q_u q}{Gt^2}\, dV \tag{8.30}$$

where $q_u = \delta q$ = shear flow due to the unit applied load and q = real shear flows which produce the deformation. Integrating Eq. (8.30) gives

$$\Delta = \frac{q_u q L_x L_y}{Gt} \tag{8.31}$$

For a structure which has N webs affecting its deflection, Eq. (8.31) becomes

$$\Delta = \sum_{i=1}^{N} \frac{q_{ui} q_i L_{xi} L_{yi}}{G_i t_i} \qquad i = 1, 2, \ldots, N \tag{8.32}$$

Equation (8.32) applies only to elastic deformations which satisfy Eq. (8.28).

8.8 TORSION OF BOX BEAMS

One of the most common applications of Eq. (8.32) is finding the angle of twist of box beams, such as that shown in Fig. 8.24. The shear flows q may result from

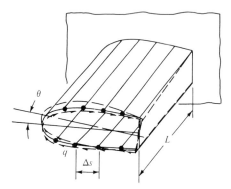

Figure 8.24 Box beam.

any condition of loading, and they are obtained by the methods used in Chap. 5. Since an angular deflection is required, a unit virtual couple (torque) will be applied as shown in Fig. 8.25 and the resulting virtual shear flows are $q_u = 1/(2A)$, where A is the enclosed area of the box. The webs are assumed to have dimensions $L_x = \Delta s$ and $L_y = L$. The angle of twist θ is obtained by substituting these values into Eq. (8.32):

$$\theta = \sum_{i=1}^{N} \frac{q_i \, \Delta s_i \, L}{2At_i G_i} \tag{8.33}$$

The summation includes all webs of the structure.

Example 8.5 The box beam shown in Fig. 8.26 has front spar-flange areas which are 3 times the rear spar-flange areas. Find the angle of twist at the free end. Assume $G = 4 \times 10^6 \ \text{lb/in}^2$.

SOLUTION The shear flows q are shown in Fig. 8.27. All these shear flows are positive except the shear flow in the right-hand web, which tends to produce

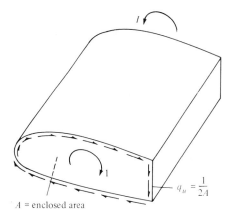

A = enclosed area

Figure 8.25 Box beam under the action of a unit torque.

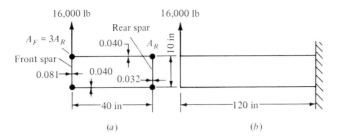

Figure 8.26

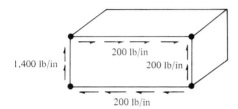

Figure 8.27

a counterclockwise rotation. Hence, from Eq. (8.33) the twist is

$$\theta = \sum \frac{q \,\Delta s \, L}{2AtG} = \frac{L}{2AG} \sum \frac{q \,\Delta s}{t}$$

$$= \frac{120}{2 \times 400 \times 4 \times 10^6} \left(\frac{1400 \times 10}{0.081} + \frac{200 \times 40 \times 2}{0.040} - \frac{200 \times 10}{0.032} \right)$$

$$= 0.020 \text{ rad}$$

8.9 ELASTIC AXIS OR SHEAR CENTER

The *elastic axis* of a wing is defined as the axis about which rotation will occur when the wing is loaded in pure torsion. For the wing shown in Fig. 8.28a, in which the cross section is uniform along the span, the elastic axis is a straight line. Points on the elastic axis do not deflect in the torsion loading, but points forward of the elastic axis are deflected upward. It is necessary to calculate the position of the elastic axis in order to make a flutter analysis of the wing.

The *shear center* of a wing cross section is defined as the point at which the resultant shear load must act to produce a wing deflection with no rotation. The shear force shown in Fig. 8.28b deflects the wing in translation, but causes no rotation of the cross section about a spanwise axis. If the wing is an elastic structure, then the shear center of a cross section must lie on the elastic axis, since a force at the shear center produces no rotation at the point of application of the couple and the couple must therefore produce no vertical deflection at the point

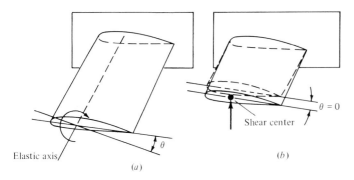

Figure 8.28

of application of the force. Practical wings deviate slightly from conditions of elasticity because the skin wrinkles and becomes ineffective in resisting compression loads but, for practical purposes, the elastic axis may be assumed to coincide with the line joining the shear centers of the various cross sections.

The shear center of a cross section may be calculated from Eq. (8.33) by finding the position of the resultant shear force which yields a zero angle of twist. The shear-center location depends on the distribution of the flange areas and the thickness of the shear webs. The procedure can be studied best by means of an illustrative example.

Example 8.6 Find the shear center for the wing cross section shown in Fig. 8.29. Web 3 has a thickness of 0.064 in, and the other webs have thicknesses of 0.040 in. Assume G is constant for all cross sections. The cross section is symmetrical about a horizontal axis.

SOLUTION The position of the shear center does not depend on the magnitude of the shear force. Thus a shear force $V = 400$ lb is assumed arbitrarily. The shear-flow increments are obtained by the methods used in Chap. 5 and are as shown in Fig. 8.30. The shear flow in web 1 is assumed to have a value of q_0, and the remaining shear flows are expressed in terms of q_0, as shown. In previous problems, the shear flow q_0 was obtained from the equilibrium of torsional moments, but the external torsional moment is not known now. So we assume that the 400-lb shear force acts at a distance $\bar{x}$ from the right side, as shown, and that this point is the shear center. The shear flow q_0 is found from the condition that the angle of twist θ be zero. From Eq. (8.33), where $L = 1$ and $2AG$ is constant for all webs,

$$\theta = \sum \frac{q \, \Delta s}{t} = 0 \tag{8.34}$$

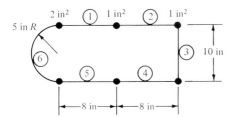

Figure 8.29

The total shear flow q is the sum of the component q_0 and the component q', where q' is the shear flow if web 1 is cut:

$$q = q_0 + q'$$

Substituting this value into Eq. (8.34) and taking q_0 outside the summation sign (because it is constant for all webs) produce:

$$q_0 \sum \frac{\Delta s}{t} + \sum q' \frac{\Delta s}{t} = 0 \qquad (8.35)$$

The numerical solution is tabulated below. Column 1 lists values of Δs, the circumferential lengths for the various webs, as shown in Fig. 8.29. These values are divided by the web thickness in column 2. The values of q', the shear flows when web 1 is cut, are tabulated in column 3. The shear flows are considered positive when they are clockwise around the outboard face of the element of Fig. 8.30. The values of $q' \Delta s/t$ are calculated in column 4.

Web	Δs (1)	$\dfrac{\Delta s}{t}$ (2)	q' (3)	$q' \dfrac{\Delta s}{t}$ (4)	$2A$ (5)	$2Aq'$ (6)	q (7)
1	8	200	0	0	80	0	−0.53
2	8	200	−10	−2000	80	−800	−10.53
3	10	156	−20	−3120	0	0	−20.53
4	8	200	−10	−2000	0	0	−10.53
5	8	200	0	0	0	0	−0.53
6	15.7	392	+20	7840	239	4780	+19.47
Total	$\cdots$	1348	$\cdots$	720	399	3980	

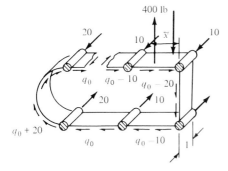

Figure 8.30

Substituting the totals of columns 2 and 4 into Eq. (8.35) yields

$$1348q_0 + 720 = 0$$

$$q_0 = -0.53 \text{ lb/in}$$

The final shear flows in the web q are calculated in column 7 by adding the value of q_0 to the value of q' in column 3.

The position of the shear center is now calculated from the equilibrium of torsional moments. The moment about any point can be obtained from the relation

$$T = \Sigma 2Aq$$

or, since $q = q_0 + q'$,

$$T = q_0\Sigma 2A + \Sigma 2Aq' \tag{8.36}$$

where A is the area enclosed by a web and the lines joining the endpoints of the web and the center of moments. The center of moments is taken as the lower right-hand corner of the box, and values of $2A$ are tabulated in column 5 and values of $2Aq'$ in column 6. The totals of columns 5 and 6, when substituted into Eq. (8.36), yield

$$400\bar{x} = -0.53 \times 399 + 3980$$

$$\bar{x} = 9.42 \text{ in}$$

This value of $\bar{x}$ determines the horizontal location of the shear center. From symmetry, the vertical location is on the line of symmetry. For cross sections which are not symmetrical about a horizontal axis, the vertical location of the shear center can be found by considering a horizontal shear force to act on the section and then proceeding in the same manner as above to find the shear flows for a zero twist. The location of the resultant of these shear flows, obtained by equating torsional moments, gives the vertical position of the shear center.

8.10 WARPING OF BEAM CROSS SECTIONS

When a rectangular box beam is subjected to torsional moments, it deforms as shown in Fig. 8.31. If the cross section is square and the web thickness is the same on all sides, the cross sections will remain plane after the box is twisted. Similarly, if the box beam is subjected to bending with no torsion, the plane cross sections will remain plane after bending. In the usual case, however, the box is rectangular and resists some torsion; therefore the cross sections do not remain plane, but warp. In the analysis of box beams in Chap. 5, we assume that torsional moments do not affect the distribution of bending stresses, or that cross sections are not restrained against warping. The shear flows computed from these assumptions are accurate for all cross sections except those which are very close to a fixed cross section.

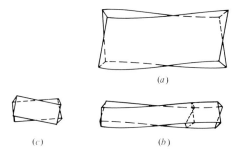

(a)

(c)   (b)   **Figure 8.31**

The amount of warping of a cross section can be measured by the angle θ between spar cross sections, as shown in Fig. 8.32a. This angle is calculated by applying unit virtual couples, as shown, and calculating the relative rotation from Eq. (8.32). The warping of cross sections of the beam of Fig. 8.26 is calculated by assuming that the shear flows q are as shown in Fig. 8.27. A unit spanwise length of the beam is considered, as shown in Fig. 8.32. The unit couples acting on the spars are represented by forces of 0.1 at distances 10 in apart, and the values of q_u must be 0.05 for all webs, as shown, in order to satisfy all conditions of static equilibrium. Substituting values from Figs. 8.27 and 8.32 into Eq. (8.32) yields

$$\theta = \sum \frac{q_u qab}{tG} = \frac{0.05 \times 200 \times 40 \times 1 \times 2}{0.040 \times 4 \times 10^6}$$

$$+ \frac{0.05 \times 200 \times 10 \times 1}{0.032 \times 4 \times 10^6} - \frac{0.05 \times 1400 \times 10 \times 1}{0.081 \times 4 \times 10^6} = 0.00362 \text{ rad}$$

This warping of the cross section is the same for all cross sections on which the shear flows are as shown in Fig. 8.27. The 1-in length along the span was selected arbitrarily, but any other length b might be used. The values of q_u shown in Fig. 8.32b would be divided by an assumed length b; then the terms in the above summation would be multiplied by b instead of the unit length, in order to yield the same final result.

At the fixed support shown in Fig. 8.26, obviously the warping of the cross

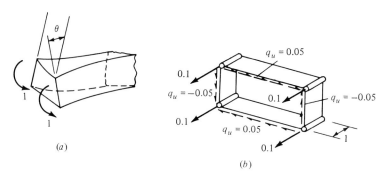

(a)

(b)

Figure 8.32

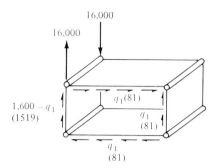

Figure 8.33

section is prevented; therefore the values of q cannot be as shown in Fig. 8.27. The values of q which are required to prevent warping are calculated now for this beam. At any cross section, the shear flows must be as shown in Fig. 8.33 in order to satisfy conditions of equilibrium. From the equilibrium of moments about a spanwise axis through one corner, and from the equilibrium of horizontal and vertical shearing forces, the shear flows q_1 must be equal and in the directions shown for three webs. For the front spar web, the shear flow is $1600 - q_1$. Substituting the values of q from Fig. 8.33 and the values of q_u from Fig. 8.32b into Eq. (8.32) gives

$$\theta = \sum \frac{q_u q a b}{t G} = \frac{0.05 q_1 \times 40 \times 1 \times 2}{0.040 \times 4 \times 10^6} + \frac{0.05 q_1 \times 10 \times 1}{0.032 \times 4 \times 10^6}$$

$$- \frac{0.05(1600 - q_1)10}{0.081 \times 4 \times 10^6} = 0.0000304 q_1 - 0.00247$$

At the fixed support, $\theta = 0$, or $q_1 = 81$ lb/in. The final shear flows are shown in parentheses in Fig. 8.33. These are seen to be considerably different from those shown in Fig. 8.27.

The shear flows of Fig. 8.27 apply a running load of 400 lb/in to the rear spar flanges and a running load of 1200 lb/in to the front spar flanges. Near the support, the shear flows apply a running load of 162 lb/in to the rear spar flanges and a running load of 1438 lb/in to the front spar flanges. Near the support, the bending stresses are therefore higher in the front spar than in the rear spar, the axial strains in the front spar flanges are greater, and the cross sections change from plane sections to warped cross sections. The spanwise distance required for the transition depends on the flange areas and web gages. In this problem, the shear flows at a section 30 in from the support have approximately the values shown in Fig. 8.27. The bending stresses outboard of this cross section are approximately equal for the two spars, since all cross sections warp the same amount. The effect of a fixed cross section is to increase the bending stresses and shear flows in the loaded spar for a spanwise distance which is approximately equal to the average of the cross-sectional dimensions.

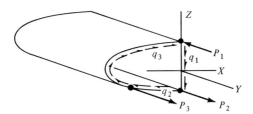

Figure 8.34

8.11 REDUNDANCY OF BOX BEAMS

The only type of box beam which is stable and statically determinate consists of three flange areas and three webs, as shown in Fig. 8.34. In this beam, the three unknown flange forces P_1, P_2, and P_3 and the three unknown shear flows q_1, q_2, and q_3 may be obtained from the six equations of static equilibrium, $\Sigma F_x = 0$, $\Sigma F_y = 0$, $\Sigma F_z = 0$, $\Sigma M_x = 0$, $\Sigma M_y = 0$, and $\Sigma M_z = 0$. As in any other statically determinate structure, the internal forces are independent of the areas of stiffness properties of the members. In any statically indeterminate structure, the areas and elastic properties affect the distribution of the internal forces in the members.

The internal bending stress distribution in all common beams is statically indeterminate, and the deformations are considered in deriving the flexure formula $\sigma = My/I$. This equation is so common that it is not customary to think of such beams as statically indeterminate. The bending stress distribution in a box beam containing more than three flanges, such as that shown in Fig. 8.35, depends on the area and elastic properties of the flanges. The shear-flow distribution, which is derived from the bending stress distribution, also depends on the areas of the flanges and is therefore statically indeterminate.

In this and the following section, we assume that the bending stress is obtained by the simple formula and that a pure torsion load shown in Fig. 8.35 produces no axial stresses in the flanges. These assumptions have been used in previous shear-flow analyses and have been shown to be accurate in most cases. The shear-flow distribution in a single-cell box then can be obtained from the conditions of statics, and such a box is considered as statically determinate for shear-flow calculations. With the assumption that a torsional moment produces no axial stresses in the flanges, the equation $q = T/(2A)$ for the shear flows is obtained from the conditions of statics.

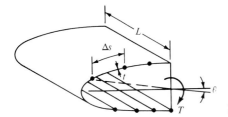

Figure 8.35

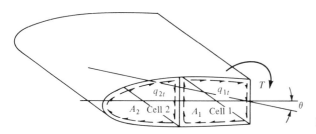

Figure 8.36

A two-cell box such as that shown in Fig. 8.36 cannot be analyzed by the equation of statics. It is assumed that the wing ribs have sufficient rigidity that the two cells deflect through the same angle θ. Then this deflection condition and the equations of statics are sufficient for the shear-flow analysis, and the structure has a single redundancy. It is, of course, assumed that the torsion produces no axial load in the flanges; hence the flanges are not shown in the sketch. A box structure with several cells has one less redundant than the number of cells, since webs in all but one cell may be cut to leave a single cell as a statically determinate base structure.

The angle of twist of a box beam was found by Eq. (8.33):

$$\theta = \sum \frac{q_i \, \Delta s_i \, L_i}{2 A t_i \, G_i} \tag{8.33}$$

where the terms are indicated in Fig. 8.35. This equation can be used for the angle of twist of a multicell structure, if the summation is evaluated around any closed path and the area A is enclosed by this closed path. Thus, for a three-cell structure, the summation can be evaluated around the entire perimeter enclosing the three cells, the entire perimeter enclosing any one cell, or the area enclosing two cells. This procedure is sometimes defined as a line integral, as follows:

$$\theta = \oint \frac{q_i \, L_i}{2 A t_i \, G_i} \, ds$$

Where the integral represents an evaluation along a closed path, returning to the starting point. The values of the summation or integral are considered positive in going clockwise around the enclosed areas.

8.12 TORSION OF MULTICELL BOX BEAMS

For the two-cell box of Fig. 8.36, the angle of twist θ_1 for cell 1 must equal the angle θ_2 for cell 2. A unit length L may be considered, since L is always the same for the two cells:

$$\sum_1 \frac{q_i \, \Delta s_i}{2 A_1 t_i \, G_i} = \sum_2 \frac{q_i \, \Delta s_i}{2 A_2 t_i \, G_i} \tag{8.37}$$

The first summation must be evaluated around the total perimeter of cell 1, including the interior web, and the second summation must include all webs of cell 2 and the interior web.

The value of q for any exterior web of cell 1 is q_{1t}, and for the interior web it is $q_{1t} - q_{2t}$. Similarly, the value of q for an exterior web of cell 2 is q_{2t}, and for the interior web it is $q_{2t} - q_{1t}$. Equation (8.37) now can be rewritten by making these substitutions, moving constant terms outside the summation signs, and assuming G to be constant in all webs:

$$\frac{q_{1t}}{A_1} \sum_1 \frac{\Delta s_i}{t_i} - \frac{q_{2t}}{A_1} \left(\frac{\Delta s}{t}\right)_{1\text{-}2} = \frac{q_{2t}}{A_2} \sum_2 \frac{\Delta s_i}{t_i} - \frac{q_{1t}}{A_2} \left(\frac{\Delta s}{t}\right)_{1\text{-}2} \tag{8.38}$$

The following abbreviations are used for the terms in Eq. (8.38):

$$\delta_{11} = \sum_1 \frac{\Delta s_i}{t_i} \qquad \delta_{22} = \sum_2 \frac{\Delta s_i}{t_i} \qquad \delta_{12} = \left(\frac{\Delta s}{t}\right)_{1\text{-}2} \tag{8.39}$$

The term δ_{11} represents a summation around the entire perimeter of cell 1, δ_{22} a summation around the entire perimeter of cell 2, and δ_{12} the value of the interior web. The terms δ_{11} and δ_{22} both include the term δ_{12} for the interior web. The δ terms do not have quite the same significance as the similar terms used in previous structures, because the constants are eliminated for simplicity and the redundants are taken as shear flows.

By utilizing Eq. (8.39), Eq. (8.38) becomes

$$\frac{1}{A_1} (q_{1t}\delta_{11} - q_{2t}\delta_{12}) = \frac{1}{A_2} (q_{2t}\delta_{22} - q_{1t}\delta_{12}) \tag{8.40}$$

The equation for equilibrium of moments about a torsional axis can be found by reference to Fig. 8.36:

$$T = 2A_1 q_{1t} + 2A_2 q_{2t} \tag{8.41}$$

Equations (8.40) and (8.41) can be solved simultaneously for the two unknowns, q_{1t} and q_{2t}.

The shear flows resulting from pure torsion may be obtained in a similar manner for a box beam with n cells. From the conditions of continuity of deformations between adjacent cells,

$$\theta_1 = \theta_2 = \theta_3 = \cdots = \theta_n$$

or

$$\frac{1}{A_1} (q_{1t}\delta_{11} - q_{2t}\delta_{12}) = \frac{1}{A_2} (q_{2t}\delta_{22} - q_{1t}\delta_{12} - q_{3t}\delta_{23})$$

$$= \frac{1}{A_3} (q_{3t}\delta_{33} - q_{2t}\delta_{23} - q_{4t}\delta_{34}) \tag{8.42}$$

$$= \frac{1}{A_n} (q_{nt}\delta_{nn} - q_{(n-1)t}\delta_{(n-1)n})$$

The terms in Eq. (8.42) are defined as

$$\delta_{nn} = \sum_n \frac{\Delta s_i}{t_i} \qquad \delta_{mn} = \delta_{nm} = \left(\frac{\Delta s}{t}\right)_{m-n}$$

and $\qquad A_n = n$th cell enclosed area $\qquad\qquad\qquad$ (8.43)

The summation of Eq. (8.43) includes all webs around the circumference of the cell, and the term δ_{mn} applies to the interior web between cells m and n. The equation for equilibrium of torsional moments is

$$T = 2A_1 q_{1t} + 2A_2 q_{2t} + 2A_3 q_{3t} + \cdots + 2A_n q_{nt} \qquad (8.44)$$

Equations (8.42) and (8.44) form a set of n simultaneous linear algebraic equations which can be solved for the n unknowns, $q_{1t}, q_{2t}, \ldots, q_{nt}$.

8.13 BEAM SHEAR IN MULTICELL STRUCTURES

Box beams usually resist transverse shearing forces in addition to the torsional moments already considered. Often it is convenient to consider the two effects separately, as a shearing force applied at the shear center and as a torsional moment about the shear center.

In a multicell box such as that shown in Fig. 8.37, the increments of flange loads ΔP can be calculated from the bending stresses at two cross sections or from the shear equations as used in Chap. 5 for a single-cell box. If one web is cut in each cell, the shear flows q' can be obtained from the equilibrium of spanwise forces on the stringers. The structure shown in Fig. 8.37 is unstable for torsional moments, but the system of shear flows q' will be in equilibrium with external shear forces acting at the shear center of the open section.

The shear flows q_{1s}, q_{2s}, and q_{3s} in the cut webs can be obtained and superimposed on the shear flows q' to give a system of shear flows which have a resultant equal to the external shearing force acting at the shear center of the closed multicell box. A superposition of the conditions shown in Figs. 8.37 and 8.38 yields the shear flows in a closed multicell box with no twist. The values of q_{1s}, q_{2s}, and q_{3s} are found from the condition that the angles of twist θ_1, θ_2, and θ_3 for each cell must be zero. After q_{1s}, q_{2s}, and q_{3s} are obtained, the equation of torsional moments yields the position of the shear center of the closed box. Then the external torque about the shear center can be computed and the shear flows resulting from this torque calculated by the methods of Sec. 8.11.

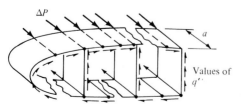

ΔP

a

Values of q'

Figure 8.37

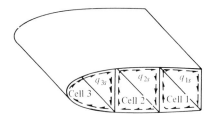

Figure 8.38

The conditions that the angles of twist for each cell be zero yield the following equation for each cell:

$$\sum \frac{q\,\Delta s}{2AGt} = 0$$

Since G usually is constant and $2A$ is always constant, these terms may be canceled from the equation. Then

$$\sum \frac{q\,\Delta s}{t} = 0 \tag{8.45}$$

For cell one Eq. (8.45) yields

$$\sum_1 \frac{q_i'\,\Delta s_i}{t_i} + q_{1s}\sum_1 \frac{\Delta s_i}{t_i} - q_{2s}\left(\frac{\Delta s}{t}\right)_{1-2} = 0 \tag{8.46}$$

in which summations are evaluated around the entire perimeter of the cell, including the interior web, and the last term applies to the interior web only. The terms in Eq. (8.46) may be abbreviated, and similar equations may be written for the other cells:

$$\delta_{10} + q_{1s}\delta_{11} - q_{2s}\delta_{12} = 0$$
$$\delta_{20} + q_{2s}\delta_{22} - q_{1s}\delta_{12} - q_{3s}\delta_{23} = 0 \tag{8.47}$$
$$\delta_{30} + q_{3s}\delta_{33} - q_{2s}\delta_{23} = 0$$

The following abbreviations are used:

$$\delta_{10} = \sum_1 \frac{q_i'\,\Delta s_i}{t_i} \qquad \delta_{20} = \sum_2 \frac{q_i'\,\Delta s_i}{t_i} \qquad \delta_{30} = \sum_{-3} \frac{q_i'\,\Delta s_i}{t_i} \tag{8.48}$$

in addition to the abbreviations given in Eq. (8.43).

Equations (8.47) may now be solved simultaneously for q_{1s}, q_{2s}, and q_{3s}. These equations are applicable to a two-cell structure if all terms containing the subscript 3 are dropped. Similar equations may also be written for any number of cells.

Example 8.7 Find the shear flows in the two-cell box of Fig. 8.39. The horizontal webs have gages of $t = 0.040$ in. Assume G is constant for all webs. The cross section is symmetrical about a horizontal centerline.

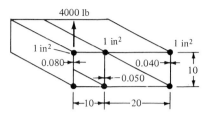

Figure 8.39

SOLUTION The shear flows may be obtained by superposition of the values of q' for the structure shown in Fig. 8.40a, and the values of q_1 and q_2 are shown in Fig. 8.40b. The shear flows q_1 and q_2 are computed as the sum of values for a load at the shear center and for a pure torsion loading. First, if we consider the shear flows q_{1s} and q_{2s} required to produce no twist of the structure, the following equation is obtained for cell 1:

$$\sum_1 \frac{q_i \, \Delta s_i}{t_i} = q_{1s}\left(\frac{20}{0.040}\right) + (q_{1s} - 100)\left(\frac{10}{0.040}\right) + q_{1s}\left(\frac{20}{0.040}\right)$$

$$+ (q_{1s} - q_{2s} + 100)\left(\frac{10}{0.050}\right) = 0$$

And so

$$1450q_{1s} - 200q_{2s} - 5000 = 0 \qquad (a)$$

A similar equation is written for cell 2:

$$\sum_2 \frac{q_i \, \Delta s_i}{t_i} = q_{2s}\left(\frac{10}{0.040}\right) + (q_{2s} + 200)\left(\frac{10}{0.080}\right) + q_{2s}\left(\frac{10}{0.040}\right)$$

$$+ (q_{2s} - q_{1s} - 100)\left(\frac{10}{0.050}\right) = 0$$

and

$$825q_{2s} - 200q_{1s} + 5000 = 0 \qquad (b)$$

Equations (a) and (b) can now be solved simultaneously, yielding $q_{1s} = 2.7$ and $q_{2s} = -5.4$ lb/in. The minus sign indicates that the shear flow q_2 is opposite to the assumed direction, or counterclockwise around the box. These values represent the shear flows for a load applied at the shear center.

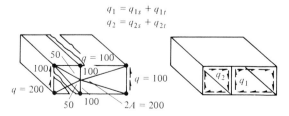

(a)   (b)   **Figure 8.40**

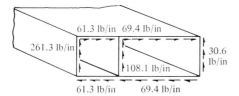

Figure 8.41

The torsional moment about the reference point O of the shear flows is

$$\Sigma 2Aq' + 2A_1 q_{1s} + 2A_2 q_{2s}$$

where A_1 and A_2 represent the enclosed areas of the cells, and

$$100 \times 200 - 100 \times 200 + 2 \times 200 \times 2.7 - 2 \times 100 \times 5.4 = 0$$

The external shearing force of 4000 lb acting at the shear center produces no torsional moment about point O, or the shear center of the cross section is at point O.

The actual load of 4000 lb acting at the left-hand web has a moment arm of 10.0 in about the shear center. The shear flows q_{1t} and q_{2t} must now be obtained for a pure torque of $T = 4000 \times 10 = 40,000$ in · lb. From Eq. (8.44)

$$40,000 = 400q_{1t} + 200q_{2t}$$

and from Eq. (8.42)

$$\frac{q_{1t}}{200}\left(\frac{2 \times 20}{0.040} + \frac{10}{0.040} + \frac{10}{0.050}\right) - \frac{q_{2t}}{200}\frac{10}{0.050}$$

$$= \frac{q_{2t}}{100}\left(\frac{2 \times 10}{0.040} + \frac{10}{0.050} + \frac{10}{0.080}\right) - \frac{q_{1t}}{100}\frac{10}{0.050}$$

These two equations are solved simultaneously and yield $q_{1t} = 66.7$ and $q_{2t} = 66.7$ lb/in. The final shear flows in the cut webs are obtained from Eqs. (a) and (b) as $q_{1s} = 69.4$ and $q_{2s} = 61.3$ lb/in. These values are now superimposed on the values of q', and the final shear flows are shown in Fig. 8.41.

PROBLEMS

8.1 Find the shear flow in each web of the beam shown in Fig. P8.1 and P8.2, and plot the

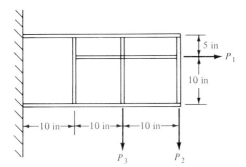

Figure P8.1 and P8.2

distribution of axial load along each stiffening member. Solve for each of the following loading conditions:

(a) $P_1 = 3000$ lb $P_2 = P_3 = 0$
(b) $P_2 = 6000$ lb $P_1 = P_3 = 0$
(c) $P_1 = 6000$ lb $P_2 = 6000$ lb $P_3 = 6000$ lb

8.2 Repeat Prob. 8.1 for the following loading conditions: $P_1 = 2400$, $P_2 = 1200$, and $P_3 = 1800$ lb.

8.3 The pulley bracket shown in Fig. P8.3 and P8.4 is attached to webs along the three sides. Find the reactions R_1, R_2, and R_3 of the webs if $P = 1000$ lb and $\theta = 45°$.

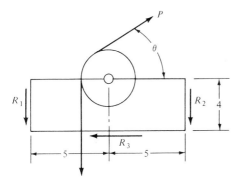

Figure P8.3 and P8.4

8.4 Repeat Prob. 8.3 for $P = 2000$ lb and $\theta = 60°$.

8.5 Find the shear flows applied by the skin to the fuselage ring shown in Fig. P8.5 to P8.8 if $P_1 = 2000$ lb and $P_2 = M = 0$.

8.6 Find the skin reactions on the fuselage ring in Fig. P8.5 to P8.8; $P_2 = 1000$ lb, and $P_1 = M = 0$.

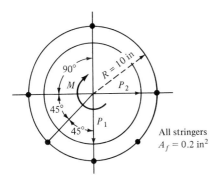

All stringers
$A_f = 0.2$ in^2

Figure P8.5 to P8.8

8.7 Find the skin reactions on the fuselage ring if $P_1 = 2000$ lb, $P_2 = 1000$ lb, and $M = 10,000$ in · lb.

8.8 Find the skin reactions on the fuselage ring if $P_1 = 1500$ lb, $P_2 = 500$ lb, and $M = 8000$ in · lb.

8.9 Find the skin reactions on the rib shown in Fig. P8.9 and P8.10 if the rib is loaded by the distributed load of 20 lb/in. Calculate the shear flows in the rib web and the axial loads in the rib flanges at vertical sections 10 and 20 in forward of the spar.

8.10 Repeat Prob. 8.9 if the rib is loaded by a concentrated upward force of 600 lb, applied at a point 20 in forward of the spar, instead of the distributed load.

8.11 Find the skin reactions on the rib shown in Fig. P8.11 to P8.16. Analyze vertical cross sections at 10-in intervals, obtaining the web shear flows and the axial loads in the rib flanges. Assume the loads $P_1 = 40,000$ lb and $P_2 = 0$. The spar flange areas are $a = b = c = d = 1$ in^2.

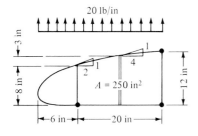

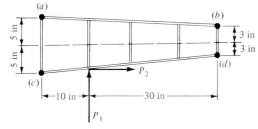

Figure P8.9 to **P8.10**

Figure P8.11 to **P8.16**

8.12 Repeat Prob. 8.11 for $P_1 = 0$ and $P_2 = 8000$ lb.

8.13 Repeat Prob. 8.11 for $P_1 = 20,000$ lb and $P_2 = 8000$ lb.

8.14 Repeat Prob. 8.11 for $P_1 = 40,000$ lb and $P_2 = 0$ if the spar flange areas are $a = 3$ and $b = c = d = 1$ in^2.

8.15 Repeat Prob. 8.11 for $P_1 = 0$ and $P_2 = 8000$ lb if the spar flange areas are $a = 3$ and $b = c = d = 1$ in^2.

8.16 Repeat Prob. 8.11 for $P_1 = 20,000$ and $P_2 = 8000$ lb if the spar flange areas are $a = 3$ and $b = c = d = 1$ in^2.

8.17 The structure of Fig. 8.16 has the dimensions in inches $h_0 = 5$, $h_i = 15$, $a = 20$, length $L = 100$. For a torque T of 40,000 in · lb, find q_0, q_1, and q_a. Find the axial loads in the corner flanges and the shear flows at a cross section 50 in from one end. Check the values by the equilibrium of torsional moments, including the in-plane components of the flange loads.

8.18 Repeat Prob. 8.17 if $h_1 = 10$ in.

8.19 Find the shear flows and the flange loads for the structure of Fig. 8.18 if only the horizontal load of 2000 lb is acting.

8.20 Find the shear flows and flange loads for the nacelle structure shown in Fig. P8.20 and P8.21.

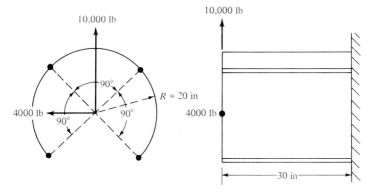

Figure P8.20 and **P8.21**

8.21 Find the shear flows and flange loads for the nacelle structure shown in Fig. P8.20 and P8.21 if a clockwise couple load of 200,000 in · lb is acting in addition to the loads shown.

8.22 Find the shear flows in all webs of the structure shown in Fig. 8.20 if a clockwise couple load of 100,000 in · lb is acting in addition to the loads shown.

8.23 A cantilever wing spar is 10 in deep between centroids of the flange areas. The bending stresses in the flanges are 30,000 lb/in^2, and the shear stresses in the web are 15,000 lb/in^2 at all points. Find the deflection resulting from shear and bending deformations as well as the percentage of the deflection contributed by the shear at (a) 20 in from the fixed support, (b) 40 in from the fixed support, and (c) 100 in from the fixed support. Use $E = 10^7$ and $G = 3,000,000$ lb/in^2.

8.24 Find the angle of twist of the wing shown in Fig. 8.26 if all four flanges have equal areas. Use $G = 4 \times 10^6$ lb/in^2.

8.25 Find the angle of twist of the wing shown in Fig. 8.26 if all webs have a thickness of 0.040 in. Use $G = 4 \times 10^6$ lb/in^2.

8.26 Calculate the location of the shear center, or elastic axis, of the wing in Fig. 8.26.

8.27 Find the shear flows in the webs of the structure shown in Fig. 8.39. The horizontal webs have gages of 0.064 in and G is constant.

8.28 Find the shear flows in the webs of the structure shown in Fig. 8.39 if all flange areas are 1 in^2. The horizontal webs have gages of 0.064 in, and G is constant.

8.29 Find the shear flows in the webs of the structure shown in Fig. P8.29 to P8.32 if all flanges have areas of 1 in^2 and all webs have gages of 0.040 in. Assume $V = 3000$ lb, $e = 8$ in, and G is constant for all webs.

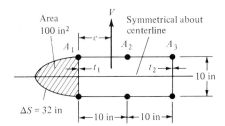

Figure P8.29 to P8.32

8.30 Repeat Prob. 8.29 if $e = 0$.

8.31 Find the shear flows in the webs of the structure shown in Fig. P8.29 to P8.32 if $A_1 = A_3 = 1$ in^2, $A_2 = 2$ in^2, $t_1 = t_2 = 0.064$ in, $V = 4000$ lb, and $e = 10$ in. The other webs have gages of 0.040 in, and G is constant for all webs.

8.32 Repeat Prob. 8.31, assuming an additional vertical web of 0.064-in gage at flange A_2.

8.33 Assume the box beam shown in Fig. 6.19a to be loaded by a torsional couple of 160,000 in · lb at the free end instead of the vertical load shown. Assume the cross section to be symmetrical about a horizontal centerline and all web gages to be $t = 0.020$ in. Calculate the warping displacements of a cross section which is free to warp, and calculate the axial flange loads and web shear flows at the wall and at cross sections at 10-in intervals along the span. Note that the stringer of 2-in^2 area resists no load and does not affect the analysis. Assume $E = 10^7$ lb/in^2 and $G = 0.4E$.

NINE

THERMAL STRESSES

9.1 INTRODUCTION

Temperature changes in structural system components are accompanied by a change in length which results in what is called thermal stresses. The subject of thermal stress analysis encompasses a wide range of structural systems and components, with applications in airframe vehicle structures, nuclear reactors, jet and rocket engines, oil refining lines, and some civil engineering structures. In this chapter, we introduce thermal stress analysis.

9.2 THERMAL STRESS PROBLEM: PHILOSOPHY

A temperature change in a given solid material causes fibers to expand and contract in different amounts. For the solid to remain continuous, a system of thermal strains and corresponding thermal stresses may be induced, depending on the characteristics of the solid and its temperature distribution. A homogeneous solid with no physical external restraints is free of thermal stresses if the temperature distribution is uniform throughout the solid. This condition is referred to as *free expansion* or *contraction* of the solid. Upon imposing external restraints, the free expansion or contraction is prevented, and thus thermal stresses are introduced to the body. To illustrate, consider the beam of Fig. 9.1 to be heated uniformly from a datum temperature T_0 to a final temperature T degrees Fahrenheit. The elongation of the beam due to the temperature change

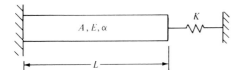

Figure 9.1

$T - T_0$ is given by

$$\delta_b = \alpha(T - T_0)L \tag{9.1}$$

where α = material thermal coefficient of expansion in/in. °F
$\quad L$ = beam length, in
$\quad T, T_0$ = final and datum temperatures, respectively, °F

The elongation of a beam due to a uniform tensile stress σ is

$$\delta_b = \frac{\sigma L}{E} \tag{9.2}$$

From compatibility conditions, the total deformations given by Eqs. (9.1) and (9.2) must equal the connecting spring deformation δ_b:

$$\alpha(T - T_0)L + \frac{\sigma L}{E} = \delta_s \tag{9.3}$$

From equilibrium conditions, it may be seen easily that

$$F_b + F_s = \sigma A + K\delta_b = 0 \tag{9.4}$$

where F_b = beam internal force = σA and F_s = spring force = $K\delta_s$. Thus

$$\delta_s = -\frac{\sigma A}{K} \tag{9.5}$$

Upon substituting Eq. (9.5) into Eq. (9.3) and solving for the stress σ, the following is obtained:

$$\sigma = -\frac{\alpha(T - T_0)L}{L/E + A/K}$$

$$= -\frac{\alpha KEL(T - T_0)}{KL + AE} \tag{9.6}$$

By examining Eq. (9.6), it can be shown that if $K = 0$, which is equivalent to the free expansion of the beam, the thermal stress is zero. In the case where $K = \infty$, which is equivalent to having the beam supported between two rigid walls, the thermal stress is

$$\sigma = -\alpha E(T - T_0) \tag{9.7}$$

In general, Eq. (9.6) may be written as

$$\sigma = -R_s \alpha E(T - T_0) \tag{9.8}$$

where R_s is a nondimensional restraint coefficient whose value ranges between 0 and 1. For $R_s = 0$, the beam is completely free to expand, whereas for $R_s = 1$, the beam is completely prevented from expansion.

From the preceding illustration, it may be concluded that the formulation of the thermal stress problem is identical to that of the isothermal stress problem in that it requires consideration of the following conditions:

1. Equilibrium of forces
2. Compatibility of deformations
3. Stress-strain and strain-displacement relationships
4. Boundary conditions

9.3 FORMULATION OF EQUATIONS FOR THERMAL STRESS ANALYSIS

Most structural problems encountered in engineering are commonly in a three-dimensional state of stress. Quite often, three-dimensional problems are difficult to solve and hence through valid simplifying assumptions are normally reduced to two- or one-dimensional problems. For instance, in the case of a plate structure, if we assume that the thickness is small compared to its other dimensions, then the stresses through the thickness direction usually are neglected, thus reducing a three-dimensional problem to a two-dimensional one. In the case of a beam, all stresses are neglected but one normal stress and one shearing stress, thus reducing the problem to a one-dimensional one. Therefore, in the rest of this book, all formulations of equations are in one or two dimensions.

Equilibrium Equations

The equilibrium equations of a two-dimensional solid shown in Fig. 9.2 are derived in Chap. 3:

$$\sigma_{xx, x} + \sigma_{xy, y} + X = 0 \tag{9.9}$$

$$\sigma_{yy, y} + \sigma_{xy, x} + Y = 0 \tag{9.10}$$

where σ_{xx}, σ_{yy} = normal stresses
$\quad\quad \sigma_{xy}$ = shearing stress
$\quad X, Y$ = body forces

In polar coordinates, Eq. (9.10) becomes

$$\sigma_{rr, r} + \frac{\sigma_{r\theta, \theta}}{r} + \frac{\sigma_{rr} - \sigma_{\theta\theta}}{r} + R = 0 \tag{9.11}$$

$$\frac{\sigma_{\theta\theta, \theta}}{r} + \sigma_{r\theta, r} + \frac{2\sigma_{r\theta}}{r} + \Theta = 0$$

where the stresses and body forces are shown in Fig. 9.3.

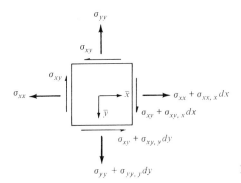

Figure 9.2

Strain-Displacement Relationships

If we consider small deformations, the strain-displacement relationships are

$$\epsilon_{xx} = q_{x,\,x} \qquad \epsilon_{yy} = q_{y,\,y} \qquad \epsilon_{xy} = q_{x,\,y} + q_{y,\,x} \tag{9.12}$$

or, in polar coordinates,

$$\epsilon_{rr} = q_{r,\,r} \qquad \epsilon_{\theta\theta} = \frac{q_r}{r} + \frac{q_{\theta,\,\theta}}{r}$$

$$\epsilon_{r\theta} = \frac{q_{r,\,\theta}}{r} + q_{\theta,\,r} - \frac{q_\theta}{r} \tag{9.13}$$

where q = generalized displacement function and ϵ denotes strain components.

Thermoelastic Strain-Stress Relationships

The thermoelastic strain-stress relationships for plane stress problems are expressed as follows:

$$\epsilon_{xx} = \frac{1}{E}(\sigma_{xx} - v\sigma_{yy}) + \alpha T$$

$$\epsilon_{yy} = \frac{1}{E}(\sigma_{yy} - v\sigma_{xx}) + \alpha T \tag{9.14}$$

$$\epsilon_{xy} = \frac{\sigma_{xy}}{G}$$

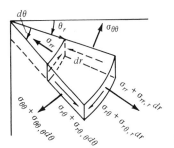

Figure 9.3

where α = coefficient of thermal expansion and T = temperature above the datum temperature. For plane strain problems Eq. (9.14) becomes

$$\epsilon_{xx} = \frac{1 - v^2}{E}\left(\sigma_{xx} - \frac{v}{1 - v}\sigma_{yy}\right) + \alpha T(1 + v)$$

$$\epsilon_{yy} = \frac{1 - v^2}{E}\left(\sigma_{yy} - \frac{v}{1 - v}\sigma_{xx}\right) + \alpha T(1 + v) \qquad (9.15)$$

$$\epsilon_{xy} = \frac{\sigma_{xy}}{G}$$

In polar coordinates, Eqs. (9.14) and (9.15) become, respectively,

$$\epsilon_{rr} = \frac{1}{E}(\sigma_{rr} - v\sigma_{\theta\theta}) + \alpha T$$

$$\epsilon_{\theta\theta} = \frac{1}{E}(\sigma_{\theta\theta} - v\sigma_{rr}) + \alpha T \qquad \text{plane stress} \qquad (9.16)$$

$$\epsilon_{r\theta} = \frac{\sigma_{r\theta}}{G}$$

$$\epsilon_{rr} = \frac{1 - v^2}{E}\left(\sigma_{rr} - \frac{v}{1 - v}\sigma_{\theta\theta}\right) + \alpha T(1 + v)$$

$$\epsilon_{\theta\theta} = \frac{1 - v^2}{E}\left(\sigma_{\theta\theta} - \frac{v}{1 - v}\sigma_{rr}\right) + \alpha T(1 + v) \qquad \text{plane strain} \qquad (9.17)$$

$$\epsilon_{r\theta} = \frac{\sigma_{r\theta}}{G}$$

Compatibility Equations

The compatibility equations can be expressed in terms of strains alone:

$$\epsilon_{xx, yy} + \epsilon_{yy, xx} = \epsilon_{xy, xy} \qquad (9.18)$$

or, in polar coordinates,

$$\epsilon_{\theta\theta, rr} + \frac{\epsilon_{rr, \theta\theta}}{r^2} + \frac{2\epsilon_{\theta\theta, r}}{r} - \frac{\epsilon_{rr, r}}{r} = \frac{\epsilon_{r\theta, r\theta}}{r} + \frac{\epsilon_{r\theta, \theta}}{r^2} \qquad (9.19)$$

In terms of stresses alone for plane stress problems, Eq. (9.18) becomes

$$\left(\frac{\partial}{\partial x^2} + \frac{\partial}{\partial y^2}\right)(\sigma_{xx} + \sigma_{yy} + E\alpha T) = -(1 + v)\left(\frac{\partial X}{\partial x} + \frac{\partial Y}{\partial y}\right) \qquad (9.20)$$

or, for plane strain problems,

$$\left(\frac{\partial}{\partial x^2} + \frac{\partial}{\partial y^2}\right)\left(\sigma_{xx} + \sigma_{yy} + \frac{\alpha E T}{1 - v}\right) = -\frac{1}{1 - v}\left(\frac{\partial X}{\partial x} + \frac{\partial Y}{\partial y}\right) \qquad (9.21)$$

Boundary Conditions

For a specific thermal stress problem, the applied surface loads S_x and S_y must be in equilibrium with the induced stresses at the boundaries of the solid:

$$S_x = \sigma_{xx}^b \, l + \sigma_{xy}^b \, m$$

$$S_y = \sigma_{yy}^b \, m + \sigma_{xy}^b \, l$$

(9.22)

where σ_{xx}^b, σ_{xy}^b and σ_{xy}^b are stresses at the boundary surfaces and l and m are direction cosines.

In polar coordinates, Eq. (9.22) becomes

$$S_r = \sigma_{rr}^b \, l + \sigma_{r\theta}^b \, m$$

$$S_{\theta\theta} = \sigma_{\theta\theta}^b \, m + \sigma_{r\theta}^b \, l$$

(9.23)

9.4 SOLUTION METHODS FOR THERMOELASTIC PROBLEMS

In general, two-dimensional thermoelastic problems involve six unknowns of stresses and strains. The method of solution quite often is dependent on the type of structure under consideration. The following are some typical techniques used in the solution of thermoelastic problems.

Direct Solution Using Equilibrium and Compatibility Conditions

A number of simple thermal stress problems can be solved by merely seeking a solution for the differential equations which describe them. Consider, for example, a circular solid plate having uniform thickness t and subjected to a temperature distribution $T = T(r)$. Equations (9.11) and (9.19) become (note that the stresses and strains are independent of θ)

$$\frac{d\sigma_{rr}}{dr} + \frac{\sigma_{rr} - \sigma_{\theta\theta}}{r} = 0$$

(9.24a)

$$\epsilon_{\theta\theta} - \epsilon_{rr} + r \frac{d\epsilon_{\theta\theta}}{dr} = 0$$

(9.24b)

Equation (9.24b) may be expressed in terms of stresses as

$$\frac{d\sigma_{\theta\theta}}{dr} - v \frac{d\sigma_{rr}}{dr} + \alpha E \frac{dT}{dr} - \frac{(1 + v)(\sigma_{rr} - \sigma_{\theta\theta})}{r} = 0$$

(9.24b')

Solving Eqs. (9.24a) and (9.24b') yields

$$r \frac{d^2\sigma_{rr}}{dr^2} + 3 \frac{d\sigma_r}{dr} = -\alpha E \frac{dT}{dr}$$

(9.25)

A direct integration of Eq. (9.24) yields

$$\sigma_{rr} = -\frac{\alpha E}{r^2} \int_0^r Tr \, dr + C_1 + \frac{C_2}{r^2} \tag{9.26}$$

For the solid plate shown in Fig. 9.4, $r = 0$ at the center; hence Eq. (9.26) becomes undefined because of the term C_2/r^2. To render a feasible solution, C_2 must be set to zero; thus Eq. (9.26) becomes

$$\sigma_{rr} = -\frac{\alpha E}{r^2} \int_0^r Tr \, dr + C_1 \tag{9.27}$$

From boundary conditions $\sigma_{rr} = 0$ at $r = R_0$. Hence the constant of integration C_1 is

$$C_1 = \frac{\alpha E}{R_0^2} \int_0^{R_0} Tr \, dr$$

and σ_{rr} becomes

$$\sigma_{rr} = -\frac{\alpha E}{r^2} \int_0^r Tr \, dr + \frac{\alpha E}{R_0^2} \int_0^{R_0} Tr \, dr$$

and

$$\sigma_{\theta\theta} = -\alpha E T + \frac{\alpha E}{r^2} \int_0^r Tr \, dr + \frac{\alpha E}{R_0^2} \int_0^{R_0} Tr \, dr \tag{9.28}$$

$$\sigma_{r\theta} = 0$$

For a hollow circular plate Eq. (9.28) becomes

$$\sigma_{rr} = \frac{\alpha E}{r^2} \left(\frac{r^2 - R_i^2}{R_0^2 - R_i^2} \int_{R_i}^{R_0} Tr \, dr - \int_{R_i}^r Tr \, dr \right)$$

$$\sigma_{\theta\theta} = \frac{\alpha E}{r^2} \left(\frac{r^2 + R_i^2}{R_0^2 - R_i^2} \int_{R_i}^{R_0} Tr \, dr + \int_{R_i}^r Tr \, dr - Tr^2 \right) \tag{9.29}$$

$$\sigma_{r\theta} = 0$$

Equations (9.26) and (9.27) are cases of generalized plane stress or plane strain problems (body forces are zero) and thus apply equally to unrestrained open-end cylindrical shells.

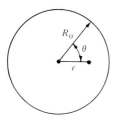

Figure 9.4

Stress Function Solution

If a stress function $\Phi = \Phi(x, y)$ is chosen such that the stresses are defined by

$$\sigma_{xx} = \frac{\partial^2 \Phi}{\partial y^2} \qquad \sigma_{yy} = \frac{\partial^2 \Phi}{\partial x^2} \qquad \sigma_{xy} = -\frac{\partial^2 \Phi}{\partial x\, \partial y} \tag{9.30}$$

then the equilibrium equations, Eqs. (9.9) and (9.10), are satisfied with the body forces set equal to zero. If Φ has to describe the true stress field, it must satisfy not only equilibrium but also compatibility and boundary conditions. Thus, substituting Eq. (9.30) into Eq. (9.20) for the plane stress problem yields

$$\frac{\partial^4 \Phi}{\partial x^4} + 2\frac{\partial^4 \Phi}{\partial x^2\, \partial y^2} + \frac{\partial^4 \Phi}{\partial y^4} + \alpha E\left(\frac{\partial^2 T}{\partial x^2} + \frac{\partial^2 T}{\partial y^2}\right) = 0 \tag{9.31}$$

In polar coordinates, Eq. (9.31) becomes

$$\left(\frac{\partial^2}{\partial r^2} + \frac{1}{r}\frac{\partial}{\partial r} + \frac{1}{r^2}\frac{\partial^2}{\partial \theta^2}\right)\left(\frac{\partial^2}{\partial r^2} + \frac{1}{r}\frac{\partial}{\partial r} + \frac{1}{r^2}\frac{\partial^2}{\partial \theta^2}\right)\Phi$$

$$+ \alpha E\left(\frac{\partial^2}{\partial r^2} + \frac{1}{r}\frac{\partial}{\partial r} + \frac{1}{r^2}\frac{\partial^2}{\partial \theta^2}\right)T = 0 \tag{9.32}$$

where the stresses are defined by

$$\sigma_{rr} = \frac{1}{r}\frac{\partial \Phi}{\partial r} + \frac{1}{r^2}\frac{\partial^2 \Phi}{\partial \theta^2}$$

$$\sigma_{\theta\theta} = \frac{\partial^2 \Phi}{\partial r^2} \tag{9.33}$$

$$\sigma_{r\theta} = -\frac{\partial}{\partial r}\left(\frac{1}{r}\frac{\partial \Phi}{\partial \theta}\right)$$

For a given specific thermal stress problem, Eq. (9.31) or (9.32) can be solved by using techniques of solving differential equations.

Equivalent Load Solution

Thermal stress problems may be handled in the same manner as isothermal problems if the actual thermal loads are converted to what is called *equivalent static loads*. In two-dimensional problems, the conversion is accomplished by making the following substitution:

$$\bar{\sigma}_{xx} = \sigma_{xx} + \frac{E\alpha T}{1 - \nu}$$

$$\bar{\sigma}_{yy} = \sigma_{yy} + \frac{E\alpha T}{1 - \nu} \tag{9.34}$$

$$\bar{\sigma}_{xy} = \sigma_{xy}$$

where $\bar{\sigma}_{xx}$, $\bar{\sigma}_{yy}$, $\bar{\sigma}_{xy}$ = equivalent stresses (displacement stresses)

σ_{xx}, σ_{yy}, σ_{xy} = actual thermal stresses

$\dfrac{E\alpha T}{1-v}$ = local uniform pressure counterbalancing free-expansion strain

Equation (9.34) may be written as

$$\sigma_{xx} = \bar{\sigma}_{xx} - \frac{E\alpha T}{1-v}$$

$$\sigma_{yy} = \bar{\sigma}_{yy} - \frac{E\alpha T}{1-v} \tag{9.35}$$

$$\sigma_{xy} = \bar{\sigma}_{xy}$$

Substituting Eq. (9.35) into Eqs. (9.9) and (9.10) and assuming that body forces X and Y do not exist yield

$$\bar{\sigma}_{xx,\,x} + \bar{\sigma}_{xy,\,y} + \bar{X} = 0$$

$$\bar{\sigma}_{yy,\,y} + \bar{\sigma}_{xy,\,x} + \bar{Y} = 0 \tag{9.36}$$

where $\bar{X}$ and $\bar{Y}$ are equivalent body forces defined as

$$\bar{X} = -\frac{E\alpha}{1-v}\frac{\partial T}{\partial x}$$

$$\bar{Y} = -\frac{E\alpha}{1-v}\frac{\partial T}{\partial y} \tag{9.37}$$

If we assume the applied surface tractions are zero, then the boundary conditions, Eq. (9.22), become

$$\bar{S}_x = \bar{\sigma}^b_{xx} l + \bar{\sigma}^b_{xy} m$$

$$\bar{S}_y = \bar{\sigma}^b_{yy} m + \bar{\sigma}^b_{xy} l \tag{9.38}$$

where $\bar{S}_x$ and $\bar{S}_y$ are equivalent surface tractions defined by

$$\bar{S}_x = \frac{E\alpha T}{1-v}\, l$$

$$\bar{S}_y = \frac{E\alpha T}{1-v}\, m \tag{9.39}$$

In polar coordinates, Eqs. (9.35) to (9.39) become

$$\sigma_{rr} = \bar{\sigma}_{rr} - \frac{E\alpha T}{1 - v}$$

$$\sigma_{\theta\theta} = \bar{\sigma}_{\theta\theta} - \frac{E\alpha T}{1 - v} \tag{9.40}$$

$$\sigma_{r\theta} = \bar{\sigma}_{r\theta}$$

$$\bar{\sigma}_{rr,r} + \frac{\bar{\sigma}_{r\theta,\theta}}{r} + \frac{\bar{\sigma}_{rr} - \bar{\sigma}_{\theta\theta}}{r} + \bar{R} = 0 \tag{9.41}$$

$$\frac{\bar{\sigma}_{\theta\theta,\theta}}{r} + \bar{\sigma}_{r\theta,r} + \frac{2\sigma_{r\theta}}{r} + \bar{\Theta} = 0$$

where

$$\bar{R} = -\frac{E\alpha T}{1 - v} \frac{\partial T}{\partial r} \tag{9.42}$$

$$\bar{\Theta} = -\frac{E\alpha T}{1 - v} \frac{\partial T}{\partial \theta}$$

$$\bar{S}_r = \bar{\sigma}_{rr}^b l + \bar{\sigma}_{r\theta}^b m \tag{9.43}$$

$$\bar{S}_\theta = \bar{\sigma}_{\theta\theta}^b m + \bar{\sigma}_{r\theta}^b l$$

and

$$\bar{S}_r = -\frac{E\alpha T}{1 - v} l \tag{9.44}$$

$$\bar{S}_\theta = -\frac{E\alpha T}{1 - v} m$$

By examining Eqs. (9.36), (9.38), (9.41), and (9.43), it is apparent that for any given temperature distribution, with the use of Eqs. (9.37), (9.39), (9.42), and (9.44) the thermal stress problem can be converted to an equivalent isothermal stress problem with conventional loads. Thus, solutions of thermal stress problems become identical to those of the isothermal ones. As an illustration, consider the unrestrained beam of Fig. 9.5 to be heated to a uniform temperature T_0. The surface tractions at the ends of the beam, if we consider axial direction only, are $\alpha E T_0$, as shown.

These surface tractions will induce a constant equivalent axial stress $\bar{\sigma}_{xx} = \alpha E T_0$ throughout the beam. Hence from Eq. (9.35) for the one-dimensional problem, the thermal stress is

$$\sigma_{xx} = \bar{\sigma}_{xx} - E\alpha T = \alpha E T_0 - \alpha E T_0 = 0$$

which is the same result as was obtained in Sec. 9.2.

$\bar{S}_x^b = \alpha E T_0$ $\alpha E T_0 = \bar{S}_x^b$

Figure 9.5

Now let us assume that the beam of Fig. 9.5 is heated according to the following temperature distribution:

$$T = T_0 \left(\frac{y}{c}\right)^3 \tag{a}$$

The surface tractions at the ends will now take the form $\bar{S}_x^b = \alpha E T_0 \, (y/c)^3$, as shown in Fig. 9.1.

These surface tractions produce equivalent end bending moments:

$$\bar{M} = \int_{-c}^{c} \frac{bE\alpha T_0}{c^3} y^3 y \, dy = \frac{3E\alpha T_0 I}{5c} \tag{b}$$

where b = beam width and I = moment of inertia of beam area. From beam theory, the equivalent stress due to the equivalent moment $\bar{M}$ is

$$\bar{\sigma}_{xx} = \frac{\bar{M}y}{I} = \frac{3E\alpha T_0}{5c} y \tag{c}$$

Thus, the actual thermal stress in the beam is

$$\sigma_{xx} = \bar{\sigma}_{xx} - \alpha E T = \alpha E T_0 \left[\frac{3}{5}\frac{y}{c} - \left(\frac{y}{c}\right)^3\right]$$

9.5 THERMAL STRESSES IN UNRESTRAINED BEAMS WITH TEMPERATURE VARIATION THROUGH THE DEPTH ONLY

Consider the beam of Fig. 9.7 to be subjected to a temperature gradient $T = T(y)$. From elementary beam theory it may be concluded that

$$\sigma_{yy} = \sigma_{zz} = \sigma_{xz} = \sigma_{yz} = 0$$

and

$$\sigma_{xx} = \sigma_{xx}(y)$$

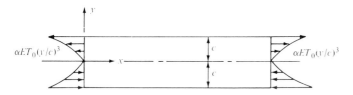

$\alpha E T_0 (y/c)^3$ $\alpha E T_0 (y/c)^3$

Figure 9.6

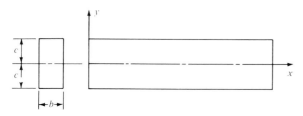

Figure 9.7

By assuming body forces are zero, it can be seen from Eqs. (9.9) and (9.10) that equilibrium is identically satisfied. The compatibility equation [Eq. (9.20)] becomes

$$\frac{d^2}{dy^2}(\sigma_{xx} + \alpha ET) = 0 \tag{9.45}$$

Therefore, the thermal stress can be obtained by simply integrating Eq. (9.45):

$$\sigma_{xx} = -\alpha ET + A_1 y + A_2 \tag{9.46}$$

The constants of integration A_1 and A_2 may be chosen such that for any temperature gradient $T(y)$, the resultant force and moment induced by σ_{xx} are zero over the ends of the beam:

$$\int_A \sigma_{xx}\, dA = \int_A \sigma_{xx} y\, dA = 0 \tag{9.47}$$

Utilizing Eq. (9.46) in Eq. (9.47) and solving for the constants yield the following expression for the thermal stress in the beam:

$$\sigma_{xx} = -\alpha ET + \frac{\alpha E}{2c}\int_{-c}^{c} T\, dy + \frac{3\alpha Ey}{2c^3}\int_{-c}^{c} Ty\, dy \tag{9.48}$$

Equation (9.48) may be written in a more conventional form as

$$\sigma_{xx} = -\alpha ET + \frac{S_T}{A} + \frac{M_T}{I}\, y \tag{9.49}$$

where

$$S_T = \alpha E\int_A T\, dA \qquad M_T = \alpha E\int_A Ty\, dA \tag{9.50}$$

with A and I being the cross-sectional area and moment of inertia of the beam, respectively.

For beams of arbitrary cross section, Eq. (9.49) becomes

$$\sigma_{xx} = -\alpha ET + \frac{S_T}{A} + \frac{I_y M_T^z - I_{yz} M_T^y}{I_y I_z - I_{yz}^2}\, y + \frac{I_z M_T^y - I_{yz} M_T^z}{I_y I_z - I_{yz}^2}\, z \tag{9.51}$$

where

$$S_T = \alpha E \int_A T \, dA \qquad M_T^y = \alpha E \int_A Tz \, dA \qquad M_T^z = \alpha E \int_A Ty \, dA \qquad (9.52)$$

Here I_y and I_z are moments of inertia about the y and z axes, respectively, and I_{yz} is the product moment of inertia.

As an illustration, consider the beam shown in Fig. 9.6, where $T = T_0(y/c)^3$. From Eq. (9.50),

$$S_T = \alpha Eb \int_{-c}^{c} T_0 \left(\frac{y}{c}\right)^3 dy = 0$$

$$M_T = \alpha Eb \int_{-c}^{c} T_0 \left(\frac{y}{c}\right)^3 y \, dy = \frac{3E\alpha T_0 I}{5c}$$

Substituting the above expressions for S_T and M_T into Eq. (9.49) yields the beam thermal stress:

$$\sigma_{xx} = \alpha E T_0 \left[\frac{3}{5}\frac{y}{c} - \left(\frac{y}{c}\right)^3\right]$$

which is the same as obtained previously by using the equivalent load method. Thus, Eqs. (9.49) and (9.51) represent the equivalent load solution for beam problems.

9.6 THERMAL STRESSES IN BUILT-UP STRUCTURES

The thermal stress of built-up structures can be analyzed easily by using the equivalent load method. Techniques such as the stiffness matrix method, energy methods, etc. that are used in the analysis of isothermal problems can be used identically for thermal problems once the equivalent thermal loads are calculated. To illustrate, consider the determinate truss structure shown in Fig. 9.8. Assume that member 1 is heated to a uniform temperature T, above a certain datum

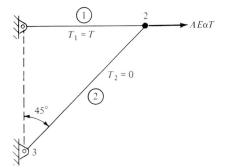

Figure 9.8

temperature. Also assume that all members are of the same material and each has a cross-sectional area A. The equivalent thermal load caused by the temperature change in member 1 is $AE\alpha T$ and is shown in Fig. 9.8. Using the equations of static equilibrium in conjunction with the joint method yields the forces in members 1 and 2:

$$\bar{F}_2 = 0 \qquad \bar{F}_1 = AE\alpha T$$

Hence from Eq. (9.35) for the one-dimensional problem, the thermal stresses are

$$\sigma_1 = \frac{\bar{F}_1}{A} - \alpha E T_1 = \frac{AE\alpha T}{A} - \alpha E T = 0$$

$$\sigma_2 = \frac{\bar{F}_2}{A} - \alpha E T_2 = \frac{0}{A} - \alpha E(0) = 0$$

It is appropriate to state at this point that determinate truss structures in which any member undergoes a uniform change in temperature are stress-free if we assume no conventional loads exist.

To illustrate the use of the stiffness matrix method in the solution of thermal problems, consider the indeterminate truss shown in Figure 9.9. Assume that member 1 has an area A and is heated to a constant temperature T. The area of members 2 and 3 is $\sqrt{2}A$ each. From Chap. 6 the element stiffness matrix relationships are:

Element 1:

$$\begin{bmatrix} \bar{F}_1^x \\ \bar{F}_1^y \\ \bar{F}_2^x \\ \bar{F}_2^y \end{bmatrix} = \frac{AE}{L} \begin{bmatrix} 0 & & \text{symmetric} & \\ 0 & 1 & & \\ 0 & 0 & 0 & \\ 0 & -1 & 0 & 1 \end{bmatrix} \begin{bmatrix} \delta_1^x = 0 \\ \delta_1^y = 0 \\ \delta_2^x \\ \delta_2^y \end{bmatrix} \qquad (a)$$

Element 2:

$$\begin{bmatrix} \bar{F}_2^x \\ \bar{F}_2^y \\ \bar{F}_3^x \\ \bar{F}_3^y \end{bmatrix} = \frac{AE}{L} \begin{bmatrix} 0.5 & & \text{symmetric} & \\ 0.5 & 0.5 & & \\ -0.5 & -0.5 & 0.5 & \\ -0.5 & -0.5 & 0.5 & 0.5 \end{bmatrix} \begin{bmatrix} \delta_2^x \\ \delta_2^y \\ \delta_3^x = 0 \\ \delta_3^y = 0 \end{bmatrix} \qquad (b)$$

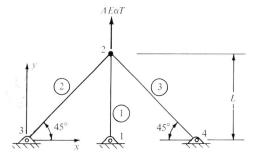

Figure 9.9

Element 3:

$$\begin{bmatrix} \bar{F}_2^x \\ \bar{F}_2^y \\ \bar{F}_4^x \\ \bar{F}_4^y \end{bmatrix} = \frac{AE}{L} \begin{bmatrix} 0.5 & & & \text{symmetric} \\ -0.5 & 0.5 & & \\ -0.5 & 0.5 & 0.5 & \\ 0.5 & -0.5 & -0.5 & 0.5 \end{bmatrix} \begin{bmatrix} \delta_2^x \\ \delta_2^y \\ \delta_4^x = 0 \\ \delta_4^y = 0 \end{bmatrix} \qquad (c)$$

The overall reduced matrix can be obtained easily, as was done in Chap. 6, and is given by

$$\begin{bmatrix} 0 \\ AE\alpha T \end{bmatrix} = \frac{AE}{L} \begin{bmatrix} 1.0 & 0 \\ 0 & 2.0 \end{bmatrix} \begin{bmatrix} \delta_2^x \\ \delta_2^y \end{bmatrix} \qquad (d)$$

Inverting and solving for the unknown thermal displacements yield

$$\begin{bmatrix} \delta_2^x \\ \delta_2^y \end{bmatrix} = \frac{L}{AE} \begin{bmatrix} 1.0 & 0 \\ 0 & 0.5 \end{bmatrix} \begin{bmatrix} 0 \\ AE\alpha T \end{bmatrix} = \begin{bmatrix} 0 \\ \dfrac{\alpha LT}{2} \end{bmatrix} \qquad (e)$$

The equivalent thermal loads on each element may be obtained from Eqs. (a) to (c):

Element 1:

$$\bar{F}_1^x = \bar{F}_2^x = 0$$

$$\bar{F}_1^y = -\bar{F}_2^y = -\frac{AE\alpha T}{2}$$

or $\qquad \bar{F}_1 = [(\bar{F}_1^x)^2 + (\bar{F}_1^y)^2]^{1/2} = \dfrac{AE\alpha T}{2} \qquad$ (tension)

Element 2:

$$\bar{F}_2^x = -\bar{F}_3^x = \frac{AE\alpha T}{4}$$

$$\bar{F}_2^y = -\bar{F}_3^y = \frac{AE\alpha T}{4}$$

or $\qquad \bar{F}_2 = \dfrac{AE\alpha T}{2\sqrt{2}} \qquad$ (tension)

Element 3:

$$\bar{F}_2^x = -\bar{F}_4^x = -\frac{AE\alpha T}{4}$$

$$\bar{F}_2^y = -\bar{F}_4^y = \frac{AE\alpha T}{4}$$

or $\qquad \bar{F}_3 = \dfrac{AE\alpha T}{2\sqrt{2}} \qquad$ (tension)

Hence the actual thermal stress in each rod is

$$\sigma_1 = \frac{F_1}{A} - \alpha E T_1 = \frac{A E \alpha T}{2A} - \alpha E T = \frac{E \alpha T}{2}$$

$$\sigma_2 = \frac{F_2}{\sqrt{2}A} - \alpha E T_2 = \frac{E \alpha A T}{2\sqrt{2}\sqrt{2}A} - 0 = \frac{E \alpha T}{4}$$

$$\sigma_3 = \sigma_2$$

The preceding problem can be solved by using the energy method. This may be accomplished by making the structure redundant, as shown in Fig. 9.10. For compatibility of deformation, the relative displacement in the direction of R_1 must be zero. Thus from Castigliano's theorem,

$$\delta_1 = \frac{\delta U}{\delta R_1} = 0$$

where U is the total strain energy stored in the structure and R_1 is the unknown equivalent internal load in member 1.

From the equations of static equilibrium in conjunction with the joint method, the equivalent internal loads in members 2 and 3 can be calculated:

$$\bar{F}_2 = \bar{F}_3 = \frac{E \alpha A T - R_1}{\sqrt{2}}$$

Therefore, the strain energy U is

$$U = \frac{1}{2} \sum_{i=1}^{3} \frac{S_i^2 L_i}{A_i E_i} = \frac{1}{2} \frac{R_1^2 L}{AE} + \frac{1}{2\sqrt{2}AE} \left(\frac{E \alpha A T - R_1}{\sqrt{2}} \right)^2 \sqrt{2}\,L$$

$$+ \frac{1}{2\sqrt{2}AE} \left(\frac{E \alpha A T - R_1}{\sqrt{2}} \right)^2 \sqrt{2}\,L$$

Performing the differentiation with respect to R_1 and setting the result equal to zero yield

$$R_1 = \frac{E \alpha A T}{2} = \bar{F}_1$$

$$\bar{F}_2 = \bar{F}_3 = \frac{E \alpha A T}{2\sqrt{2}}$$

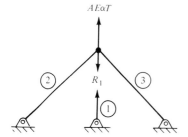

Figure 9.10

or

$$\sigma_1 = \frac{E\alpha A T}{2A} - E\alpha T = \frac{E\alpha T}{2}$$

$$\sigma_2 = \sigma_3 = \frac{E\alpha A T}{2\sqrt{2}\sqrt{2A}} - 0 = \frac{E\alpha T}{4}$$

which are the same results obtained by using the stiffness matrix technique. As may be seen from the two preceding problems, the advantage of the stiffness matrix method is that it yields not only the internal equivalent thermal loads but also the actual thermal deflections.

Rigid-frame structures under the action of thermal loads are analyzed in the same manner as truss structures. First the equivalent thermal loads are calculated, and then any convenient technique employed in conventionally loaded structures can be used.

PROBLEMS

9.1 Find the radial displacement of a thin, circular, solid plate which is subjected to a temperature distribution $T = T(r)$.

9.2 Find the radial displacement of a circular cylinder whose ends are held between two rigid walls. Assume a temperature distribution $T = T(r)$.

9.3 Find the thermal stresses in strips 1 and 2 of the composite beam shown in Fig. P9.3. Assume uniform temperatures T_1 and T_2 for strips 1 and 2, respectively. Also assume that no slippage takes place at the bond line.

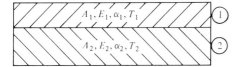

Figure P9.3

9.4 Find the thermal stresses in the idealized structure shown in Fig. P9.4. Assume the skin and the stringer to be heated to uniform temperatures T_s and T_b, respectively.

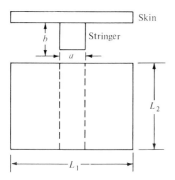

Figure P9.4

9.5 The inner and outer surface temperatures of a circular cylindrical shell (inner radius $= R_i$ and outer radius $= R_o$) are kept constant at T_i and T_o, respectively. The end surfaces are perfectly insulated. Find the steady-state temperature distribution in the cylinder, and plot the results. Assume the datum temperature to be T_o.

   Hint: See Gatewood, Jr., *Aero. Sci.*, vol. 21, no. 9, 1954, pp. 645–646.

$$\frac{d^2T}{dr^2} + \frac{1}{r}\frac{dT}{dr} = 0$$

9.6 Refer to Prob. 9.5. Find the stresses in the cylinder. Assume unrestrained conditions. Plot the results.

9.7 For the riveted structure shown in Fig. P9.7, find the loads on each rivet. The critical rivet load

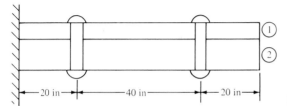

Figure P9.7

and deflection are 1200 lb and 0.006 in, respectively. Assume steady-state temperature distribution. The following data are given:

$$E_1 = E_2 = 10^7 \ \text{lb/in}^2$$

$$2A_1 = A_2 = 2.0 \ \text{in}^2$$

$$\alpha_1 = \alpha_2 = 10^{-7} \ \text{in/(in} \cdot {}^\circ\text{F)}$$

$$2T_1 = T_2 = 400^\circ\text{F above datum temperature}$$

$$T_0 = 80^\circ\text{F}$$

   Hint: See Gatewood, Jr., *Aero. Sci.*, vol. 21, no. 9, 1954, pp. 645–646.

9.8 The general equation which describes the general state of stress in a one-dimensional thermally loaded beam is

$$\sigma_{xx} = \alpha E[-T(y) + Ay + B]$$

where A and B are arbitrary constants which depend on the end conditions of the beam. Find the stress in an unrestrained beam whose cross section is shown in Fig. P9.8 and subjected to a temperature distribution given by $T = T_0 \, y e^{-\beta y}$.

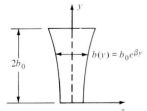

Figure P9.8

9.9 In Prob. 9.8, consider a beam whose cross section is shown in Fig. P9.9. The skin is made out of aluminum and is at a constant temperature T_s, while the web and flange are made out of different material and at a constant temperature T_w. Find the stresses in the skin, web, and flange.

(*a*) Assume that the beam is completely unrestrained.

(*b*) Assume that the beam is restrained in axial compression.

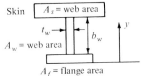

Figure P9.9

9.10 The temperature of a beam of rectangular cross section, as shown in Fig. P9.10, is $T = T_0 e^{(zx + \beta y)}$. Using the Airy stress function $\psi = \rho(x)\beta(y)$, find the stress distribution throughout the beam. Assume the beam is unrestrained.

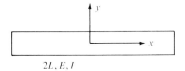

Figure P9.10

9.11 Design the truss structure shown in Fig. P9.11. Assume 2024-T42 aluminum-alloy tubing construction is used.

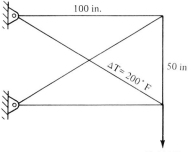

20,000 lb   **Figure P9.11**

9.12 Find the thermal stresses and deflections for each of the structures shown in Fig. P9.12.

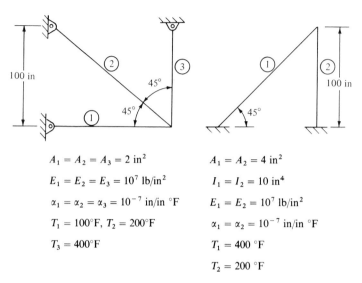

$$A_1 = A_2 = A_3 = 2 \text{ in}^2$$

$$E_1 = E_2 = E_3 = 10^7 \text{ lb/in}^2$$

$$\alpha_1 = \alpha_2 = \alpha_3 = 10^{-7} \text{ in/in } °F$$

$$T_1 = 100°F, \ T_2 = 200°F$$

$$T_3 = 400°F$$

$$A_1 = A_2 = 4 \text{ in}^2$$

$$I_1 = I_2 = 10 \text{ in}^4$$

$$E_1 = E_2 = 10^7 \text{ lb/in}^2$$

$$\alpha_1 = \alpha_2 = 10^{-7} \text{ in/in } °F$$

$$T_1 = 400 \ °F$$

$$T_2 = 200 \ °F$$

Figure P9.12

9.13 Find the displacements for the structure shown in Fig. P9.13. Assume the beam is subjected to a temperature distribution given by $T = 400y/h$, as shown. Assume $E = 10^7 \text{ lb/in}^2$ and $\alpha = 10^{-7}$ in/(in · °F) for both rod and beam.

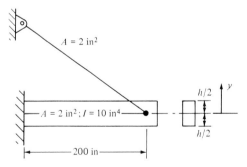

Figure P9.13

DESIGN OF MEMBERS IN TENSION, BENDING, OR TORSION

10.1 TENSION MEMBERS

Tension members are analyzed and designed more readily than other types of members. The stress conditions existing in tension members at the ultimate load condition are accurately known and are not subject to the uncertainties which exist in joints, fittings, and other types of structural members. The allowable tensile stress for a structural material is easy to determine, and a single value of the allowable stress applies to members of any shape. It is shown that the allowable stresses for structural members in bending, torsion, or compression depend on the shapes of the members and on other factors which are not considered for tension members.

For a concentric tension load P on a member with a net area A, the tensile stress is found from the equation $\sigma_t = P/A$. The allowable tension stress σ_{tu} is the minimum guaranteed value for the material. The margin of safety may be calculated in the usual manner as $\sigma_{tu}/\sigma_t - 1$.

Tension members frequently must resist bending and compression stresses under other loading conditions, and these other conditions often determine the shape of a member, even when the tension load is the largest load.

The primary tension structure in a semimonocoque wing consists of the skin, stringers, and spar caps on the under surface. Although the positive bending moments in a wing are about twice as large as the negative bending moments, the compression loads from negative bending determine the design of most of the

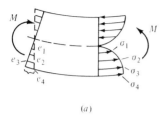

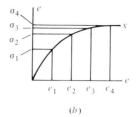

(a)   (b)

Figure 10.1

structure on the underside of the wing. The wing skin resists tension stress, but buckles and becomes ineffective for compression stress. Thus the compressive area is less than the tensile area, and the allowable compressive stress is considerably less than the allowable tensile stress. Even though the compressive loads are smaller, they must always be considered when the shape of the stiffening members is determined and frequently they determine the required areas.

10.2 PLASTIC BENDING

In the previous calculations of bending stresses, we assume that the stresses are below the elastic limit. In most types of machine design and structural design, the strength at the yield stress is the important criterion for design, and the conventional elastic stress distribution is satisfactory for use in design. In airframe structures, however, the ultimate strength of a member is the design criterion. Before failure, the stress exceeds the elastic limit and is said to be in the *plastic range*. The assumptions used in deriving the flexure formula $\sigma_b = My/I$ no longer apply.

The initially straight beam shown in Fig. 10-1a has bending stresses exceeding the elastic limit. Plane sections remain plane after bending, and thus the strain distribution is proportional to the distance from the neutral axis, as in elastic beams. If the beam is deflected so that the extreme fiber has a strain e_4, there will be a stress σ_4 at this point, as shown by the stress-strain curve of Fig. 10.1b. For other strains e_1, e_2, and e_3, the corresponding stresses σ_1, σ_2, and σ_3 do not vary linearly with the strains above the elastic limit, and the stress distribution on the beam cross section varies as shown in Fig. 10.2.

The ultimate resisting moment of a beam depends on both the shape of the

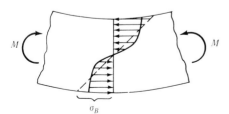

Figure 10.2

beam cross section and the shape of the stress-strain curve. Since there is no simple theoretical relationship which applies to a general case, frequently an empirical method is used to determine the ultimate bending strength. A fictitious stress σ_b, termed the *bending modulus of rupture*, is defined by the equation $\sigma_b = Mc/I$, where M is the ultimate bending moment, as determined from tests of similar beams, I is the moment of inertia of the cross section, and c is the distance from the beam neutral axis to the extreme fiber. The true stress distribution is shown in Fig. 10.2, and the fictitious straight-line stress distribution which yields an equal bending moment and has a maximum value of σ_B is shown by the dotted line. For geometrically similar sections such as round tubes with the same ratio of outside diameter to wall thickness D/t, the bending modulus of rupture may be found for any material by means of tests.

The bending modulus of rupture for round tubes of chrome-molybdenum steel is shown in Fig. 10.3 for various values of D/t and of σ_{tu}, the ultimate tensile

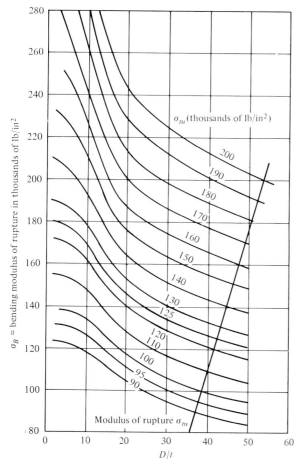

Figure 10.3

stress to which the material is heat-treated. For the larger values of D/t, the tube walls are thin and tend to cripple locally. The local crippling stress of the tube wall, which in itself is difficult to compute theoretically, corresponds to the stress σ_4 shown in Fig. 10.1. Thus the tests take into consideration the effects of local crippling, as well as the effects of the shape of cross section and of the stress-strain curve. The bending modulus of rupture is proportional to the bending moment, and the margin of safety may be computed from the usual relation, $\sigma_B/\sigma_b - 1$. The true maximum stress σ_4 of Fig. 10.1 is not proportional to the bending moment and cannot be used in obtaining the margin of safety.

Example 10.1 A $1\frac{1}{2}$ by 0.083-in steel tube resists a bending moment of 25,000 in · lb. What is the margin of safety if the material is heat-treated to an ultimate tensile stress σ_{tu} of 180,000 lb/in²?

SOLUTION The properties of a $1\frac{1}{2}$ by 0.083-in steel tube are $D/t = 18.08$ and $I/c = 0.1241$ in³. From Fig. 10.3, $\sigma_B = 220,000$ lb/in². The fictitious bending stress σ_b is obtained from the simple flexure formula:

$$\sigma_b = \frac{Mc}{I} = \frac{25,000}{0.1241} = 201,000 \text{ lb/in}^2$$

The margin of safety (MS) is now obtained in the usual manner:

$$MS = \frac{220,000}{201,000} - 1 = 0.09$$

It is also necessary to determine a yield margin of safety. For this material, the yield stress is 165,000 lb/in². The applied or limit bending moment is $\frac{2}{3} \times 25,000 = 16,670$ in · lb and so

$$\sigma_b = \frac{16,670}{0.1241} = 134,000 \text{ lb/in}^2$$

or the margin of safety for yielding is

$$MS = \frac{165,000}{134,000} - 1 = 0.23$$

10.3 CONSTANT BENDING STRESS

For some materials, the stress-strain curve remains almost horizontal after the elongation exceeds a value corresponding to the yield point. If a beam of such a material is subjected to bending beyond the yield stress, the bending stresses will approximate those shown in Fig. 10.4. Both the tension and the compression stresses may be assumed to have constant values of σ_0 over the entire area. The bending moment is obtained by taking the sum of the moments of infinitesimal forces $\sigma_0 \, dA$ about the neutral axis:

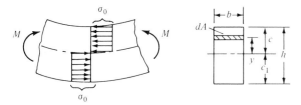

Figure 10.4

$$M = \sigma_0 \int_{-c_1}^{c} y \, dA \tag{10.1}$$

In the case of a cross-sectional area which is symmetrical with respect to a horizontal axis, the bending moment becomes

$$M = 2Q\sigma_0 \tag{10.2}$$

where

$$Q = \int_0^c y \, dA \tag{10.3}$$

For the symmetrical area, the neutral axis corresponds with the axis of symmetry, as in the case of elastic bending. For an unsymmetrical area, the neutral axis is not at the centroid, but is located so that the cross-sectional area above the neutral axis is equal to the area below it, since the total tension force must equal the total compression force.

For a rectangular beam of width b and depth h, $Q = bh^2/8$. Substituting in Eq. (10.2) yields

$$M = \frac{\sigma_0 \, bh^2}{4} \tag{10.4}$$

The bending modulus of rupture σ_b can be found by equating the bending moment of Eq. (10.4) to the expression which defines σ_b, or $M = \sigma_b I/c$. For the rectangular section, $I/c = bh^2/6$, and $\sigma_b = 1.5\sigma_0$.

The parabolic shear stress distribution for a rectangular beam in which the stresses are below the elastic limit was obtained from the bending stress distribution, and does not apply for other distributions of bending stress. For a rectangular cross section, or for other similar cross sections in which the bending modulus of rupture is considerably larger than the actual stress, the shearing stresses are seldom very high and may be approximated with sufficient accuracy.

The plastic bending of a beam in which the cross section is not symmetrical about the neutral axis is considered by analyzing a numerical example. The area shown in Fig. 10.5 has its centroidal axis 0.3 in above the base. For plastic bending with a constant stress σ_0, the neutral axis will be 0.2 in above the base, in order for the tension area to be equal to the compression area. The tension and compression forces will be equal to $0.12\sigma_0$ and will resist a bending moment

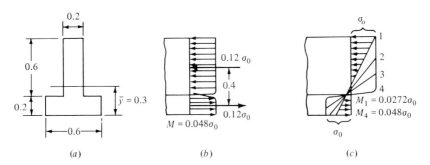

Figure 10.5

of $0.048\sigma_0$, as shown in Fig. 10.5b. The elastic bending stress for this area is obtained from the equation $M = \sigma_b I/c = 0.027\sigma_b$. The bending modulus of rupture is therefore equal to $\sigma_b = (0.048/0.0272)\sigma_0 = 1.765\sigma_0$. The stress distribution of various bending moments is shown in Fig. 10.5c. For bending moments less than $M = 0.0272\sigma_0$, the stresses are below the elastic limit and have a straight-line distribution with the neutral axis at the centroid of the area, as shown by curve 1. For larger values of the bending moment, the stresses will exceed the elastic limit at the upper side of the beam, but remain below the elastic limit on the lower side, with the neutral axis shifting downward, as shown by curve 2. For further increases in bending moment, the stresses approach the constant values shown by curve 4, with the neutral axis between the two rectangles.

10.4 TRAPEZOIDAL DISTRIBUTION OF BENDING STRESS

The stress-strain curves for most aircraft materials can be approximated accurately by a trapezoidal curve, as shown in Fig. 10.6a. The idealized bending stress distribution is shown in Fig. 10.6b. This approximation for obtaining the bending strength in the plastic range was proposed by Cozzone.[7] The bending moment for the trapezoidal stress distribution is readily obtained as the sum of the bend-

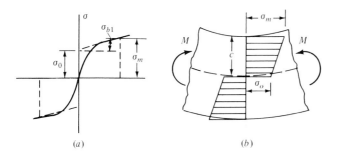

Figure 10.6

ing moment for a constant stress σ_0, as given by Eq. (10.2), and the bending moment for a linear stress distribution varying from 0 to σ_{b1} :

$$M = 2Q\sigma_0 + \frac{\sigma_{b1} I}{c} \tag{10.5}$$

The term σ_m may be introduced instead of σ_{b1} by substituting $\sigma_{b1} = \sigma_m - \sigma_0$ into Eq. (10.5). Making this substitution and dividing by I/c give

$$\sigma_B = \frac{Mc}{I} = \sigma_m + \sigma_0 \left(\frac{2Q}{I/c} - 1 \right) \tag{10.6}$$

The term in parentheses depends on the shape of the cross section and may vary from 0 for concentrated flange areas to 1.0 for a diamond shape. If this term is designated by K, Eq. (10.6) becomes

$$\sigma_B = \sigma_m + K\sigma_0 \tag{10.7}$$

where

$$K = \frac{2Q}{I/c} - 1 \tag{10.8}$$

Some values of K for various cross sections are shown in Fig. 10.7.

The value of σ_0 should be determined in such a way that the bending moment resisted by the assumed trapezoidal stress distribution is equal to the bending moment resisted by the actual stresses. Therefore, the correct value of σ_0 would depend somewhat on the cross-sectional area. If the value of σ_0 were calculated for each area, there would be no advantage in assuming a trapezoidal stress distribution, since it would be necessary to calculate the true resisting moment of the beam in order to calculate σ_0. Cozzone has shown that it is sufficiently accurate to calculate σ_0 for a rectangular cross section and to use this value for all cross sections.

Example 10.2 A beam with the cross section shown in Fig. 10.8 is made of aluminum-alloy forging. The true shape of the forging is shown by the dotted lines, but the trapezoids shown by the solid lines are assumed. Calculate the

Section	K	Section	K
	0		0.25 to 0.7
	0.5		1.0
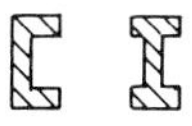	0 to 0.5		$\frac{1}{3}$

Figure 10.7

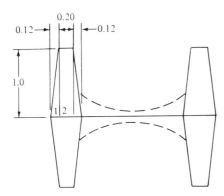

Figure 10.8

ultimate bending strength about a horizontal axis if $\sigma_m = 65,000$, $\sigma_0 = 60,000$, and the yield stress $\sigma_{ty} = 50,000$ lb/in².

SOLUTION The values of I and Q are calculated for the assumed area, which is composed of eight of the triangles (1) and four of the rectangles (2):

$$I = \frac{8 \times 0.12 \times 1^3}{12} + \frac{4 \times 0.20 \times 1^3}{3} = 0.347 \text{ in}^4$$

$$2Q = 8 \times 0.06 \times 0.333 + 4 \times 0.20 \times 0.5 = 0.56 \text{ in}^3$$

The bending modulus of rupture is now calculated from Eqs. (10.7) and (10.8):

$$K = \frac{2Q}{I/c} - 1 = \frac{0.56}{0.347} - 1 = 0.61$$

$$\sigma_B = \sigma_M + K\sigma_0 = 65,000 + 0.61 \times 60,000 = 101,600 \text{ lb/in}^2$$

The ultimate bending strength is

$$M = \frac{\sigma_B I}{c} = 101,600 \, \frac{0.347}{1.0} = 35,300 \text{ in} \cdot \text{lb}$$

In this case, it would not be possible to utilize the full ultimate bending strength because the stress at the applied load condition would exceed the yield stress. The exact amount of permanent set permitted at the applied or limit load is not specified clearly for a member in bending, but depends somewhat on the judgment of the designer. In some cases, the bending modulus of rupture is not permitted to exceed the yield stress; or, for this problem, the value of Mc/I would not exceed 50,000 lb/in² at the applied load and consequently could not exceed 75,000 lb/in² at the ultimate or design load. Even at the yield stress, however, some plastic bending effects may be considered. The stress-strain diagram for the material of Fig. 10.8, in which the stress does not exceed the yield stress shown in Fig. 10.9, may be represented by the trapezoid with $\sigma_0 = 21,200$ and $\sigma_m = 50,000$ lb/in². The

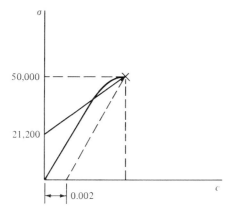

Figure 10.9

bending modulus of rupture at the limit load can be calculated from Eq. (10.7):

$$\sigma_B = \sigma_m + K\sigma_0 = 50,000 + 0.61 \times 21,200 = 63,000 \ \text{lb/in}^2$$

The allowable value of Mc/I would then be 63,000 for the limit load and $1.5 \times 63,000 = 94,500 \ \text{lb/in}^2$ at the ultimate or design load.

10.5 CURVED BEAMS

Most beam structures are analyzed by the methods previously considered, in which any initial curvature of the axis of the beam is neglected. However, when the radius of curvature is of the same order of magnitude as the depth of the beam, the stress distribution differs considerably from that for straight beams. The stresses on the concave side of the beam are higher than those for a similar straight beam, and the stresses on the convex side are lower. When the maximum stresses exceed the elastic limit, local yielding occurs, which permits a re-distribution of stress. At the ultimate bending moment, the stresses approach the same distribution as for the plastic bending of a straight beam. Thus the beam curvature has the effect of reducing the yield strength but of not appreciably changing the ultimate bending strength.

The beam shown in Fig. 10.10 has an initial radius of curvature R measured to the centroid of the cross section. A plane cross section pp remains plane after bending, and its relative position after bending is shown by nn. A longitudinal fiber of the beam of initial length L is extended a distance δ. Since δ is measured between the straight lines pp and nn, it varies linearly with the distance y from the centroid to the fiber:

$$\delta = k_1 + k_2 y$$

The terms k_1, k_2, and k_3 are determined constants. The length of the fiber L is proportional to its distance from the center of curvature:

$$L = k_3(R + y)$$

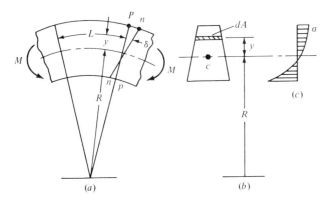

Figure 10.10

The unit stress is now obtained as the product of the unit strain δ/L and the elastic modulus E:

$$\sigma = E \frac{\delta}{L} = E \frac{k_1 + k_2 y}{k_2(R + y)}$$

This expression may be simplified by dividing the numerator by the denominator and grouping the constants into two new, undetermined constants a and b:

$$\sigma = a + \frac{b}{R + y} \tag{10.9}$$

This stress distribution is pictured in Fig. 10.10c.

If a resultant tension force P acts at the centroid of the area, it must equal the sum of the internal forces, that is, $\int \sigma \, dA$. From Eq. (10.9) we have

$$P = \int_\sigma dA = aA + b \int \frac{dA}{R + y} \tag{10.10}$$

where A is the total cross-sectional area and a and b are undetermined constants. For the case of pure bending, the force P vanishes.

The external bending moment M about the centroidal axis must equal the moment of the internal forces $\int fy \, dA$:

$$M = \int_\sigma y \, dA = a \int y \, dA + b \int \frac{y \, dA}{R + y}$$

The first integral on the right side of the equation is zero because y is measured from the centroidal axis. The second integral may be separated into two terms by division:

$$\int \frac{y \, dA}{R + y} = \int \left(1 - \frac{R}{R + y}\right) dA = A - R \int \frac{dA}{R + y}$$

The bending-moment equation may now be written as

$$M = bA - bR \int \frac{dA}{R + y} \tag{10.11}$$

The unknown constant b can be found from Eq. (10.11) since all other terms are known from the geometry and loading of the beam. The other unknown constant, a, is obtained from Eq. (10.10). The stress distribution is then found from Eq. (10.9). The effect of the axial load P is to change the constant a and the stress σ by an amount P/A. Hence the same stress distribution can be obtained by superimposing the bending stresses for $P = 0$ and the stresses P/A resulting from the axial load P at the centroid of the area.

The extreme fiber stresses for various curved beams can be determined as ratios of the stresses computed by the flexure formula. These ratios have been computed for various cross sections. The terms K for the equation $\sigma = KMc/I$ are plotted in Fig. 10.11 for a few common cross sections. It is observed that the stresses always become infinite on the concave side when $R/c = 1$, which corresponds to a sharp reentrant angle on the concave surface of the beam. Such reentrant angles should be avoided in any structure or machine part.

Another effect of beam curvature which cannot be analyzed by simple theory occurs in beams with thin flanges, as indicated in Fig. 10.12. If the concave side of a beam is in compression, the flanges tend to deflect toward the neutral axis, as shown in Fig. 10.12b. The bending stress is not distributed uniformly along the

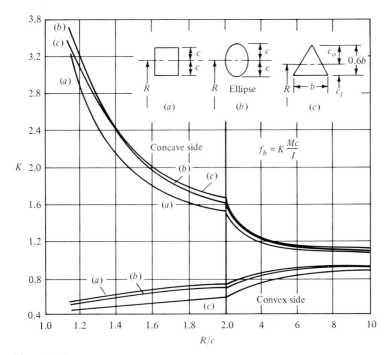

Figure 10.11

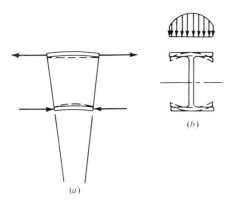

Figure 10.12

horizontal beam flange, but is much higher near the web than it is in the outstanding legs, as shown in Fig. 10.12b. When the bending produces compression on the convex side of the beam, the flanges deflect away from the neutral axis, but the stress distribution along the horizontal width of the flange is essentially the same as shown.

10.6 TORSION OF CIRCULAR SHAFTS

The stresses resulting from torsional moments acting on elastic cylindrical members of circular cross section can be obtained readily. It has been found experimentally that there is no distortion of any cross section of the shaft either in the direction normal to the plane of the cross section or in the plane of the cross section. Any two cross sections of the shaft, such as those shown in Fig. 10.13b, have a relative rotation about the axis of the shaft. Since the two cross sections have no relative displacement radially, or along the axis of the shaft, the only stresses are the shearing stresses in the circumferential and axial directions, as shown in Fig. 10.13b.

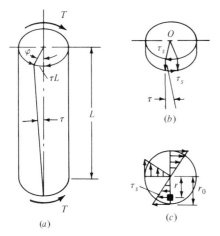

Figure 10.13

The shearing strain γ and consequently the shearing stress τ_s must vary linearly in proportion to the distance r from the center of the shaft:

$$\tau_s = Kr \tag{10.12}$$

The term K is a constant of proportionality which is determined later. The external torsional moment T must equal the sum of the moments of the internal shearing forces on the cross section:

$$T = \int \tau_s r \, dA = K \int r^2 \, dA \tag{10.13}$$

The integral of Eq. (10.13) represents the polar moment of inertia of the cross-sectional area, and usually it is designated as J or I_P. The value of K may be found from Eq. (10.13) as $K = T/J$. Substituting this value into Eq. (10.12) yields the formula for torsional shear stresses in circular shafts:

$$\tau_s = \frac{Tr}{J} \tag{10.14}$$

The angle of twist of a circular shaft may be determined from the angle of shearing strain γ. The shearing modulus of elasticity G is defined as the ratio of shear stress to shear strain:

$$G = \frac{\tau_s}{\gamma} \tag{10.15}$$

And G is related to E for an isotropic material only by the following equation:

$$G = \frac{E}{2(1 + \mu)} \tag{10.16}$$

The shaft of radius r_0 shown in Fig. 10.13 twists through an angle ϕ in length L. A point on the circumference of the upper cross section moves a distance ϕr_0 during the deformation. This point is also displaced a distance γL, as shown:

$$\phi r_0 = \gamma L$$

The angle of twist ϕ may be expressed in other forms by substituting values from Eqs. (10.15) and (10.14):

$$\phi = \frac{\gamma L}{r_0} = \frac{\tau_0 L}{G r_0} = \frac{TL}{JG} \tag{10.17}$$

These equations apply only to torsion members with solid circular cross sections, or to tubular members with hollow circular cross sections.

10.7 TORSION OF A NONCIRCULAR SHAFT

The stresses in a torsion member of arbitrary cross section cannot be obtained by means of a simple, general equation. A few special cases may be analyzed, how-

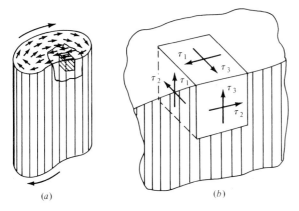

(a)   (b)

Figure 10.14

ever, and some general properties of the shear stress distribution are examined. A cylindrical torsion member of arbitrary cross section is shown in Fig. 10.14a. The small cubical element at the surface of the member is enlarged in Fig. 10.14b. In general, three pairs of shearing stresses τ_1, τ_2, τ_3 exist on any such element; but in the case where one face of this element is a free surface of the member, the shear stresses τ_2 and τ_3 on the free surface must be zero. Thus the element at the boundary has only the shear stresses τ_1, which are parallel to the boundary. Hence the shear stresses on a cross section near any boundary are parallel to the boundary, as shown in Fig. 10.14a. The resultant of all shearing forces on the cross section must be equal to the external torque.

The exaggerated deformations of a square shaft in torsion are shown in Fig. 10.15a. An element at a corner of the cross section may be compared to the element shown in Fig. 10.14b, and since two perpendicular faces of the element can have no shear stresses, all the shear stresses τ_1, τ_2, and τ_3 must be zero at a corner of the member. The shear stresses on the cross section have maximum values at the center of each side and have directions approximately as shown in

(b)

(a)

Figure 10.15

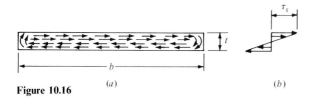

Figure 10.16   (a)   (b)

Fig. 10.15b. The cubical elements at the corners remain cubical, as indicated in Fig. 10.15a, but elements near the centers of the sides have rather large shearing deformations. The cross sections therefore do not remain plane, but warp as shown.

One type of member which is used frequently on flight vehicle structures has a narrow, rectangular cross section. While such cross sections are inefficient for torsion members, often they must resist some torsional stresses. For the cross section shown in Fig. 10.16 of length b and width t, the shear stresses must be parallel to the boundary. If the length b is large compared to the thickness t, the end effects are small, and the shear stresses may be assumed to be distributed as shown in Fig. 10.16b for the entire length b. It can be shown that the shear stress has the following value:

$$\tau_s = \frac{3T}{bt^2} \tag{10.18}$$

Equation (10.18) is accurate when the width b is large compared to the thickness t. For rectangular cross sections in which the dimensions are of the same order, the maximum stress, which occurs at the middle of the longest side, is found by

$$\tau_s = \frac{T}{\alpha bt^2} \tag{10.19}$$

Values of α are given in Table 10.1. These values have been calculated by theoretical methods which are not within the scope of the present discussion. For large ratios of b/t, the value of α is 0.333, which corresponds to Eq. (10.18). For smaller values of b/t, the effects of the ends are more noticeable, and the values of α are smaller than 0.333.

The angle of twist of a rectangular shaft of length L can be obtained from

$$\phi = \frac{TL}{\beta bt^3 G} \tag{10.20}$$

where ϕ is in radians and β is a constant which is given in Table 10.1.

Table 10.1 Constants for Eqs. (10.19) and (10.20)

b/t	1.00	1.50	1.75	2.00	2.50	3.00	4	6	8	10	∞
α	0.208	0.231	0.239	0.246	0.258	0.267	0.282	0.299	0.307	0.313	0.333
β	0.141	0.196	0.214	0.229	0.249	0.263	0.281	0.299	0.307	0.313	0.333

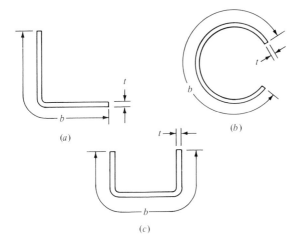

(a)

(b)

(c)

Figure 10.17

The torsional properties of a rectangular plate are not appreciably affected if the plate is bent to some cross section, such as those shown in Fig. 10.17, provided that the end cross sections of the member are free to warp. Thus these sections may be analyzed by Eqs. (10.19) and (10.20). The angle of twist for a shaft of any cross section may be expressed in the form

$$\phi = \frac{TL}{KG} \tag{10.21}$$

where K is a constant which depends on only the cross-sectional area. By comparing Eqs. (10.20) and (10.21), the value of K for a rectangular cross section is found to be βbt^3. For a cross section made up of several rectangular elements, such as that shown in Fig. 10.18, the value of K is given approximately by

$$K = \beta_1 b_1 t_1^3 + \beta_2 b_2 t_2^3 + \beta_3 b_3 t_3^3 \tag{10.22}$$

Example 10.3 The round tube shown in Fig. 10.19a has an average radius R and a wall thickness t. Compare the torsional strength and rigidity of this tube with that of a similar tube which is slit for its entire length, as shown in Fig. 10.19b and c. Assume $R/t = 20$.

SOLUTION Approximate values of the area and moment of inertia are satis-

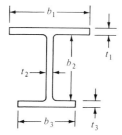

Figure 10.18

factory for thin-walled tubes. The area is equal to the product of the circumference $2\pi R$ and the wall thickness t. The polar radius of gyration is approximately equal to R, and the polar moment of inertia is then obtained as follows:

$$J = 2\pi R^3 t \tag{10.23}$$

It is sufficiently accurate to use the average radius R in place of the outside radius in computing the maximum shearing stress. The values of the closed tube are obtained from Eqs. (10.14) and (10.17):

$$\tau_s = \frac{Tr}{J} = \frac{T}{2\pi R^2 t} \tag{10.24}$$

$$\phi = \frac{TL}{JG} = \frac{TL}{2\pi R^3 tG} \tag{10.25}$$

The slit tube is analyzed by Eqs. (10.18) and (10.20), assuming $b = 2\pi R$ and $\alpha = \beta = 0.333$:

$$\tau_s = \frac{3T}{2\pi Rt^2} \tag{10.26}$$

$$\phi = \frac{3TL}{2\pi Rt^3 G} \tag{10.27}$$

The ratio of shearing stresses for the two members is obtained by dividing Eq. (10.26) by Eq. (10.24). The values of the stress and angle for the closed tube are designated τ_{s0} and ϕ_0, respectively:

$$\frac{\tau_s}{\tau_{s0}} = \frac{3R}{t} = 60$$

The shearing stresses are therefore 60 times as high in the slit tube as in the closed tube, or the closed tube would be 60 times as strong if the allowable shearing stresses were equal.

The torsional stiffness of the two members may be compared by dividing Eq. (10.27) by Eq. (10.25):

$$\frac{\phi}{\phi_0} = \frac{3R^2}{t^2} = 1200 \tag{10.28}$$

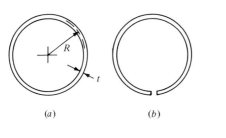 

(a)   (b)   (c)   **Figure 10.19**

For a given angle of twist, the closed tube resists 1200 times the torsion of the open section; or, for a given torque, the open section twists through an angle 1200 times as great as the closed tube. We assumed that the open tube was free to distort as shown in Fig. 10.19c, or that the end cross sections were not restrained against warping.

10.8 END RESTRAINT OF TORSION MEMBERS

In Sec. 10.7 we assume that the end cross sections of the torsion members are free to warp from their original plane and that there are no stresses normal to the cross sections. It has been pointed out that many aircraft structural members must be constructed with thin webs and that such members are very inefficient in resisting torsional loads unless they form a closed box. In some cases, it is necessary to use open sections with thin webs; and in most of these cases, the ends should be restrained to provide additional torsional rigidity and strength.

The *I* beam shown in Fig. 10.20 resists part of the torsion by means of the shear stresses distributed for the individual rectangles, as shown in Fig. 10.16. The remainder of the torsion is resisted by horizontal bending of the beam flanges, as shown in Fig. 10.20b. The proportion of the torsion which is resisted by each of the two ways depends on the dimensions of the cross section and the length of the member. This proportion also varies along the member, as more of the torsion is resisted by flange bending near the fixed end than near the free end.

For members in which the webs are thin and the length is not great, all the torsion may be assumed to be resisted by flange bending. In the case of long members with thick webs, all the torsion may be assumed to be resisted by the torsional resistance of the rectangular elements. In some cases, however, it may be necessary to calculate the proportion of the torsion resisted by each method.

A member which is not restrained at the ends twists as shown in Fig. 10.21. The angle of twist varies uniformly along the length and may be computed from Eq. (10.21). For a member resisting torsion by means of flange bending, as shown in Fig. 10.20b, the flange bending stresses vary from zero at the free end to maximum values at the fixed end. The angle of twist and the amount of cross section warping vary along the span. For the *I*-beam cross section, Timoshenko and Gere[9] give the following equations for the maximum flange bending moment

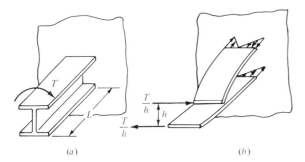

(a) (b) **Figure 10.20**

Figure 10.21

$M_{\max}$, for the maximum torsion resisted by web shears $T'_{\max}$, and for the angle of twist θ in length L (the constant a is a ratio of the relative flange bending rigidity to the torsional rigidity):

$$M_{\max} = \frac{T}{h} a \tanh \frac{L}{a} \tag{10.29}$$

$$T'_{\max} = T\left(1 - \operatorname{sech} \frac{L}{a}\right) \tag{10.30}$$

$$\theta = \frac{T}{KG}\left(L - a \tanh \frac{L}{a}\right) \tag{10.31}$$

$$a = \frac{h}{2}\sqrt{\frac{2I_f E}{KG}} \tag{10.32}$$

The term I_f is the moment of inertia of one flange of the beam about a vertical axis, K is defined by Eq. (10.22), and h is the beam depth between centers of flanges. The analysis for the I beam shown in Fig. 10.20 also applies for a beam of length $2L$ which has a torque of $2T$ applied at the center and which resists half

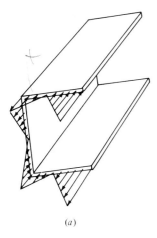

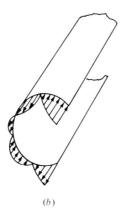

(a)   (b)

Figure 10.22

the torque at each end, if the ends are not restrained against warping. In this case, a cross section at the center of the beam is prevented from warping because of the symmetry of loading.

A channel cross section, such as that shown in Fig. 10.22a, frequently is used as an aircraft structural member. If a channel member is restrained at the end, it will resist torsion by bending of the flanges in much the same manner as the *I*-beam flanges shown in Fig. 10.20b. The analysis for flange bending is slightly more complicated than for the *I* beam, because the vertical web of the channel acts with the flanges in resisting flange bending, as shown in Fig. 10.22a. The cross section shown in Fig. 10.22b acts in the same way as the channel section, but the stress distribution is still more difficult to analyze.

10.9 TORSIONAL STRESSES ABOVE THE ELASTIC LIMIT

It has been assumed that the stresses are below the elastic limit in the previous analysis of torsional stresses. In many design applications, the ultimate torsional strength is desired. While there is not much published information concerning stress-strain curves for specimens in pure shear, these curves will have the same general shape as the tension stress-strain curves and will have ordinates approximately 0.6 of those for the tension curves. Thus the torsional stresses in a round bar are distributed as shown in Fig. 10.23 when the stresses exceed the elastic limit.

As in the plastic bending of beams, it is convenient to work with a fictitious stress instead of the exact stress distributions. This stress is designated as the torsional modulus of rupture τ_T, which is defined by

$$\tau_T = \frac{Tr}{J} \tag{10.33}$$

where T is the ultimate torsional strength of the member. For steel tubes, the value of τ_T depends on the proportions of the cross section. The values of the ratio τ_T/σ_{tu} are shown in Fig. 10.24 for various values of the ratio of outside diameter to wall thickness D/t. These curves are taken from MIL-HDBK-5.

Equation (10.33) applies to only circular or hollow circular cross sections,

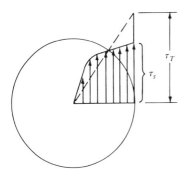

Figure 10.23

and Fig. 10.24 supplies information for allowable stresses on all such cross sections for various aircraft steels. The plastic torsional stress distribution in noncircular sections usually cannot be obtained by any simple analysis. Often it may be necessary to make static tests on torsion members of noncircular cross sections to determine the allowable torsional moments to use for design.

Example 10.4 A round steel tube with 1-in outer diameter (OD) 0.065 in thick resists a design torsional moment of 5000 in · lb. Find the margin of safety if the ultimate tensile stress $\sigma_{tu} = 100,000$ lb/in^2.

SOLUTION The values $D/t = 15.38$ and $I/y = 0.04193$ in^3 are obtained for this tube. From Fig. 10.24, the ratio $\tau_T/\sigma_{tu} = 0.6$ is obtained. The margin of safety is therefore calculated as follows:

$$\tau_T = 0.6 \times 100,000 = 60,000 \text{ lb/in}^2$$

$$\tau_t = \frac{Tr}{J} = \frac{T}{2I/y} = \frac{5000}{2 \times 0.04193} = 59,600 \text{ lb/in}^2$$

$$\text{MS} = \frac{\tau_T}{\tau_t} - 1 = \frac{60,000}{59,600} - 1 = 0.007$$

Example 10.5 Design a round tube to resist a torsional moment of 8000 in · lb. The minimum permissible wall thickness is 0.049 in, and the material has an ultimate tensile stress σ_{tu} of 100,000 lb/in^2.

SOLUTION A torsion tube must be designed by a trial-and-error process, because allowable stress depends on the D/t ratio and cannot be determined exactly until the tube is selected. For tubes of approximately the same weight, the larger I/y values are obtained for larger diameters, but the higher allowable moduli of rupture are obtained for thicker tube walls. Thus several tubes may have about the same weight and strength, although tubes with higher D/t ratios usually have a strength-weight advantage. Assume an average ratio

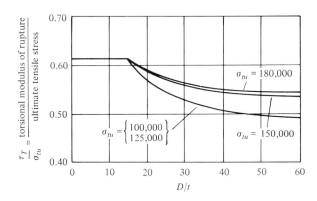

Figure 10.24

Table 10.2

Tube, in	$\dfrac{\omega t}{100\ \text{in}}$	$\dfrac{D}{t}$	$\dfrac{I}{y}$	τ_T	τ_t	MS
$1\frac{1}{2} \times 0.049$	6.32	30.60	0.07847	53,000	51,000	0.04
$1\frac{3}{8} \times 0.049$	5.78	28.05	0.06534	54,000	61,500	-0.12
$1\frac{1}{4} \times 0.058$	6.15	21.55	0.06187	56,000	64,700	-0.12

τ_T/σ_{tu} of 0.60 as a first approximation, or $\tau_T = 60{,}000$ lb/in^2. The required tube properties are now obtained as follows:

$$\text{Required } \frac{J}{r} = \frac{2I}{y} = \frac{T}{\tau_T} = \frac{8000}{60{,}000}$$

or

$$\text{Required } \frac{I}{y} = 0.0667 \text{ in}^3$$

The lightest tube which meets the requirement of 0.0667 in^3 is a $1\frac{1}{2}$- by 0.049-in tube with $D/t = 30.60$, $I/y = 0.07847$, and a weight of 6.32 lb/100 in. From Fig. 10.24, $\tau_T/\sigma_{tu} = 0.53$, or $\tau_T = 53{,}000$ lb/in^2. The margin of safety is

$$\tau_t = \frac{T}{2I/y} = \frac{8000}{2 \times 0.07847} = 51{,}000 \text{ lb/in}^2$$

$$\text{MS} = \frac{\tau_T}{\tau_t} - 1 = \frac{53{,}000}{51{,}000} - 1 = 0.04$$

Other tubes are compared in Table 10.2. The $1\frac{1}{2}$- by 0.049-in tube is the only one which has a positive margin of safety. Any other tubes which are lighter than $1\frac{1}{2}$ by 0.049 in and which have a minimum wall thickness of 0.049 in are obviously under the required strength.

10.10 COMBINED STRESSES AND STRESS RATIOS

The design of members resisting tension, bending, or torsion is discussed in earlier sections. Many structural members, however, must resist the simultaneous action of two or more of these loading conditions. If stresses are below the elastic limit of the material, the normal and shear stresses at any point may be combined by utilizing Mohr's circle. When working stresses based on the elastic limit are used, as in most types of structural design, it is customary to determine the principal stresses and the maximum shearing stresses at a point and to compare these to the allowable working stresses.

When members are designed on the basis of ultimate strength, it is not feasible to calculate the true principal stresses in the case of plastic bending or torsion. Even if the true stresses are known, it is difficult to predict the loads at which failure would occur under combined loading conditions. A tension member fails when the average stress reaches the ultimate tensile strength σ_{tu} for the material, but a member resisting combined stresses may fail before the maximum principal stress reaches the value of σ_{tu}. Failure in pure shear, for example, occurs when the principal tension and compression stresses are about $0.6\sigma_{tu}$. Various theories of failure of materials under combined loading have been developed, but none gives a simple method of predicting the failure of all materials.

Shanley and Ryder have proposed a method which provides a practical means of considering the combination of the fictitious stresses of bending or torsion and of obtaining the allowable ultimate loads for combined loadings. This method consists of using stress ratios and has been extensively adopted in the analysis of flight vehicle structures. The stress ratio method may be applied to almost any combination of two or more types of loading, although in some cases it may be necessary to test some specimens in order to apply the method. The method of stress ratios is applied first to some special cases of loading and is stated later in a general form.

One of the simplest types of combined loadings is that of tension and bending, as shown in Fig. 10.25. The stresses may be added algebraically, and for small loads the stress is $P/A + My/I$ at any point in the cross section, as shown in Fig. 10.25a. When the stresses exceed the elastic limit, however, the distribution becomes similar to that shown in Fig. 10.25b. The true stress distribution is difficult to calculate, and it is convenient to use the method of stress ratios in predicting the strength. For pure bending with no torsion, failure will occur when the ratio of the applied bending stress σ_b to the allowable bending stress σ_B approaches unity ($R_b = \sigma_b/\sigma_B = 1$). Similarly, for tension with no bending, failure will occur when the tension-stress ratio $R_t = \sigma_t/\sigma_{tu}$ approaches unity. Since the stresses below the elastic limit add directly, it seems logical to add the stress ratios, and tests substantiate this method. The failure under combined tension and bending therefore occurs under the following condition:

$$R_b + R_t = 1 \tag{10.34}$$

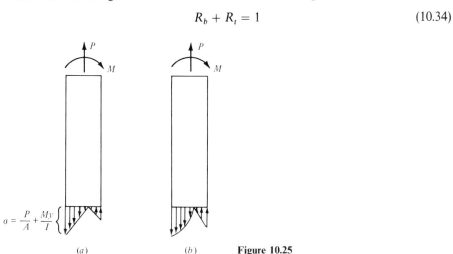

$$\sigma = \frac{P}{A} + \frac{My}{I}$$

(a)   (b)   **Figure 10.25**

The margin of safety is defined by the following equation, which corresponds with previously used expressions when either of the stress ratios is zero:

$$MS = \frac{1}{R_b + R_t} - 1 \tag{10.35}$$

For round tubes in combined bending and torsion, the stresses do not add algebraically. For stresses below the elastic limit, the maximum stress at any point may be obtained readily by using Mohr's circle. The maximum tension and shearing stresses in the tube shown in Fig. 10.26a will occur at the support and on the upper surface, as shown. The tension stress in the direction of the axis of the tube σ_b is found from the bending moment M as follows:

$$\sigma_b = \frac{My}{I} \tag{10.36}$$

The shearing stress τ_s on the planes shown is obtained from the torsional moment T:

$$\tau_s = \frac{Ty}{2I} \tag{10.37}$$

The small element at the top of the tube is enlarged in Fig. 10.26b. The principal stresses and the maximum shearing stresses at this point on the tube are obtained from the stresses σ_b and τ_s by Mohr's circle, as shown in Fig. 10.26c. The maximum shearing stress τ_{max} is equal to the radius of the circle

$$\tau_{max} = \sqrt{\tau^2 + \left(\frac{\sigma_b}{2}\right)^2} \tag{10.38}$$

Substituting values σ_b and τ_s from Eqs. (10.36) and (10.37) into Eq. (10.38) yields

$$\tau_{max} = \frac{\sqrt{M^2 + T^2}}{2I/y} = \frac{T_e}{2I/y} \tag{10.39}$$

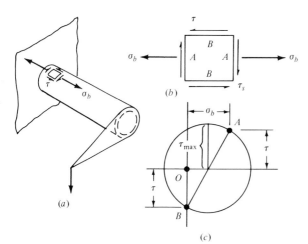

(a)

(b)

(c)

Figure 10.26

where T_e is an equivalent torque defined by Eq. (10.39). If the torsion is larger than the bending moment, the maximum shearing stress may be used to predict the strength of the tube. If the bending moment is large in comparison to the torsion, the principal stresses are more important than the shearing stresses. In machine design practice, shafts in bending and torsion are designed so as to keep the shearing stress of Eq. (10.39) and the principal stresses smaller than the corresponding allowable working stresses, which are a certain fraction of the yield stress.

When a tube is stressed beyond the elastic limit, Eqs. (10.36) and (10.37) do not yield the true stresses, but yield the fictitious stresses defined as the bending modulus of rupture and the torsional modulus of rupture. Equations (10.38) and (10.39), therefore, are not exact when the stresses are beyond the elastic limit. When the torsion is large in comparison to the bending moment, however, Eq. (10.39) may be used to predict the ultimate for the allowable stress. A more accurate method of designing tubes in torsion or bending is by stress ratios. The bending and torsional moduli of rupture σ_b and τ_t, respectively, are calculated from Eqs. (10.36) and (10.37), and the allowable values σ_B and τ_T are found by the same methods as for tubes in bending or torsion only. The stress ratio in bending $R_b = \sigma_b/\sigma_B$ is combined with the stress ratio in torsion $R_{st} = \tau_t/\tau_T$ in the same manner as the loads are combined in Eq. (10.39). Failure occurs for the following condition:

$$R_b^2 + R_{st}^2 = 1 \tag{10.40}$$

The margin of safety is

$$MS = \frac{1}{\sqrt{R_b^2 + R_{st}^2}} - 1 \tag{10.41}$$

Where either the bending moment or the torsional moment is zero, Eqs. (10.40) and (10.41) yield values which were previously obtained for bending or torsion only.

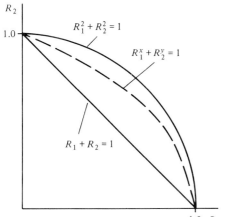

Figure 10.27

Equations (10.34) and (10.40) represent two ways in which stress ratios may be combined. These equations may be plotted with the two stress ratios as coordinates, as shown in Fig. 10.27. The graph for Eq. (10.34) is shown by the straight line, and that for Eq. (10.40) by the circle. Other combinations of loading conditions may be represented by a more general equation:

$$R_1^x + R_2^y = 1 \tag{10.42}$$

where the exponents x and y usually must be found experimentally by plotting test results, as shown in Fig. 10.27, and by writing an equation for a curve passing through the points.

Example 10.6 A 2- by 0.095-in steel tube is heat-treated to an ultimate tensile strength $\sigma_{tu} = 180,000$ lb/in².
(a) Find the margin of safety if the tube resists a design tension load of 50,000 lb and a design bending moment of 30,000 in · lb.
(b) Find the margin of safety if the tube resists a bending moment of 30,000 in · lb and a torsional moment of 50,000 in · lb.

SOLUTION The tube properties are $A = 0.5685$ in², $D/t = 21.05$, and $I/y = 0.2486$ in³.
(a) From Fig. 10.3, $\sigma_B = 211,000$ lb/in². The stresses and stress ratios are

$$\sigma_b = \frac{My}{I} = \frac{30,000}{0.2586} = 116,000 \text{ lb/in}^2$$

$$R_b = \frac{\sigma_b}{\sigma_B} = \frac{116,000}{211,000} = 0.550$$

$$\sigma_t = \frac{P}{A} = \frac{50,000}{0.5865} = 88,000 \text{ lb/in}^2$$

$$R_t = \frac{\sigma_t}{\sigma_{tu}} = \frac{88,000}{180,000} = 0.488$$

The margin of safety is obtained from Eq. (10.35):

$$MS = \frac{1}{R_b + R_t} - 1 = -0.035$$

The negative margin of safety indicates that the tube is unsatisfactory.
(b) The stress ratio for bending is the same as computed in part (a). The torsional modulus of rupture is obtained from Fig. 10.24, for $D/t = 21.05$ and $\tau_T = 0.58 \times 180,000 = 104,000$ lb/in². The stress ratio for torsion is

$$\tau_t = \frac{Ty}{2I} = \frac{50,000}{2 \times 0.2586} = 96,500$$

$$R_{st} = \frac{\tau_t}{\tau_T} = \frac{96,500}{104,000} = 0.925$$

The margin of safety is now calculated from Eq. (10.41):

$$MS = \frac{1}{\sqrt{R_b^2 + R_{st}^2}} - 1 = -0.07$$

The tube is unsatisfactory.

10.11 FAILURE THEORIES IN STRUCTURAL DESIGN

Since the stress-ratio technique was developed by Shanley and Ryder[35] years ago, many theories of failure in structural design have been presented. The three most commonly used are the maximum stress theory, maximum shear theory, and maximum distortion energy theory.[37]

In the maximum stress theory, failure is said to occur in a structural member under the action of combined stresses when one of the principal stresses reaches the failure value (yield stress, ultimate stress, etc.) in simple tension σ_0. In a two-dimensional state of stress, failure is defined as follows:

$$\text{Failure occurs when } \sigma_1 = \pm\sigma_0 \qquad (\sigma_1 > \sigma_2)$$
$$\text{Failure occurs when } \sigma_2 = \pm\sigma_0 \qquad (\sigma_2 > \sigma_1)$$

or

$$\frac{\sigma_1}{\sigma_0} = \pm 1 \qquad (\sigma_1 > \sigma_2)$$

$$\frac{\sigma_2}{\sigma_0} = \pm 1 \qquad (\sigma_2 > \sigma_1)$$

(10.43)

In the maximum shear theory, failure is said to occur in a structural member under the action of combined stresses when the maximum shear stress reaches the value of shear failure stress in tension, $\tau_{max} = \sigma_0/2$. Under combined stresses in two dimensions, three possible shear values exist:

$$\tau_1 = \frac{\sigma_0}{2} = \pm\frac{\sigma_1 - \sigma_2}{2}$$

$$\tau_2 = \frac{\sigma_0}{2} = \pm\frac{\sigma_2}{2}$$

(10.44)

$$\tau_3 = \frac{\sigma_0}{2} = \pm\frac{\sigma_1}{2}$$

Therefore, failure is defined by one of the following:

$$\sigma_1 - \sigma_2 = \pm\sigma_0$$

$$\sigma_2 = \pm\sigma_0$$

$$\sigma_1 = \pm\sigma_0$$

or
$$\frac{\sigma_1}{\sigma_0} - \frac{\sigma_2}{\sigma_0} = \pm 1 \tag{10.45}$$

$$\frac{\sigma_2}{\sigma_0} = \pm 1$$

$$\frac{\sigma_1}{\sigma_0} = \pm 1$$

If the principal stresses σ_1 and σ_2 have opposite signs, then the maximum shear is defined by one of Eqs. (10.44). However, if σ_1 and σ_2 have the same sign, then failure is defined by

$$\tau_{\max} = \pm \frac{\sigma_1}{2} \quad (\sigma_1 > \sigma_2)$$

$$\tau_{\max} = \pm \frac{\sigma_2}{2} \quad (\sigma_2 > \sigma_1) \tag{10.46}$$

In the distortion energy theory, failure is said to occur in a structural member under the action of combined stresses when the energy of distortion reaches the same energy of failure in tension. The energy of distortion for a two-dimensional state of stress is given by

$$U_D = \frac{1+v}{3E}(\sigma_1^2 + \sigma_2^2 - \sigma_1\sigma_2) \tag{10.47}$$

In tension only,

$$U_D = \frac{1+v}{3E}\sigma_0^2 \tag{10.48}$$

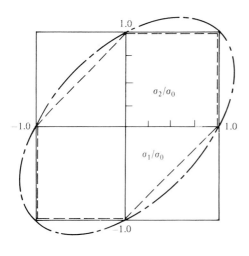

— · — · — Distortion energy theory

— — — — — Maximum shear theory

————————— Maximum stress theory

Figure 10.28

Thus, setting Eq. (10.47) equal to Eq. (10.48) yields the distortion energy theory of failure as

$$\sigma_1^2 + \sigma_2^2 - \sigma_1\sigma_2 = \sigma_0^2$$

or
$$\left(\frac{\sigma_1}{\sigma_0}\right)^2 + \left(\frac{\sigma_2}{\sigma_0}\right)^2 - \frac{\sigma_1\sigma_2}{\sigma_0^2} = 1 \tag{10.49}$$

A graphical representation of the three failure theories is shown in Fig. 10.28.

PROBLEMS

10.1 What is the tensile strength of a $1\frac{1}{2}$- by 0.065-in steel tube with ends welded so that the welds make angles less than $30°$ with the axis of the tube? Assume (a) $\sigma_{tu} = 95{,}000$ lb/in^2, (b) $\sigma_{tu} = 150{,}000$ lb/in^2, heat-treated after welding, and (c) $\sigma_{tu} = 150{,}000$ lb/in^2, welded after heat treatment. Obtain allowable stresses near welds from MIL-HDBK-5.

10.2 What is the ultimate tensile strength of a 2024-T3 aluminum-alloy tube that is $1\frac{1}{2}$ by 0.054 in? The ends are connected by one line of AN4 bolts. Assume $\sigma_{tu} = 64{,}000$ lb/in^2 for the tube.

10.3 Find the ultimate bending moments which may be resisted by steel tubes of dimensions $1\frac{3}{4}$ by 0.049 in, $1\frac{3}{4}$ by 0.058 in, and $1\frac{3}{4}$ by 0.083 in, investigating each tube for heat treatments of $\sigma_{tu} = 125{,}000, 150{,}000$ and $180{,}000$ lb/in^2.

10.4 Find the lightest standard steel tube which can resist a bending moment of 10,000 in · lb if $\sigma_{tu} = 180{,}000$ lb/in^2.

10.5 Design a round steel tube to resist a bending moment of 30,000 in · lb, with $\sigma_{tu} = 125{,}000$ lb/in^2. Calculate the margin of safety for each trial size.

10.6 Find the ultimate resisting moment of a beam with a rectangular cross section, with $b = 0.5$ and $h = 2$ in. The material is a 2014-T6 aluminum-alloy forging. Assume a trapezoidal stress-strain curve with $\sigma_m = 65{,}000$ and $\sigma_0 = 60{,}000$ lb/in^2.

10.7 A 2- by 0.083-in tube of 2024-T3 aluminum alloy is 20 in long and resists a torsional moment T of 8000 in · lb. Find the maximum shearing stress and the angle of twist.

10.8 An elevator torque tube is made of 2024-T3 aluminum alloy and has dimensions of 2 by 0.083 in. Find the angle of twist if the shearing stress has a value of 10,000 lb/in^2 for a length of 80 in.

10.9 An extrusion of 2024-T3 aluminum alloy of the shape shown in Fig. 10.18 has dimensions $b_1 = b_2 = b_3 = 2$ and $t_1 = t_2 = t_3 = 0.2$ in. Find the shearing stress and the angle of twist if the section resists a torque of 100 in · lb and has a length of 10 in. Assume (a) both ends are free to warp and (b) one end is restrained against warping.

10.10 Repeat Prob. 10.9 for a length of 20 in.

In the following problems, all loads are ultimate or design loads.

10.11 Design a round steel tube to resist a torsional moment of 10,000 in · lb. Assume material properties of (a) $\sigma_{tu} = 100{,}000$, (b) $\sigma_{tu} = 125{,}000$, and (c) $\sigma_{tu} = 180{,}000$ lb/in^2.

10.12 Design a round steel tube to resist a bending moment of 6000 and a torsional moment of 8000 in · lb. Use Eq. (10.39) for a preliminary trial, but employ the method of stress ratios for the final design. Assume (a) $\sigma_{tu} = 125{,}000$ and (b) $\sigma_{tu} = 180{,}000$ lb/in^2.

BUCKLING DESIGN OF
STRUCTURAL MEMBERS

11.1 BEAM-DEFLECTION EQUATIONS

The methods used in the design of compression members, or columns, are based on beam-deflection equations. Columns do not fail as a result of the direct compression stresses, only as a result of the combined compression and bending stresses. Since the magnitudes of the bending stresses depend on the bending deflections, it is necessary to derive the column equations from beam-deflection equations.

The equations for beam deflections are derived from the customary assumptions that stress is proportional to strain and that the deflections are small in comparison to the original dimensions. Only deformations resulting from bending stresses is usually considered, but if shearing deformations are appreciable, they may be computed separately and superimposed. An initially straight beam is shown with exaggerated deflections in Fig. 11.1. The two cross sections a distance dx apart are parallel to the unstressed condition, but have a relative angle $d\theta$ in the stressed condition. The angle $d\theta$ may be obtained by considering the small triangle between the neutral axis and a point a distance c below the neutral axis. The stress σ at this point is obtained from the flexure formula:

$$\sigma = \frac{Mc}{I} \tag{11.1}$$

The longitudinal fiber a distance c from the neutral axis has an elongation $\sigma \, dx/E$

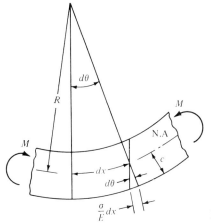

Figure 11.1

in the length dx. The angle $d\theta$ is obtained by dividing this elongation by the distance c, as shown in Fig. 11.1:

$$d\theta = \frac{\sigma \, dx}{Ec} \qquad (11.2)$$

From Eqs. (11.1) and (11.2),

$$d\theta = \frac{M}{EI} \, dx \qquad (11.3)$$

The deflection curve of the beam can be represented by x and y coordinates, as shown in Fig. 11.2. The beam is assumed to be initially straight and parallel to the x axis. The deflections are small, and the angle θ between a tangent to the deflection curve and the x axis is small enough that it is sufficiently accurate to assume that the angle in radians, the sine of the angle, and the tangent of the angle are all equal:

$$\theta = \sin \theta = \tan \theta \qquad (11.4)$$

$$\theta = \frac{dy}{dx} \qquad (11.5)$$

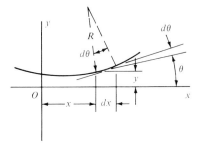

Figure 11.2

From Eqs. (11.3) and (11.5),

$$\frac{d^2y}{dx^2} = \frac{M}{EI} \tag{11.6}$$

All conventional methods of obtaining beam deflections are based on Eq. (11.6). It should be noted that the deflections y are measured positive upward, so that a positive bending moment M produces a positive curvature d^2y/dx^2. Equation (11.6) often is derived from the assumption that y is positive downward, and a minus sign is introduced because a positive bending moment would then produce a negative curvature.

11.2 LONG COLUMNS

Compression members tend to fail as a result of the lateral bending induced by the compression load, an action which is commonly termed *buckling*. In the case of columns which are long in comparison to their other dimensions, elastic buckling occurs, or the columns buckle when the compressive stresses are below the elastic limit. Such columns are termed *long columns*.

The initially straight column shown in Fig. 11.3 is assumed to be held in the deflected position by means of the compressive forces P. The bending moment at any cross section is found from

$$M = -Py \tag{11.7}$$

It is assumed that the material does not exceed the elastic limit at any point, and therefore Eq. (11.6) is applicable. The differential equation of the deflection curve is obtained from Eqs. (11.6) and (11.7):

$$\frac{d^2y}{dx^2} + \frac{P}{EI}\, y = 0 \tag{11.8}$$

The general solution of Eq. (11.8) is

$$y = C_1 \sin\sqrt{\frac{P}{EI}}\, x + C_2 \cos\sqrt{\frac{P}{EI}}\, x \tag{11.9}$$

This solution may be verified by substitution and must be a general solution of the second-order differential equation because it contains the two arbitrary constants C_1 and C_2. In order to satisfy the end conditions shown in Fig. 11.3, the deflection curve must pass through the points ($x = 0$, $y = 0$) and ($x = L$,

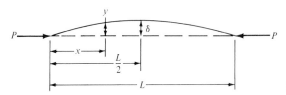

Figure 11.3

$y = 0$). Substituting the first condition into Eq. (11.9) yields $C_2 = 0$. The second condition, that $y = 0$ when $x = L$, may be satisfied when $C_1 = 0$, which is a trivial solution, corresponding to the conditions of small loads when the column remains straight. The only solution of interest in column analysis is that for which the column is deflected and C_1 is not zero. This solution, in which the column is buckled, is obtained when the value of P satisfies the condition

$$\sqrt{\frac{P}{EI}} \, L = \pi, \, 2\pi, \, 3\pi, \, \ldots, \, n\pi$$

or
$$P = \frac{n^2 \pi^2 EI}{L^2} \tag{11.10}$$

The value of P is obviously a minimum when $n = 1$, and higher values of n have no significance in this case, since the column will fail at the smallest value of P that will produce buckling. The critical, or buckling, load P_{cr} is therefore defined as follows:

$$P_{cr} = \frac{\pi^2 EI}{L^2} \tag{11.11}$$

Often it is more convenient to work with a buckling stress $\sigma_{cr} = P_{cr}/A$. This may be found by introducing the radius of gyration of the cross-sectional area, $\rho = \sqrt{I/A}$, into Eq. (11.11):

$$\sigma_{cr} = \frac{\pi^2 E}{(L/\rho)^2} \tag{11.12}$$

The analysis of long columns was first published by the Swiss mathematician Euler. Equation (11.11) [or Eq. (11.12)] is commonly called the *Euler equation*, and the buckling load is often called the *Euler load*.

The value of C_1 cannot be obtained at the critical load. This value is equal to the maximum deflection δ at the center of the column, which is indeterminate for the assumed conditions. For loads smaller than P_{cr}, the deflection C_1 or δ must be zero, or else the column remains straight. At the critical load, any deflection δ for which the maximum stress is below the elastic limit will satisfy conditions of equilibrium. This may be shown experimentally by loading a long column in a standard testing machine. As the ends of the column are moved together, the column remains straight until the Euler load is obtained. As the ends continue to move together, the load remains constant at the Euler load, but the lateral deflection δ increases. If the elastic limit is not exceeded, the column returns to its initial shape when the load is removed.

11.3 ECCENTRICALLY LOADED COLUMNS

In an actual structure, it is not possible for a column to be perfectly straight or to be loaded exactly at the centroid of the area. The action of a practical long column may be approximated by the member shown in Fig. 11.4, in which the

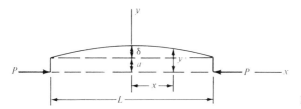

Figure 11.4

column is initially straight but the loads both have eccentricity a. The axes of coordinates are taken as shown. The equation of the deflection curve is still represented by Eq. (11.9) since Eqs. (11.7) and (11.8) are applicable, but constants C_1 and C_2 must be found from the condition that the deflection curve satisfy the two conditions $x = 0$, $y = \delta + a$ and $x = 0$, $dy/dx = 0$. Substituting these two conditions into Eq. (11.9) produces

$$y = (\delta + a) \cos \sqrt{\frac{P}{EI}}\, x \qquad (11.13)$$

The value of δ can be found now from the condition that $y = a$ for $x = L/2$. By substituting these values into Eq. (11.13) the following value of δ is obtained:

$$\delta + a = a\left(\sec \sqrt{\frac{P}{EI}}\frac{L}{2}\right) \qquad (11.14)$$

The deflection δ of an eccentrically loaded column thus increases with an increase in the load P. As the value of P reaches the Euler load P_{cr}, as defined by Eq. (11.11), the deflection becomes infinite, since $\sec(\pi/2) = \infty$. Figure 11.5 shows the relationship between P and δ for various eccentricities a, as determined from Eq. (11.14). All curves are asymptotic to the line $P = P_{cr}$, for this is the theoretical buckling load regardless of the eccentricity of loading. A large deflection may stress the material beyond the elastic limit and cause failure before the Euler load is obtained, since then the long-column equations would no longer apply.

11.4 SHORT COLUMNS

Columns of any specific material are classified according to their slenderness ratio L/ρ. For a slenderness ratio greater than a certain critical value, the column is a long column and is analyzed by Eq. (11.12). Short columns have a slenderness ratio less than this critical value. The critical L/ρ corresponds to the value for which the maximum compressive stress in the column is equal to the stress at which the compressive stress-strain curve deviates from a straight line, as shown by point B in Fig. 11.6. Usually this stress is considerably smaller than the yield stress, point C of Fig. 11.6 at which the material has a permanent unit elongation of 0.002.

Most flight vehicle materials have stress-strain curves similar to that shown in Fig. 11.6, in which the stress-strain curve has a positive slope at all points. The

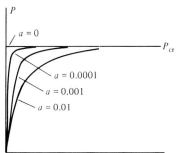

P

$a = 0$

P_{cr}

$a = 0.0001$

$a = 0.001$

$a = 0.01$

δ **Figure 11.5**

constant slope of this stress-strain curve below the elastic limit is equal to the modulus of elasticity E, and the variable slope above the elastic limit is termed the *tangent modulus of elasticity* E_t. Ductile materials, such as mild steel, may have a zero or negative value of E_t near the yield point. If E_t is positive at all points, a short column may remain perfectly straight when loaded to stresses beyond the yield point. If such a column has a slight lateral deflection, the internal resisting moment is found from an equation similar to Eq. (11.6), except that E is replaced by E_t, the tangent modulus for the compressive stress:

$$M = E_t I \frac{d^2 y}{dx^2} \tag{11.15}$$

If this internal resisting moment is greater than the bending moment produced by the load P, the column will remain straight when loaded. If the internal resisting moment is not as large as the external bending moment, the deflection will increase and the column will probably fail. When the bending moment of the load P is equal to the resisting moment defined by Eq. (11.15), P may be obtained in the same manner as in the Euler equation, but with E_t substituted for E:

$$P = \frac{\pi^2 E_t I}{L^2} \tag{11.16}$$

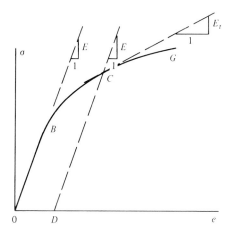

σ

E_t

E E

1

G

1 1

C

B

0 D e **Figure 11.6**

This equation is called the *tangent modulus equation*, or the *Engesser equation*.

The tangent modulus equation does not quite represent the true conditions for short columns. At point C of the stress-strain diagram in Fig. 11.6, a small increase in the compressive strain produces an increase in compressive stress, as determined by the portion CG of the curve, which has slope E_t. A small decrease in the compressive strain, however, produces a decrease in stress, as indicated by line CD, which has slope E. If a short column deflects laterally in such a way that the compressive strain on the convex side is decreased, the resisting moment will be greater than that given by Eq. (11.15) because the modulus of elasticity for part of the cross section is E rather than E_t. Thus the correct modulus of elasticity should be a value between E and E_t. Values of an effective modulus of elasticity should be derived on the assumption that the column is supported laterally and remains straight until the ultimate load is applied and then buckles with no change in axial load. The column formula obtained by substituting this modulus into the Euler equation is termed the *reduced modulus equation* and frequently is referred to in the literature.

Shanley[8] has shown that the correct load resisted by a short column is between the values given by the tangent modulus equation and by the reduced modulus equation. The tangent modulus equation yields values which are slightly low, since some strain reversal must take place before the ultimate column load is reached. The reduced modulus equation always yields values which are too high, since the column is not laterally supported when the load is applied. The tangent modulus equation is used frequently, because it corresponds closely to test results and is always conservative.

It is customary to represent column equations by plotting the average compressive stress $\sigma_c = P/A$ against the slenderness ratio L/ρ. Figure 11.7 shows such curves. The slenderness ratio beyond which the material acts as a long column is about 115. The stress at this point corresponds to the stress at point B of Fig. 11.6. For slenderness ratios less than 115, the compressive stress is higher, and the tangent modulus of elasticity E_t is smaller than E. Thus the points on the column curve are below the Euler curve in the short-column range. The test points are seen to follow the tangent modulus curve very closely. Such test loads are always slightly lower than theoretical loads because of unavoidable eccentricities of loading. The curves shown in Fig. 11.7 represent values for an actual specimen, whereas similar design curves are based on minimum guaranteed properties of the material and give somewhat lower stresses.

11.5 COLUMN END FIXITY

In the previous analysis, we assume that the column is hinged at both ends so that it can rotate freely. In most cases, however, compression members are connected in such a way that they are restrained against rotation at the ends. In order to have the means of determining the buckling load for a column with various end fixities, Eq. (11.8) is written as follows:

$$\frac{d^4y}{dx^4} + k^2 \frac{d^2y}{dx^2} = 0 \tag{11.17}$$

where $k^2 = P/(EI)$. The solution of Eq. (11.17) can be found easily:

$$y = C_1 \sin kx + C_2 \cos kx + C_3 x + C_4 \tag{11.18}$$

Constants C_1 through C_4 are determined from the boundary conditions at the ends of the column. For instance, in the case of pin-ended columns, the boundary conditions are

$$y = 0 \qquad d^2y/dx^2 = 0 \qquad \text{at } x = 0$$

$$y = 0 \qquad d^2y/dx^2 = 0 \qquad \text{at } x = L$$

For columns which are completely fixed at both ends, the boundary conditions are

$$y(0) = 0 \qquad \frac{dy(0)}{dx} = 0$$

$$y(L) = 0 \qquad \frac{dy(L)}{dx} = 0$$

Columns which are fixed at $x = 0$ and pinned at $x = L$ have the boundary conditions

$$y(0) = 0 \qquad\qquad y(L) = 0$$

$$\frac{dy(0)}{dx} = 0 \qquad \frac{d^2y(L)}{dx^2} = 0$$

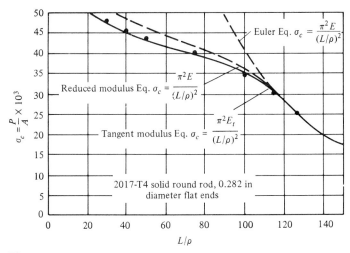

Figure 11.7

Upon substituting the boundary conditions in Eq. (11.18), a set of four algebraic homogeneous equations is obtained. For nontrivial solutions, the determinant of the undetermined coefficients C_1 to C_4 must vanish. This will yield a transcendental equation from which the buckling load is determined. For example, substituting the last set of boundary conditions (column fixed at one end and pinned at the other) yields

$$C_2 + C_4 = 0$$

$$kC_1 + C_3 = 0$$

$$(\sin kL)C_1 + (\cos kL)C_2 + LC_3 + C_4 = 0$$

$$(k^2 \sin kL)C_1 + (k^2 \cos kL)C_2 = 0$$

For nontrivial solutions of the above set of equations, the determinant must vanish, i.e.,

$$\begin{vmatrix} 0 & 1 & 0 & 1 \\ k & 0 & 1 & 0 \\ \sin kL & \cos kL & L & 1 \\ k^2 \sin kL & k^2 \cos kL & 0 & 0 \end{vmatrix} = 0$$

Expanding yields the following:

$$k^2(kL \cos kL - \sin kL) = 0$$

Since $k^2 = P/(EI)$, k cannot be zero. Therefore

$$kL \cos kL - \sin kL = 0$$

or

$$\tan kL = kL \qquad \text{(transcendental equation)}$$

The critical buckling load occurs at the smallest value of kL which satisfies the transcendental equation above. This value can be shown to be equal to $\pi/0.7$. Hence the critical buckling load is

$$P_{er} = \frac{2.04\pi^2 EI}{L^2} \tag{11.19}$$

If a compression member is rigidly fixed against rotation at both ends, the deflection curve for elastic buckling may be determined by using the above procedure and will have the shape shown in Fig. 11.8b. At the quarter points of the fixed column, there will be points of reverse curvature, or points of contraflexure. At points of contraflexure there is no curvature and hence no bending moment. The portion of the column between points of contraflexure thus may be treated as a pin-ended column. The length L' between the points of contraflexure is used in place of L in the column equations previously derived, and the slenderness ratio is defined as L'/ρ. An end-fixity term c is used often and is defined in the following equation:

$$\sigma_{\text{cr}} = \frac{\pi^2 E}{(L'/\rho)^2} = \frac{c\pi^2 E}{(L/\rho)^2} \qquad \text{for long columns only} \qquad (11.20)$$

or
$$L' = \frac{L}{\sqrt{c}} \qquad \text{for all columns} \qquad (11.21)$$

For the fixed-end condition of Fig. 11.8*b*, $L' = L/2$ and $c = 4$. Thus the fixed-end column will resist 4 times the load of a similar pin-ended column, if both are in the long-column range. This same relation does not hold in the short-column range, because the value of E_t in Eq. (11.16) is smaller for the smaller values of L'. This fact is evident from Fig. 11.7, where it may be seen that a reduction in L'/ρ has a much smaller effect on σ_c in the short-column range than it has in the Euler column range.

In order to obtain complete end fixity, the compression member must be attached to a structure of infinite rigidity at both ends. This condition is approached less frequently in practice than the condition of hinged ends. Most practical columns have end conditions somewhere between hinged and fixed, as shown in Fig. 11.8*c*. The ends are rigidly attached to a structure which deflects and permits the ends to rotate slightly. The true end-fixity conditions seldom can be determined exactly, and so conservative assumptions must be made. Fortunately, short columns usually are used, and the effect of end fixity on the allowable compressive stress is much smaller than it would be for long columns.

Other common end conditions for columns are shown in Fig. 11.9. For the column fixed at one end and free to both rotate and move laterally at the other end, as shown in Fig. 11.9*a*, length L' is twice length L, since the column is similar to one-half of the column with two hinged ends. The column with one end fixed and the other end free to rotate but not free to move laterally has an effective length $L' \cong 0.7L$, as shown in Fig. 11.*b*.

Welded trusses made of steel tubes frequently are utilized in vehicle struc-

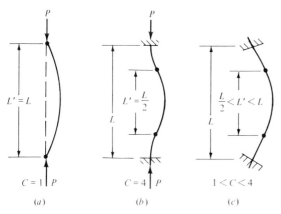

Figure 11.8

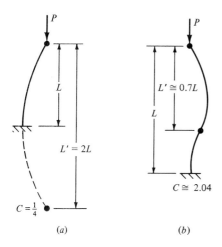

(a)   (b)

tures. The ends of a compression member in such a truss cannot rotate without bending all the other members at the end joints. Such a member is shown in Fig. 11.10. The problem of obtaining the true end fixity of such a compression member is difficult, since the member may buckle either horizontally or vertically and is restrained by the torsional and bending rigidities of many other members. For a steel-tube fuselage truss, usually it is conservative to assume $c = 2.0$ for all members. If a very heavy compression member is restrained by comparatively light members, a smaller end fixity might be obtained. Similarly, a light compression member restrained by heavy members may approach the fixity condition $c = 4$. If all the members at a joint are compression members, they may all have a tendency to rotate in the same direction, so that none helps restrain the others, and all should be designed as pin-ended. Where this rare case exists with members in any plane, the members perpendicular to this plane probably would supply torsional restraint to the joint. Tension members that connect to the ends of compression members supply greater restraint than similar compression members. Often steel-tube engine mounts are designed with the conservative assumption of pin-ended member with $c = 1.0$.

Stringers which act as compression members in semimonocoque wing or

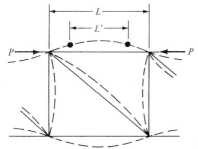

Figure 11.10

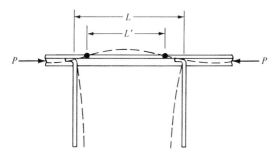

Figure 11.11

fuselage structures often are supported by comparatively flexible ribs or bulk-heads. Such a stringer is shown in Fig. 11.11. Since the ribs or bulkheads are free to twist as shown, their restraining effect is neglected and the effective column length L' is assumed equal to the length L between bulkheads. Where the bulk-heads are rigid enough to provide restraint and clips are provided to attach the stringers to the bulkheads, a value $c = 1.5$, corresponding to an effective length L' of $0.815L$, is sometimes used.

11.6 EMPIRICAL FORMULAS FOR SHORT COLUMNS

One disadvantage of the tangent modulus formula for short columns is that the relation between the allowable column stress σ_c and L'/ρ cannot be expressed by a simple equation. It is often more convenient to express this relationship by a simple approximate equation which is reasonably close to the points obtained directly from column tests or from the tangent modulus equation. The short-column curves for many materials approximate the parabola

$$\sigma_c = \sigma_{c0} - K\left(\frac{L'}{\rho}\right)^2 \tag{11.22}$$

The constants σ_{c0} and K must be chosen so that the parabola fits the test data and is tangent to the Euler curve. By equating the slope of this parabola to that of the Euler curve at the point of tangency and substituting the resulting value of K into Eq. (11.22), the following equation is obtained:

$$\sigma_c = \sigma_{c0}\left[1 - \frac{\sigma_{c0}(L'/\rho)^2}{4\pi^2 E}\right] \tag{11.23}$$

The term σ_{c0} is called the *column yield stress*. It has little physical significance, since very short columns ($L'/\rho < 12$) fail by block compression rather than column action, and Eq. (11.23) is not applicable in this range. The value of σ_{c0} is determined so that Eq. (11.23) will fit short-column test data for values of L'/ρ above the block compression range.

The general second-degree parabola equation is shown in Fig. 11.12 with the corresponding Euler equation. The value of σ_{c0} represents the intercept of this

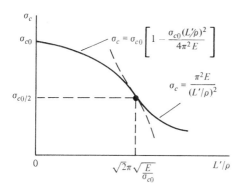

Figure 11.12

curve at the point $L'/\rho = 0$. The parabola is always tangent to the Euler curve when $\sigma_{c0} = \sigma_{c0}/2$, as is found by solving Eq. (11.23) simultaneously with the Euler equation. In the same way, the critical slenderness ratio, which divides the long-column and short-column ranges, is found to be $L/\rho = \sqrt{2}\,\pi\,\sqrt{E/\sigma_{c0}}$.

The short-column curves for most aluminum alloys and for several other materials are represented more accurately by straight lines. A straight line tangent to the Euler curve has the equation:

$$\sigma_c = \sigma_{c0}\left(1 - \frac{0.385 L/\rho}{\pi\sqrt{E/\sigma_{c0}}}\right) \tag{11.24}$$

The coordinates of the point of tangency of this curve and the Euler curve are $L/\rho = \sqrt{3}\,\pi\,\sqrt{E/\sigma_{c0}}$ and $\sigma_c = \sigma_{c0}/3$ (see Fig. 11.13). The value of L/ρ at this point is the critical value dividing the short-column and long-column ranges.

Other materials have column curves which may be represented by a semi-cubic equation. A 1.5-degree equation which is tangent to the Euler curve has the form

$$\sigma_c = \sigma_{c0}\left[1 - 0.3027\left(\frac{L/\rho}{\pi\sqrt{E/\sigma_{c0}}}\right)^{1.5}\right] \tag{11.25}$$

The coordinates of the point of tangency are again found by solving the short-

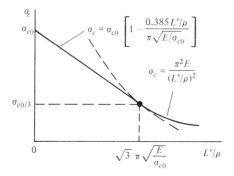

Figure 11.13

column and long-column equations simultaneously. The critical slenderness ratio is $L/\rho = 1.527\pi\sqrt{E/\sigma_{c0}}$ corresponding to a stress of $\sigma_c = 0.429\sigma_{c0}$.

Equations (11.23) to (11.25) and the Euler equation may be expressed in dimensionless form by using coordinates B and R_a as defined by

$$B = \frac{L/\rho}{\pi\sqrt{E/\sigma_{c0}}} \tag{11.26}$$

$$R_a = \frac{\sigma_c}{\sigma_{c0}} \tag{11.27}$$

The Euler equation, Eq. (11.12), then becomes

$$R_a = \frac{1}{B^2} \tag{11.28}$$

Equations (11.23) to (11.25) will have the following forms, respectively:

$$R_a = 1.0 - 0.25B^2 \tag{11.29}$$

$$R_a = 1.0 - 0.3027B^{1.5} \tag{11.30}$$

$$R_a = 1.0 - 0.385B \tag{11.31}$$

These equations are plotted in Fig. 11.14. The dimensionless form of expressing column curves has the advantage of showing column curves for all materials on a single graph

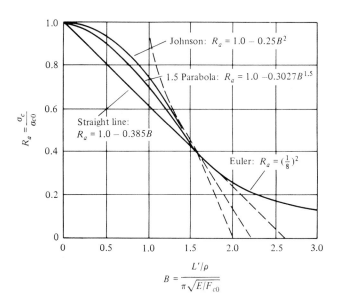

Figure 11.14

11.7 DIMENSIONLESS FORM OF TANGENT MODULUS CURVES

The materials used in flight vehicle construction are improved frequently. Designers must adopt new materials and processes which save structural weight, even though the new materials are more expensive than standardized ones. When a new or improved material is introduced, it is difficult to make extensive column tests and crippling tests in order to establish new design allowable stresses. It would be much better to obtain simple compressive stress-strain curves for the new material and to base new column and crippling allowable stresses on these tests than to test numerous built-up column specimens. The Ramberg-Osgood[19] equation for the stress-strain curve, discussed in Sec. 4.3, provides the necessary data for comparing similar materials.

The Ramberg-Osgood equation of the stress-strain curve is

$$\bar{\epsilon} = \bar{\sigma} + \tfrac{3}{7}\bar{\sigma}^n \tag{11.32}$$

where $\bar{\sigma}$ and $\bar{\epsilon}$ are dimensionless functions of the stress σ and the strain ϵ and the modulus of elasticity E:

$$\bar{\epsilon} = \frac{E\epsilon}{\sigma_1} \tag{11.33}$$

$$\bar{\sigma} = \frac{\sigma}{\sigma_1} \tag{11.34}$$

The stress σ_1 is approximately equal to the yield stress at a permanent strain of 0.002, but it must be defined as the stress at a secant modulus of elasticity of $0.7E$, in order for stress-strain curves with equal values of n to be geometrically similar.

The tangent modulus of elasticity $E_t = d\sigma/d\epsilon$ is readily obtained from Eqs. (11.32) to (11.34):

$$\frac{E}{E_t} = E\frac{d\epsilon}{d\sigma} = \frac{d\bar{\epsilon}}{d\bar{\sigma}} = 1 + \tfrac{3}{7}n\bar{\sigma}^{n-1} \tag{11.35}$$

$$\frac{E_t}{E} = \frac{1}{1 + \tfrac{3}{7}n\bar{\sigma}^{n-1}} \tag{11.36}$$

See Fig. 11.15. Now the tangent modulus equation can be written as a single expression which includes both the long- and short-column ranges, since Eq. (11.36) represents the modulus of elasticity below as well as above the elastic limit:

$$\sigma_c = \frac{\pi^2 E_t}{(L'/\rho)^2} = \frac{\pi^2 E}{(L'/\rho)^2}\left(\frac{1}{1 + \tfrac{3}{7}n\bar{\sigma}^{n-1}}\right) \tag{11.37}$$

For low values of σ, the expression in brackets is approximately unity, and Eq. (11.37) corresponds to the Euler equation. The expression in brackets, corresponding to E_t/E of Eq. (11.36), is plotted in Fig. 11.15 for various values of n.

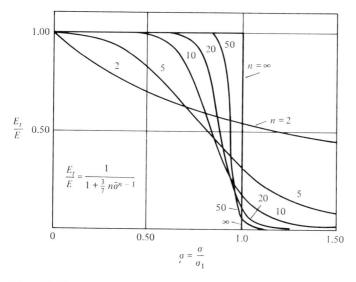

Figure 11.15

In order to plot column curves given by Eq. (11.37) in a dimensionless form similar to that shown in Fig. 11.14, the stress σ_1 must be used rather than σ_{c0} in Eqs. (11.26) and (11.27):

$$B = \frac{L/\rho}{\pi\sqrt{E/\sigma_1}}$$

(11.38)

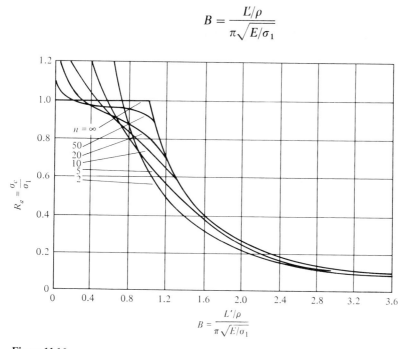

Figure 11.16

$$R_a = \frac{\sigma_c}{\sigma_1} \tag{11.39}$$

By substituting these values from Eqs. (11.38) and (11.39) into Eq. (11.37), the following column equation is obtained:

$$R_a = \frac{1}{B^2} \frac{E_t}{E} \tag{11.40}$$

This equation is plotted in Fig. 11.16 for various values of n.

The form of dimensionless column curve given by Eq. (11.40) and Fig. 11.16 was proposed by Cozzone and Melcon[7]. They also use this same basic diagram for local crippling, initial buckling of sheet in compression and shear, and buckling of sheet between rivets. These further applications are discussed later. These column curves have a very distinct advantage when structures of new materials are analyzed. It is necessary only to obtain the basic compression stress-strain diagram of the material. The shape factor n and the stress σ_1 corresponding to the yield stress supply all the necessary information on the new material. All the information obtained from tests of columns of one material are immediately applicable to a new material.

11.8 BUCKLING OF ISOTROPIC FLAT PLATES IN COMPRESSION

A flat plate, in which the thickness is small compared to the other dimensions, does not act as a number of parallel narrow beams when resisting bending stresses. The initially flat plate shown in Fig. 11.17a may be compared to the narrow beam shown in Fig. 11.17b. The initially rectangular cross section of the narrow beam distorts to the trapezoidal cross section, because the compression stresses on the upper face of the beam produce a lateral elongation, while the tensile stresses on the lower face of the beam produce a lateral contraction. The cross sections of the flat plate, however, must remain rectangular.

If the shaded element of Fig. 11.17a, shown to a larger scale in Fig. 11.17c, is considered, it will have unit elongations ϵ_x and ϵ_y as follows:

$$\epsilon_x = \frac{\sigma_x}{E} - v\frac{\sigma_y}{E} \tag{11.41}$$

and

$$\epsilon_y = \frac{\sigma_y}{E} - v\frac{\sigma_x}{E} \tag{11.42}$$

where v is Poisson's ratio. In the plate, the elongation in the y direction must be zero if the plate is assumed to have no curvature in the y direction. Substituting $\epsilon_y = 0$ into Eqs. (11.41) and (11.42) yields

$$\sigma_y = v\sigma_x \tag{11.43}$$

$$\epsilon_x = \frac{\sigma_x}{E}(1 - v^2) \tag{11.44}$$

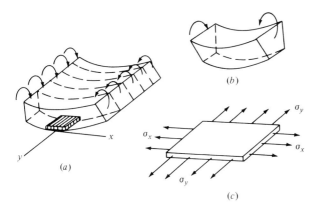

(b)

(c)   **Figure 11.17**

Thus, when the flat plate is deflected with single curvature in the x direction, the stresses in the y direction are equal to Poisson's ratio times the stresses in the x direction. Similarly, the unit elongations in the x direction have the ratio of $1 - v^2$ to the corresponding elongations in a narrow beam.

Since a flat plate has smaller unit elongations than the corresponding narrow beam, the curvature resulting from an equivalent bending moment will be smaller by the ratio $1 - v^2$. Similarly, if the term $M/(EI)$ of the general beam-deflection relation, Eq. (11.6), is replaced by $M(1 - v^2)/(EI)$, the Euler formula for a flat plate may be obtained as follows:

$$P_{\text{cr}} = \frac{\pi^2 EI}{(1 - v^2)L^2} \tag{11.45}$$

This equation applies for the condition shown in Fig. 11.18 where the unloaded edges are free and the loaded edges are simply supported, or free to rotate but not free to deflect normal to the plane of the plate. Substituting $I = bt^3/12$, $L = a$,

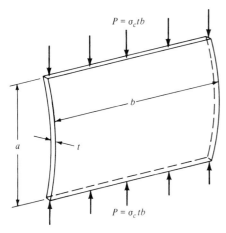

$P = \sigma_c t b$

b

a

t

$P = \sigma_c t b$ **Figure 11.18**

and $P_{cr} = \sigma_{cr} tb$ into Eq. (11.45) gives

$$\sigma_{cr} = \frac{\pi^2 E}{12(1 - v^2)} \left(\frac{t}{a}\right)^2 \tag{11.46}$$

For a plate simply supported on all four edges, as shown in Fig. 11.19, the buckling compression load is considerably higher. As the plate deflects, both vertical and horizontal strips must bend. The supporting effect of the horizontal strips may be sufficient to cause a vertical strip to deflect into two or more waves, as shown in Fig. 11.19. It can be shown that the buckling stress is[15]

$$\sigma_{cr} = \frac{\pi^2 E}{12(1 - v^2)} \left(\frac{bm}{a} + \frac{a}{bm}\right)^2 \left(\frac{t}{b}\right)^2 \tag{11.47}$$

where m is the number of waves in the buckled sheet. The value of v is approximately 0.3 for all metals. Since a large error in v produces only a small error in σ_{cr}, it is seldom necessary to consider the variation of Poisson's ratio. Equation (11.47) may be written as follows:

$$\sigma_{cr} = KE\left(\frac{t}{b}\right)^2 \tag{11.48}$$

where K is a function of a/b and is plotted in Fig. 11.20 for $v = 0.3$. Only the curve of Fig. 11.20 which gives the minimum value of K is significant, since the sheet will buckle into the number of waves that requires the smallest load. It is seen from Fig. 11.20 that the wavelength of the buckles is approximately equal to the width b, or that $m = 1$ for $a/b = 1$, $m = 2$ for $a/b = 2$, etc. The ratio a/b at which the number of waves changes from m to $m + 1$ is obtained from Eq. (11.47) as $a/b = \sqrt{m(m + 1)}$. The buckling stress obtained from Eq. (11.47) for a square plate with four edges simply supported is 4 times that obtained from Eq. (11.46) for the plate with sides free and ends simply supported.

Figure 11.19

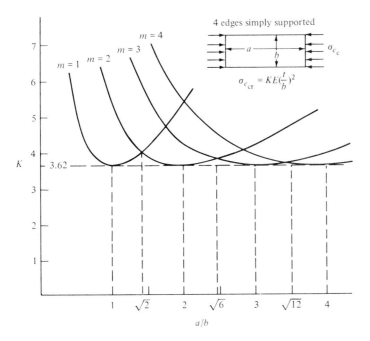

Figure 11.20

The buckling loads for rectangular plates with other edge conditions also can be found from Eq. (11.48) by using the correct values of K. Values of K are shown in Fig. 11.22 for various conditions. The loaded edges are termed ends, and the unloaded edges termed sides, as designated on the curves. A free edge may rotate or deflect in a direction normal to the plate. A fixed edge, as shown in Fig. 11.21a, is prevented from rotating or deflecting. The simply supported edge shown in Fig. 11.21b is free to rotate, but not to deflect normal to the plane of the plate.

The true edge-fixity conditions for flat plates in an airplane structure cannot be calculated in most cases. It is necessary to estimate the edge fixity after the supporting structure has been considered, in a manner similar to that for estimating column end-fixity conditions. The upper skin of an airplane wing, for

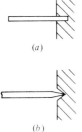

(a)

(b)   **Figure 11.21**

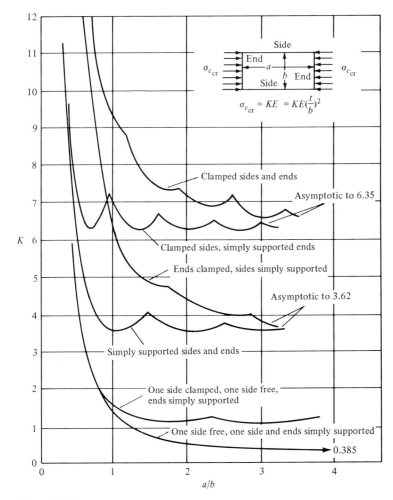

Figure 11.22

example, is compressed in a spanwise direction. If the stringers are flexible torsionally, they will rotate as the sheet buckles and will act almost as simple supports for the sheet between the stringers, as shown in Fig. 11.23a. If the stringers have considerable torsional rigidity, as do the "hat" sections and the

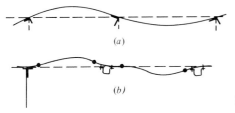

Figure 11.23

spar flange shown in Fig. 11.23*b*, they will rotate only slightly and will provide almost clamped edge conditions. In most structures, it is necessary to assume a value for the term *K* of Eq. (11.48) which will represent a conservative mean between simply supported and clamped edge conditions.

11.9 ULTIMATE COMPRESSIVE STRENGTH OF ISOTROPIC FLAT SHEET

The buckling of sheets in compression does not cause the collapse of a semi-monocoque structure, because the stiffening members usually can resist stresses which are much higher than those at which the initial sheet buckling occurs. We showed that a long column may resist a compression load when in the buckled condition and that the load is the same for a small lateral deflection as for a large deflection, provided that the stress does not exceed the elastic limit. The compression load resisted by a flat sheet with the sides free also remains constant for any lateral deflection. If the sides are supported, however, the compression load resisted by the sheet will increase as the lateral deflection increases, because the sides of the sheet must remain straight and consequently must be stressed in proportion to the strain in the direction of loading.

The plate shown in Fig. 11.24 is simply supported at all four edges and is loaded by a rigid block. The compression stresses are uniformly distributed as shown in Fig. 11.19 if the load is smaller than the buckling load. The stress distribution over the width of the plate is indicated in Fig. 11.25 by lines 1 and 2,

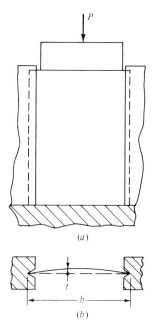

(*a*)

(*b*)   **Figure 11.24**

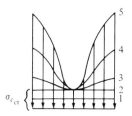

Figure 11.25

with line 2 indicating the stress at initial buckling. As the load is increased beyond the buckling load, the stress distribution is indicated by lines 3, 4, and 5. Near the middle of the cross section, the compressive stress remains approximately equal to the buckling stress, or a vertical strip acts in a similar manner to a long column. At the sides of the sheet, buckling is prevented, and the stress increases in proportion to the vertical motion of the loading block. The load may be increased until failure occurs by crushing of the sheet at the sides, although in common aircraft structures the stiffening members supporting the sheet usually fail before the sheet fails.

The curve representing the distribution of compressive stress over the width of a sheet is difficult to obtain, and even if it were known, it would be tedious to use in analysis. It is more convenient to obtain the total compression load corresponding to a given compression stress at the side of the sheet. It is customary to work with effective widths w, shown in Fig. 11.26, which are defined in such a way that the constant stresses σ_c acting over the effective widths will yield the total compression load. Thus w is selected so that the area under the two rectangles in Fig. 11.26a is equal to the area under the curve of the actual stress distribution. The total compression load P and the edge stress σ_c can be found experimentally, and the widths w may be calculated from

$$2tw\sigma_c = P \tag{11.49}$$

An approximate value of w may be obtained by assuming that a long sheet of total width $2w$ will have a buckling stress of σ_c. From Eq. (11.48) and Fig. 11.20,

$$\sigma_c = 3.62E\left(\frac{t}{2w}\right)^2$$

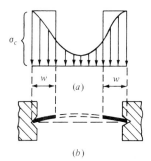

Figure 11.26

or

$$w = 0.95t \sqrt{\frac{E}{\sigma_c}}$$

Test results indicate that this value is too high and that it is more accurate to use

$$w = 0.85t \sqrt{\frac{E}{\sigma_c}} \tag{11.50}$$

In obtaining Eq. (11.50), we assume that the sheet is free to rotate at all four edges. In actual structures some degree of restraint always exists, and the effective widths may be much greater in many cases. Tests indicate that stringers provide considerable edge fixity at low stresses, but do not provide much restraint at stresses approaching the ultimate strength of the stringers. Numerous other equations have been used in place of Eq. (11.50); no equations provide accurate correlation with test results under all conditions. Uncertainties regarding the effects of edge restraints in the actual structure, accidental eccentricities in the sheet, and the effects of stresses beyond the elastic limit further complicate the problem. Equation (11.50) yields a smaller effective width than do most other equations and is conservative for use in design. For normal aircraft structures in which the sheet is relatively thin, the weight penalty introduced by using Eq. (11.50) is small; for high-speed aircraft in which the skin is relatively thick, a more accurate analysis may be justified.

The buckling stress for a flat sheet with a large ratio of length to width, with one side simply supported and the other free, can be obtained from Eq. (11.48) and Fig. 11.21. For $K = 0.385$, $\sigma_{cr} = 0.385E(t/b)^2$. The ultimate load resisted by such a sheet when the supported side is stressed by a value σ_c is found by considering that an effective width w_1 resists the stress σ_c and by obtaining w_1 as b from Eq. (11.50):

$$\sigma_c = 0.385E \left(\frac{t}{w_1} \right)^2$$

or

$$w_1 = 0.62t \sqrt{\frac{E}{\sigma_c}}$$

A more conservative value is recommended:

$$w_1 = 0.60t \sqrt{\frac{E}{\sigma_c}} \tag{11.51}$$

These effective sheet widths w and w_1 shown in Fig. 11.27 are obtained from Eqs. (11.50) and (11.51).

Example 11.1 The sheet stringer panel shown in Fig. 11.28 is loaded in compression by means of rigid members. The sheet is assumed to be simply supported at the loaded ends and at the rivet lines and to be free at the sides.

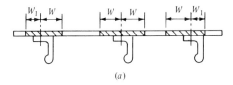

(a)

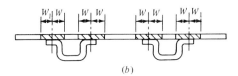

(b)

Figure 11.27

Each stringer has an area of 0.1 in^2. Assume $E = 10,300,000$ lb/in^2 for the sheet and stringers. Find the total compressive load P:
(a) When the sheet first buckles
(b) When the stringer stress σ_c is 10,000 lb/in^2
(c) When the stringer stress σ_c is 30,000 lb/in^2

SOLUTION (a) The sheet between the stringers is simply supported on all four edges and has dimensions of $a = 10$, $b = 5$, and $t = 0.040$ in. From Fig. 11.20, for $a/b = 2.0$ the value $K = 3.62$ is obtained. The buckling stress is $\sigma_c = KE(t/b)^2 = 3.62 \times 10,300,000(0.040/5)^2 = 2390$ lb/in^2. The edge of the sheet has dimensions of $a = 1$ and $b = 10$ in and is simply supported on three

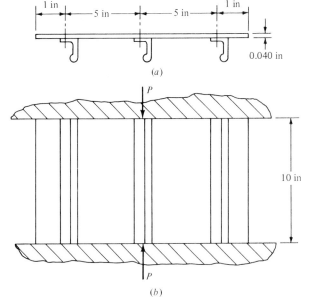

(a)

(b)

Figure 11.28

edges and free on the fourth edge. From Fig. 11.21, $K = 0.385$. The buckling stress is $\sigma_c = KE(t/b)^2 = 0.385 \times 10{,}300{,}000(0.040/1)^2 = 6200 \text{ lb/in}^2$.

The sheet therefore buckles initially between the stringers. The total area of the sheet is assumed to be effective before buckling occurs. The buckling of a flat sheet in compression is a gradual process, and the load does not drop appreciably when buckling occurs. The load is therefore calculated as follows:

$$A = 3 \times 0.1 + 12 \times 0.040 = 0.78 \text{ in}^2$$

$$P = \sigma_c A = 2390 \times 0.78 = 1865 \text{ lb}$$

(b) The effective sheet widths are obtained from Eqs. (11.50) and (11.51):

$$w = 0.85\sqrt{\frac{E}{\sigma_c}} = 0.85 \times 0.040\sqrt{\frac{10{,}300{,}000}{10{,}000}} = 1.09 \text{ in}$$

$$w_1 = 0.60t\sqrt{\frac{E}{\sigma_c}} = 0.77 \text{ in}$$

The effective sheet area is

$$A_1 = (4w + 2w_1)t = (4 \times 1.09 + 2 \times 0.77) \times 0.040 = 0.236 \text{ in}^2$$

The total compressive load is

$$P = \sigma_c A = 10{,}000(0.3 + 0.236) = 5360 \text{ lb}$$

(c) The solution is similar to that of part (b):

$$w = 0.85t\sqrt{\frac{E}{\sigma_c}} = 0.85 \times 0.040\sqrt{\frac{10{,}300{,}000}{30{,}000}} = 0.63 \text{ in}$$

$$w_1 = 0.60t\sqrt{\frac{E}{\sigma_c}} = 0.44 \text{ in}$$

$$A = 0.3 + (4 \times 0.63 + 2 \times 0.44) \times 0.040 = 0.436 \text{ in}^2$$

$$P = \sigma_c A = 30{,}000 \times 0.436 = 13{,}080 \text{ lb}$$

11.10 PLASTIC BUCKLING OF FLAT SHEET

In the discussion of buckling of sheet elements, we assume that the stress does not exceed the proportional elastic limit for the material. This elastic buckling action for flat sheets is similar to the elastic buckling of long columns in that the modulus of elasticity is the only significant material property. Equation (11.48) for sheet buckling is similar to the Euler equation for columns, and in each case the buckling stress is proportional to the modulus of elasticity of the material.

In the case of sheet elements for which the thickness is greater in comparison to the other dimensions, the compressive stresses will exceed the elastic limit before buckling will occur, as is the case for short columns. Equation (11.48) will be valid in this case if the tangent modulus of elasticity E_t is substituted for the modulus E:

$$\sigma_{cr} = KE_t\left(\frac{t}{b}\right)^2 \tag{11.52}$$

This equation may be written as

$$\sigma_{cr} = \frac{KE_t}{(b/t)^2} \tag{11.53}$$

Equation (11.53) is similar to the tangent modulus equation for short columns:

$$\sigma_c = \frac{\pi^2 E_t}{(L'/\rho)^2} \tag{11.54}$$

The tangent modulus curve and other curves for short columns were plotted with values of σ_c as ordinates and values of L'/ρ as abscissas. Values of σ_{cr} and b/t could be similarly plotted from Eq. (11.53) for a known value of K. In fact, the column curves can be used for plastic sheet buckling if the values of b/t are multiplied by a constant which is obtained by equating the right side of Eq. (11.53) and the right side of Eq. (11.54) as follows:

$$\text{Equivalent } \frac{L'}{\rho} = \frac{\pi}{\sqrt{K}}\frac{b}{t} \tag{11.55}$$

A typical column curve for an aluminum-alloy material is shown in Fig. 11.29. The allowable column stress is obtained from the curve for a known value

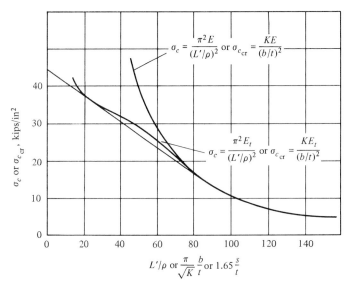

Figure 11.29

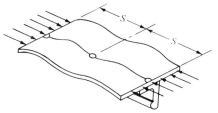

Figure 11.30

of L/ρ. In the short-column range, the theoretical tangent modulus curve may be replaced by the more conservative straight line, in order to account for accidental eccentricities or other unknown conditions. Similarly, the curves in Fig. 11.29 yield the allowable buckling stress for a flat plate in the plastic, or short-column, range. The value of σ_{cr} may be obtained for any known value of $(\pi/\sqrt{K})(b/t)$. Either of the short-column curves may be used, depending on the possible initial eccentricities of the sheet element and the degree of conservatism desired. The value of K is obtained from Fig. 11.20 or 11.21.

One common application of plastic buckling is the buckling of compressive skin between rivets attaching the skin to the stringers or spar caps. A skin element of this type is shown in Fig. 11.30. The rivets have a uniform spacing s along the stringer, and the restraint is such that the skin element of length s and indefinite width has clamped ends and free sides. The element therefore resists 4 times the load of a similar element with hinged ends, which was analyzed by Eq. (11.46). Substituting $a = s$ and $E = E_t$ into Eq. (11.46) and multiplying the right-hand side by 4 (to account for the end fixity) yield

$$\sigma_c = \frac{\pi E_t}{3(1 - v^2)} \left(\frac{t}{s}\right)^2$$

or, for $v = 0.3$,

$$\sigma_c = \frac{3.62 E_t}{(s/t)^2} \tag{11.56}$$

The tangent modulus short-column curve may be used in solving Eq. (11.56). An equivalent slenderness ratio may be obtained by equating the right-hand sides of Eqs. (11.54) and (11.55):

$$\text{Equivalent } \frac{L'}{\rho} = \frac{\pi}{\sqrt{3.62}} \frac{s}{t} = 1.65 \frac{s}{t} \tag{11.57}$$

Example 11.2 Find the compression buckling stress for a sheet 4 by 4 by 0.125 in with all four edges simply supported, assuming that the tangent modulus column curve for the material is represented by Fig. 11.29.

SOLUTION For this sheet $a = b = 4$ and $t = 0.125$. From Fig. 11.21 for

$a/b = 1$ and simply supported edges, $K = 3.62$. From Eq. (11.55) the equivalent L/ρ is

$$\frac{\pi}{\sqrt{K}}\frac{b}{t} = \frac{\pi}{\sqrt{3.62}} \times \frac{4}{0.125} = 52.8$$

From Fig. 11.29, $\sigma_{cr} = 28{,}000 \text{ lb/in}^2$. If this point had been on the right-hand portion of Fig. 11.29, corresponding to the long-column or elastic range, the buckling stress would correspond to that given by Eq. (11.48).

Example 11.3 The angle extrusion shown in Fig. 11.31 is loaded in compression. Each leg of the angle buckles as a plate simply supported on the ends and on one side and free on the other side. Find the stress at which this buckling occurs. Assume that Fig. 11.29 represents properties of this material.

SOLUTION For each leg, $b = 1$, $a = 8$, and $t = 0.072$. For $a/b = 8$ the value of K from Fig. 11.21 is approximately 0.385. The equivalent L/ρ is

$$\frac{\pi}{\sqrt{K}}\frac{b}{t} = \frac{\pi}{\sqrt{0.385}} \times \frac{1}{0.072} = 70.4$$

From Fig. 11.29, $\sigma_{c_{cr}} = 20{,}500 \text{ lb/in}^2$.

The type of failure indicated for this section is typical of crippling failures for aluminum-alloy extrusions. The ordinary short-column curves apply only to round tubes or to stable cross sections which do not cripple locally. Since light extrusions are used extensively as column members in aircraft structures, the subject of crippling failure is very important and is discussed in detail later.

Example 11.4 An 0.040 sheet is riveted to an extrusion by rivets spaced 1 in apart. What compression stress in the extrusion will produce buckling of the

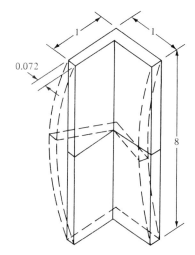

Figure 11.31

sheet between rivets, as shown in Fig. 11.30, if the sheet has column properties as represented by Fig. 11.29?

SOLUTION From Eq. (11.57) the equivalent L/ρ is

$$1.65 \frac{s}{t} = 1.65 \frac{1}{0.040} = 41.2$$

From Fig. 11.29, $\sigma_c = 31{,}300 \text{ lb/in}^2$.

11.11 NONDIMENSIONAL BUCKLING CURVES

The plastic buckling stresses in Sec. 11.10 are obtained from column curves of the
type shown in Fig. 11.29. A column curve of this type is applicable to only one
material, since the column curve is affected by the shape of stress-strain curve, the
modulus of elasticity, and the yield stress of the material. There are numerous
advantages to plotting column curves in dimensionless form, as shown in Fig.
11.16 and as discussed in Sec. 11.7. When several materials have stress-strain
curves of the same general shape, as indicated by the value of n, a single-column
curve presents the data for all these materials. Test information for any one of the
materials is therefore applicable to all of them.

Cozzone and Melcon propose that the nondimensional curves of Fig. 11.16
be used for all problems of plastic sheet buckling, interrivet buckling, and local
crippling of compression members. The curves of Fig. 11.16 are presented by

$$\frac{\sigma_a}{\sigma_1} = \frac{E_t}{E} \frac{1}{B^2} \tag{11.58}$$

where σ_a is the allowable average stress for a column for sheet buckling, or for
crippling, and σ_1 is the secant yield stress corresponding to the stress at the
intersection between the stress-strain curve and a line through the origin having
slope $0.7E$.

For columns, the term B is defined by Eq. (11.38):

$$B = \frac{L/\rho}{\pi \sqrt{E/\sigma_1}} \tag{11.38}$$

For plastic sheet buckling, the value of B is obtained from Eqs. (11.53) and
(11.58):

$$B = \frac{b/t}{\sqrt{EK/\sigma_1}} \tag{11.59}$$

For interrivet buckling, the value of B is obtained from Eqs. (11.56) and (11.58):

$$B = \frac{0.525s/t}{\sqrt{E/\sigma_1}} \tag{11.60}$$

Thus the values of B may be calculated from Eqs. (11.38), (11.59), or (11.60), and then the value of σ_a/σ_1 may then be read from the proper curve of Fig. 11.16.

11.12 COLUMNS SUBJECT TO LOCAL CRIPPLING FAILURE

The column equations previously derived are applicable to closed tubular sections with comparatively thick walls or to other cross sections which are not subject to local crippling failure. Many of the columns used in semimonocoque flight vehicle structures are made of extruded sections or of bent sheet sections and fail by local crippling. The assumed column curve is that shown by line A of Fig. 11.32, where the σ_{cc} is the crippling stress. Tests of columns of extrusions or bent sheet with thin walls subject to local crippling yield values represented by curve B of Fig. 11.32 and indicate that sections subject to crippling failure should be analyzed by different column equations than stable cross sections of the same material.

Usually it is desirable to make tests which will cover a range of slenderness ratios for each thin-walled section which is to be used as a column. This procedure is not always practical for preliminary design, since the designer, having a wide choice of cross sections, must be able to select some sections and predict their strength at an early stage of the design. Tests on aluminum-alloy columns subject to crippling failures show that the short-column curve closely approximates a second-degree parabola, as represented by Eq. (11.23) or (11.29). The crippling stress σ_{cc} is substituted for the stress σ_{c0} as follows:

$$\sigma_c = \sigma_{cc}\left[1 - \frac{\sigma_{cc}(L/\rho)^2}{4\pi^2 E}\right] \tag{11.61}$$

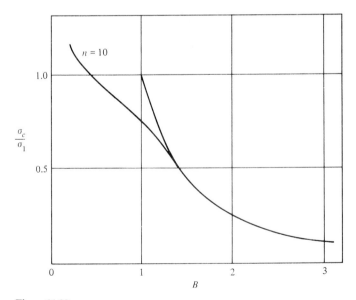

Figure 11.32

As in the case of other short-column curves, Eq. (11.61) does not apply for very short columns ($L/\rho < 12$) because the end supports increase the crippling stress. Thus the crippling stress σ_{cc} can be obtained by testing a column with an L/ρ of about 12. An approximate value of the crippling stress may be derived by finding the sum of the plastic buckling strengths of the rectangular elements of the cross section.

The column cross sections shown in Fig. 11.33 may be considered as made of rectangular plates of width b, thickness t, and length a, which is large in comparison to b. The plates with widths designated b' are assumed to be simply supported on both sides, and those with widths designated b are assumed to be free on one side and restrained on the other side. In the case of the angles shown in Fig. 11.33a and e, the plates are assumed to be simply supported on one side, since the two plates buckle at the same stress and neither plate supplies any edge restraint for the other. In the case of the other cross sections, however, the plates which have one side free have edge conditions between the clamped and simply supported cases for the other side. This difference on edge restraint is seen by comparing the buckled form of the angle shown in Fig. 11.31 to the buckled form of the channel shown in Fig. 11.34. The legs of the angle buckle in one half-wave regardless of the length of the column, as is the case for a flat plate with one side free and the other side simply supported. The legs of the channel buckle into the same number of half-waves as the back of the channel which buckles in approximately square panels, as shown in Fig. 11.34.

The initial buckling stress of the plates may be smaller than the stress at which collapse of the member occurs, since the corner resists load after the initial buckling. This effect is considered empirically by assuming the effective width b to be less than the total width, as shown in Fig. 11.33e. The extrusions resist a greater load at the corners than the bent sheet sections, as indicated by the

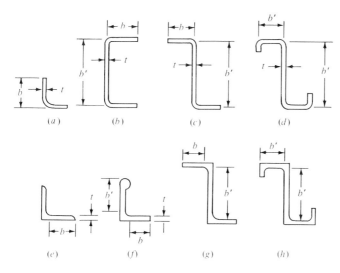

Figure 11.33

Figure 11.34

widths b in Fig. 11.33. The edge conditions for the plates in a bulb angle extrusion of the type shown in Fig. 11.33 f depend on the bending stiffness of the bulb, but usually it is assumed that the bulb supports the plate as indicated.

After the plastic buckling stress is determined for each element of area by the method of Secs. 11.10 and 11.11, the total crippling load on the cross section is found as the sum of the loads on the individual areas. If the areas have dimensions $b_1 t_1$, $b_2 t_2$, and $b_3 t_3$ and buckling stresses σ_1, σ_2, and σ_3, then the total crippling stress is

$$\sigma_{cc} = \frac{\sigma_1 b_1 t_1 + \sigma_2 b_2 t_2 + \sigma_3 b_3 t_3}{b_1 t_1 + b_2 t_2 + b_3 t_3} = \frac{\Sigma \sigma b t}{\Sigma b t} \tag{11.62}$$

The denominator of Eq. (11.62) may not be equal to the total area because the corner areas are not included. The crippling load is obtained by multiplying the stress σ_{cc} by the total area, and it may be greater than the numerator of Eq. (11.62) because of the load on the corners.

Example 11.5 Find the equation of the short-column curve for the extrusion shown in Fig. 11.35 given that $E = 10,700,000$, $n = 10$, and $\sigma_1 = 37,000$.

SOLUTION Assume the column curve for this material is represented by Fig. 11.29. For plastic buckling, the equivalent L'/ρ is obtained from Eq. (11.55) with $K = 3.62$ from Fig. 11.21. For area 1, $b/t = 1.564/0.05 = 31.3$; for area 2, $b/t = 0.70/0.093 = 7.52$. From Eq. (11.55),

$$\text{Equivalent } \frac{L'}{\rho} = \frac{\pi}{\sqrt{K}} \frac{b}{t} = 1.65 \frac{b}{t}$$

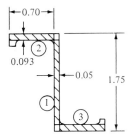

Figure 11.35

For area 1, the equivalent L'/ρ is $1.65 \times 31.3 = 51.6$; for area 2, it is $1.65 \times 7.52 = 12.4$. From Fig. 11.29, $\sigma = 29,000$ for area 1 and $\sigma = 45,000$ lb/in² for area 2. From Eq. (11.62),

$$\sigma_{cc} = \frac{\Sigma \sigma bt}{\Sigma bt} = \frac{29,000 \times 1.564 \times 0.05 + 45,000 \times 0.70 \times 0.093 \times 2}{1.564 \times 0.05 + 2 \times 0.70 \times 0.093}$$

$$= 39,000 \text{ lb/in}^2$$

The crippling stresses for the individual areas also can be determined from the nondimensional curve of Fig. 11.16. From Eq. (11.59),

$$B = \frac{b/t}{\sqrt{10,700,000 \times 3.62/37,000}} = 0.0308 \frac{b}{t}$$

For area 1, $B = 0.0308 \times 31.3 = 0.965$. From Fig. 11.16, $\sigma/\sigma_1 = 0.77$, or $\sigma = 28,500$ lb/in².

For area 2, $B = 0.0308 \times 7.52 = 0.232$. From Fig. 11.16, $\sigma/\sigma_1 = 1.20$, or $\sigma = 45,000$ lb/in². These check the values obtained from Fig. 11.29. The short-column curve is now obtained from Eq. (11.61):

$$\sigma_c = \sigma_{cc} \left[1 - \frac{\sigma_{cc}(L'/\rho)^2}{4\pi^2 E} \right]$$

$$= 39,000 \left[1 - \frac{39,000(L'/\rho)^2}{4\pi^2 \, 10,700,000} \right]$$

$$= 39,000[1 - 0.0000923(L'/\rho)^2]$$

Example 11.6 Given the section shown in Fig. 11.36, assume that $n = 10$, $E = 9700$, and $\sigma_1 = 46,000$. Find the crippling stress for the cross section.

SOLUTION The web is assumed to be simply supported on both sides, with $K = 3.62$, and buckles into approximately rectangular panels in a manner similar to the channel section shown in Fig. 11.34. The half-waves are approximately 1.12 in long; therefore, the flanges may be considered as simply supported at ends 1.12 in apart and on one side. From Fig. 11.21 for

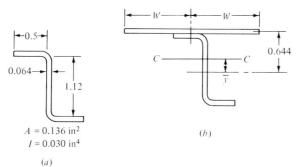

$A = 0.136 \text{ in}^2$
$I = 0.030 \text{ in}^4$

(a)

(b)

Figure 11.36

$a/b = 2.000$, $K = 0.60$. From Eq. (11.59),

$$B = \frac{b/t}{\sqrt{EK/\sigma_1}}$$

For the flanges,

$$B = \frac{0.5/0.064}{\sqrt{9,700,000 \times 0.6/46,000}} = 0.693$$

For the web,

$$B = \frac{1.12/0.064}{\sqrt{9,700,000 \times 3.62/46,000}} = 0.633$$

From Fig. 11.16, $\sigma/\sigma_1 = 0.88$, or $\sigma = 40,500$ lb/in^2, for the flanges, and $\sigma/\sigma_1 = 0.905$, or $\sigma = 41,600$ lb/in^2 for the web.

From Eq. (11.62),

$$\sigma_{cc} = \frac{\Sigma \sigma b t}{\Sigma b t} = \frac{\Sigma \sigma b}{\Sigma b} = \frac{2 \times 40,500 \times 0.5 + 41,600 \times 1.12}{2 \times 0.5 + 1.12}$$

$$= 41,000 \text{ lb/in}^2$$

11.13 NEEDHAM AND GERARD METHODS FOR DETERMINING CRIPPLING STRESSES

More recent semiempirical methods than that of Sec. 11.12 for the determination of crippling stresses of columns were developed by Needham[45] and Gerard.[46–49] In the Needham method, the structural member section is divided into angle elements, as shown in Fig. 11.37. The crippling strength of these elements can be established by theory and/or tests. The crippling failure strength of the member section then can be determined by summing the crippling strengths of each angle element that makes up the total section. Through extensive tests Needham arrived at the following semiempirical equation for the crippling stress of angle sections:

$$\sigma_c = \frac{k_e(E_c \sigma_{cy})^{0.5}}{(b'/t)^{0.75}} \tag{11.63}$$

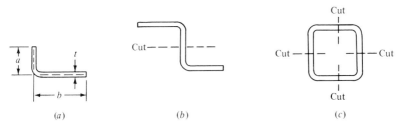

Figure 11.37 (a) Two edges free; (b) one edge free; (c) no edge free.

where σ_c = crippling stress
E_c = compressive modulus of elasticity
σ_{cy} = material compressive yield stress

$$\frac{b'}{t} = \frac{a+b}{2t}$$

k_e = constant coefficient whose magnitude depends on support condition of angle edges: $k_e = 0.316$ for the two edges free, 0.342 for one edge free, and 0.366 with no edge free

The crippling stress for the actual member section is obtained by utilizing the following equation:

$$\sigma_{cs} = \frac{\Sigma \sigma_{ci} A_i}{\Sigma A_i} \tag{11.64}$$

where σ_{cs} = member section crippling stress
σ_{ci} = ith angle crippling stress, calculated from Eq. (11.63)
A_i = ith angle cross-sectional area

Gerard's method[49] for calculating crippling stresses is a generalization of Needham's method. His extensive investigation led to the formulation of three semiempirical equations for determining the crippling stresses in various shapes of structural members.

For sections with straight unloaded edges such as plates, tee, cruciform, and H sections, the following crippling stress equation applies:

$$\sigma_{cs} = 0.67\sigma_{cy}\left[\frac{gt^2}{A}\left(\frac{E_c}{\sigma_{cy}}\right)^{1/2}\right]^{0.85} \tag{11.65a}$$

For sections with distorted unloaded edges such as tubes, V-groove plates, angles, stiffened panels, and multicorner sections, the following crippling equation applies:

$$\sigma_{cs} = 0.56\sigma_{cy}\left[\frac{gt^2}{A}\left(\frac{E_c}{\sigma_{cy}}\right)^{1/2}\right]^{0.4} \tag{11.65b}$$

For sections such as two corner sections, J, Z, and channel sections, the following equation applies:

$$\sigma_{cs} = 3.2\sigma_{cy}\left[\frac{t^2/A}{(E_c/\sigma_{cy})^{1/3}}\right]^{0.75} \tag{11.65c}$$

where A = section area and g = number of flanges which make up the section plus the number of cuts required to divide the section into a number of flanges. See Fig. 11.38. The maximum crippling stress σ_{cs} must not exceed those specified in Table 11.1 unless it is verified experimentally.

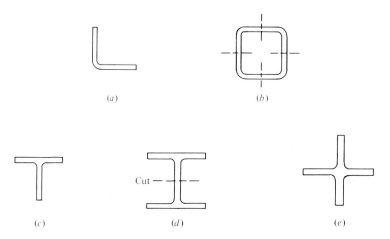

Figure 11.38 (*a*) Basic angle section; g = 2; (*b*) tube; g = 4 cuts + 8 flanges = 12; (*c*) basic T section; g = 3; (*d*) H section; g = 1 cut + 6 flanges = 7; (*e*) cruciform section; g = 0 cut + 4 flanges = 4.

Table 11.1

Section shape	Maximum σ_{cs}
L	$0.7\sigma_{cy}$
T, +, H	$0.8\sigma_{cy}$
Z, J, ⊔	$0.9\sigma_{cy}$
□, multicorner	$0.8\sigma_{cy}$

11.14 CURVED SHEET IN COMPRESSION

A thin-walled circular cylinder loaded in compression parallel to its axis may fail by local instability of the thin walls. This type of failure is similar to that which occurs in the compression skin of semimonocoque wing and fuselage structures. The compression buckling of flat sheets was considered, but most actual structures are made of curved sheets, and the curvature has a considerable effect on the buckling and ultimate strengths. A cylinder which is loaded in compression will assume a buckled form similar to that shown in Fig. 11.39. The number of circumferential waves depends on the ratio of R/t, where R is the radius and t is

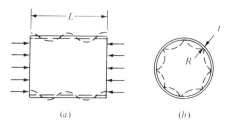

(*a*)   (*b*)   **Figure 11.39**

the wall thickness of the cylinder. A large number of waves develop for a large value of R/t. The length of the longitudinal waves is the same magnitude as the length of the circumferential waves. For the high ratios of R/t which are common in semimonocoque wing and fuselage skins, the wavelengths are so small that a sector of a cylinder, with simply supported edges as shown in Fig. 11.40a resists approximately the same buckling stress as the complete cylinder. This sector corresponds to the skin between adjacent stringers, as shown in Fig. 11.40b. For smaller values of R/t, the length of the circumferential waves is greater, and the stringers or edge supports of Fig. 11.40b prevent the formation of the waves and thus increase the buckling stress.

The compression buckling stress for a thin-walled cylinder may be determined theoretically in a manner similar to that used in obtaining the buckling stresses of flat plates. The classic analysis of cylinders, which is based on the assumption of small displacements, yields

$$\sigma_{c_{cr}} = 0.606E\ \frac{t}{R} \tag{11.66}$$

if Poisson's ratio is 0.3. Test values, however, are much lower than those given by Eq. (11.66), and test results show considerable scatter. This is in contrast to the excellent correlation between theoretical and experimental values for the buckling stresses for flat sheet.

Von Karman, Dunn, and Tsien[6] have shown that the assumptions made in the analysis by the classic theory are in error. In the case of buckling of a flat plate, a longitudinal strip of the plate is supported elastically by lateral strips which exert restraining forces in proportion to their deflection. When the flat plate buckles, it may buckle in either direction, and the load after buckling remains equal to the buckling load, as in a Euler column. In the case of a compressed cylinder, however, the longitudinal strips are supported by circumferential rings which exert restraining forces that are not proportional to their radial deflection. The stiffness of a circular ring increases as it is deflected outward and decreases as it is deflected inward. Thus the thin walls of a compressed cylinder buckle inward much more readily than they buckle outward. The buckling is accompanied by a sudden decrease in both the load and the length if there are no eccentricities of the walls. The buckling load is considerably reduced by small eccentricities of the walls. The buckling stress depends on the rigidity of the testing machine, since any testing machine has some elasticity, and the plates of

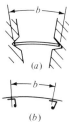

(a)

(b)   **Figure 11.40**

the machine move together slightly as the resistance of the specimen decreases. The large effects of specimen eccentricity and of testing machine elasticity explain the large scatter of test results.

If the compression load on a cylinder P is plotted against the axial compressive deformation e, curves similar to those shown in Fig. 11.41 are obtained. Curve 1 represents a theoretical curve for an ideal cylinder in which the walls are perfectly cylindrical and homogeneous. The point A_1 corresponds to the theoretical buckling stress, obtained from Eq. (11.66), which cannot be determined experimentally by the most careful testing because the upper branch of the curve is so close to the lower branch. At a deformation e corresponding to point B_1, the cylinder assumes a buckled form and the load drops. If the deformation has exceeded that corresponding to B_1 before buckling occurs, the cylinder suddenly decreases in length when buckling occurs. Because of the elasticity of the testing machine, the plates of the machine move together and the cylinder decreases in length when the load drops, even for test specimens with small eccentricities, as represented by curve 2.

Buckling loads obtained experimentally are represented by points A_2 and A_3 of Fig. 11.41. In the case of unstiffened cylinders, these buckling loads represent the ultimate strength of the cylinder in compression. Several empirical equations have been derived from experimental results, and the various equations yield widely divergent values of buckling stress, as might be expected because of the scatter of test values. Kanemitsu and Nojima propose the following equations:

$$\frac{\sigma_{c_{cr}}}{E} = 9\left(\frac{t}{R}\right)^{1.6} + 0.16\left(\frac{t}{L}\right)^{1.2} \tag{11.67}$$

where L is the length of the cylinder. This equation appears to give satisfactory agreement with test values within the ranges of $500 < R/t < 3000$ and $0.1 < L/R < 2.5$.

Another equation which yields reasonable values of the buckling stress for smaller values of R/t is obtained as approximately one-half of the value of Eq. (11.66):

$$\sigma_{c_{cr}} = 0.3E\,\frac{t}{R} \tag{11.68}$$

This equation yields results which are much higher than experimental values in

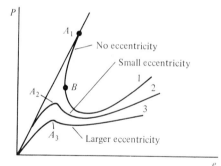

Figure 11.41

cases where R/t is large. Perhaps it is reasonable to use Eq. (11.68) for values of R/t less than 500 and to use Eq. (11.67) for the range in which it applies.

In the case of curved sheet which is stiffened by longitudinal members, as is common in semimonocoque construction, the sheet would resist a buckling stress as given by Eq. (11.48) if there were no curvature and an additional stress, as given by Eq. (11.67), because of the curvature. While there is little theoretical justification for adding these buckling stresses, this procedure is substantiated reasonably well by tests.

The compression buckling stress for curved skin on the upper surface of a wing is increased considerably by the negative air pressure on the sheet. Since the curved sheet has a tendency to buckle inward, the aerodynamic forces reduce this tendency. Equation (11.67) is very conservative in this case. It is very important to prevent the buckling of the wing skin of high-speed aircraft because of the aerodynamic drag of the irregular airfoil section.

The ultimate strength of a stiffened, curved sheet panel may be found in a similar manner to that used in obtaining the ultimate strength of a flat sheet panel in Sec. 11.9. In addition to the compression load resisted by the stringers and by the effective widths of skin acting with the stringers, the sheet between stringers resists load because of its curvature, even though it has buckled. The load resisted by a buckled, curved sheet is indicated by the right-hand portion of the curves of Fig. 11.41. While this load depends on the elongation e of the stringers, many other unknown factors are involved. The method for calculating this load is to assume that a skin width of $b - 2W$ between stringers resists a stress of $0.25Et/R$, as shown in Fig. 11.42. As an alternative method, this stress might be calculated by Eq. (11.67). Where this buckling stress for the curved sheet exceeds the stringer stress σ_c, the entire sheet area is assumed to resist a stress σ_c.

Example 11.7 For the wing shown in Fig. 11.43, $R = 50$, $t = 0.064$, $b = 6$, and the rib spacing is $L = 18$ in. Find the compressive stress in the skin at which the buckling occurs if $E = 10^7$ lb/in^2.

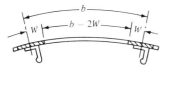

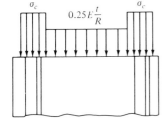

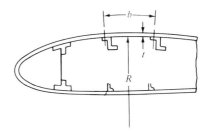

Figure 11.42   **Figure 11.43**

SOLUTION The buckling stress is obtained as the sum of the buckling stress for a flat sheet simply supported on four sides, as obtained from Eq. (11.48), and the buckling stress for a cylinder, as obtained from Eq. (11.67). From Eq. (11.48),

$$\sigma_{c_{cr}} = KE\left(\frac{t}{b}\right)^2 = 3.62 \times 10^7 \times \left(\frac{0.064}{6}\right)^2 = 4110 \text{ lb/in}^2$$

From Eq. (11.67),

$$\frac{\sigma_{c_{cr}}}{E} = 9\left(\frac{t}{R}\right)^{1.6} + 0.16\left(\frac{t}{L}\right)^{1.2}$$

$$= 9\left(\frac{0.064}{50}\right)^{1.6} + 0.16\left(\frac{0.064}{18}\right)^{1.2}$$

$$\sigma_{c_{cr}} = 2130 + 1560 = 3690 \text{ lb/in}^2$$

The total buckling stress is the sum of these two values:

$$\sigma_{c_{cr}} = 4110 + 3690 = 7800 \text{ lb/in}^2$$

11.15 ELASTIC SHEAR BUCKLING OF FLAT PLATES

The buckling of rectangular plates which resist direct compression stresses is discussed in Sec. 11.8. Other types of stresses, such as shear stresses and bending stresses, also may produce elastic buckling of thin plates. Only loads in the plane of the plate are discussed here, and components normal to the plane of the plate are assumed to be zero.

The elastic buckling stresses for thin rectangular plates in shear can be calculated theoretically. The analysis is beyond the scope of this book, but the results may be expressed in the same form as Eq. (11.48) if Poisson's ratio v is assumed constant for all materials:

$$\tau_{cr} = KE\left(\frac{t}{b}\right)^2 \tag{11.69}$$

The values of K are plotted in Fig. 11.44 for $v = 0.3$ and for the two conditions of all four edges clamped and all four edges simply supported. The term t is the plate thickness, and E is the elastic modulus of the plate material. For the compressed plate discussed in Sec. 11.8, the width b is perpendicular to the direction of loading and the length a is parallel to the loads; but since the plate in shear is loaded on all four sides, the dimension b is considered as the smaller of the two plate dimensions. The critical shearing stress τ_{cr} is uniformly distributed along all four sides of the plate.

The rectangular plate which is loaded in pure shear has principal tension and compressive stresses at 45° to the edges. These principal stresses are equal to the shearing stresses. The diagonal compression stresses cause the sheet buckling, and

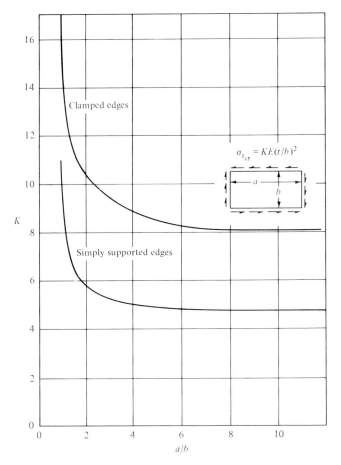

Figure 11.44

when buckling occurs, the wrinkles form at approximately 45° angles to the edges. The buckling shearing stresses τ_{cr} are considerably higher than the buckling compression stresses $\sigma_{c_{cr}}$ for plates with equal dimensions. This is a result of the restraining effect of the diagonal tension in the plate which is loaded by shearing forces.

The critical buckling stresses in a thin plate loaded in bending as shown in Fig. 11.45 also can be calculated theoretically and expressed in the same form as the equations for compression and shear buckling:

$$\sigma_{b_{cr}} = KE\left(\frac{t}{b}\right)^2 \tag{11.70}$$

where $\sigma_{b_{cr}}$ is the critical maximum bending stress shown in Fig. 11.45 and K is given by the curve of Fig. 11.45 if all four edges are simply supported.

In the case of buckling of thin plates under the combined action of two of the

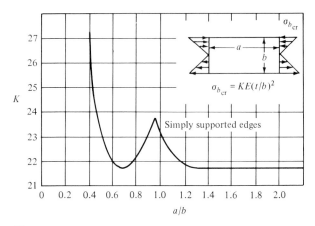

Figure 11.45

conditions of compression, shear or bending, the initial buckling stresses have been determined empirically by the method of stress ratios. The initial buckling occurs when one of the following equations is satisfied:

Compression and bending:

$$R_b^{1.75} + R_c = 1 \qquad (11.71)$$

Compression and shear:

$$R_s^{1.5} + R_c = 1 \qquad (11.72)$$

Bending and shear:

$$R_b^2 + R_s^2 = 1 \qquad (11.73)$$

where R_b, R_s, and R_c represent the ratios of the stresses in the plate to the critical buckling stresses $\sigma_b/\sigma_{b_{cr}}$, $\sigma_c/\sigma_{c_{cr}}$, and τ/τ_{cr}.

11.16 ELASTIC BUCKLING OF CURVED RECTANGULAR PLATES

A large part of the structure of a semimonocoque airplane consists of the outer shell, or skin. This skin usually is curved to provide the necessary aerodynamic shape, and it must resist tension, compression, shear, and bending stresses. In addition to the conditions of ultimate strength and yield strength, which must be considered in the design of flight vehicle structural members, often the skin must be designed so that it will not wrinkle under normal flight conditions. Skin wrinkles or other surface irregularities seriously affect the airflow in the case of high-speed aircraft, but may be permissible for slower-speed aircraft. Unfortunately, both the buckling strength and the ultimate strength of curved plates depend on many uncertain factors and are difficult to predict accurately. Initial plate eccentricities, air pressure normal to the plate, and conditions of the supports are difficult to evaluate; yet they may have a considerable effect on buckling loads.

The buckling stress for a curved plate in shear, such as shown in Fig. 11.46, is higher than the buckling stress for a flat plate with corresponding dimensions. The buckling stresses obtained experimentally usually are smaller than those calculated theoretically for an ideal plate with small deflections. A condition similar to that described in Sec. 11.14 for plates in compression exists for plates in shear; the theoretical buckling stresses for flat plates correspond closely with test results for practical plates, but theoretical buckling stresses for curved plates usually are higher than values obtained experimentally.

The theoretical shear buckling stresses for curved plates have been calculated by Batdorf, Stein, and Schildcrout.[50] For a constant value of Poisson's ratio, $v = 0.3$, the shear buckling stress τ_{cr} may be expressed in the form of previous buckling equations:

$$\tau_{cr} = K_s E \left(\frac{t}{b}\right)^2 \tag{11.74}$$

The term K_s is a function of the ratios a/b and $b^2/(rt)$ and is plotted in Figs. 11.46 and 11.47. When the circumferential length is greater than the axial length, Fig.

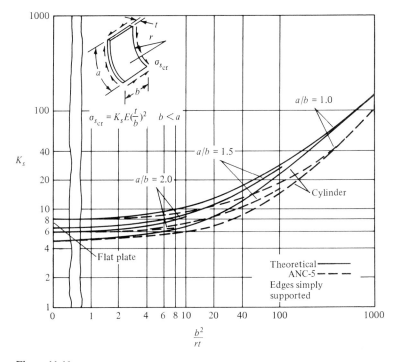

Figure 11.46

11.47 is applicable; when the axial length is greater, Fig. 11.46 must be used. The dimension b is smaller than the dimension a in either case. Both figures apply only to plates for which all four edges are simply supported. The points at the left side of the charts, for $b^2/(rt) = 0$, correspond to the buckling stress coefficients for flat plates, as given in Fig. 11.44.

For design purposes, it is necessary to consider the effects of initial accidental eccentricities, which always cause the buckling stresses to be smaller than the theoretical values. Often the designer must use judgment in evaluating these effects for a particular structure. An empirical equation is proposed:

$$\tau_{cr} = KE\left(\frac{t}{b}\right)^2 + K_1 E \frac{t}{r} \tag{11.75}$$

where the first term represents the buckling stress for a flat plate, as given by Fig. 11.44, and the last term represents the additional stress which can be resisted because of the curvature. The value $K_1 = 0.10$ is recommended. By rewriting Eq. (11.75) and comparing it to Eq. (11.74), the following relations are obtained:

$$\tau_{cr} = \left(K + K_1 \frac{b^2}{rt}\right)E\left(\frac{t}{b}\right)^2$$

$$K_s = K + K_1 \frac{b^2}{rt} \tag{11.76}$$

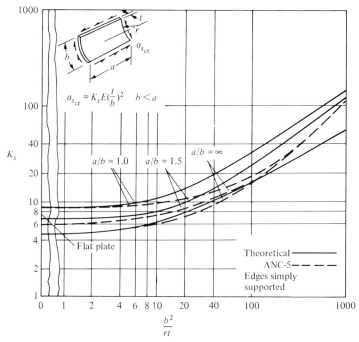

Figure 11.47

The values of K_s from Eq. (11.76) are plotted as the dotted lines in Figs. 11.46 and 11.47, assuming $K_1 = 0.10$. The values obtained from Eq. (11.75) are seen to represent conservative approximations for all values shown on the chart, except for the case of large values of a/b and $b^2/(rt)$ shown in Fig. 11.47. Except for this range, Eq. (11.75) approximates most of the available test information closely and conservatively and may be used in practical design. While the theoretical curves of Figs. 11.46 and 11.47 apply only to plates with simply supported edges, Eq. (11.75) may be used with Fig. 11.44 for plates with clamped edges or for other edge conditions, by interpolation of Fig. 11.44.

11.17 PURE TENSION FIELD BEAMS[51]

The ultimate strength of thin webs in shear is much greater than the initial buckling strength. In the case of structural members which are not exposed to the airstream, such as wing spars, the shear webs may be permitted to wrinkle at a small fraction of their ultimate loads. To describe the manner in which loads are resisted by shear webs after buckling has occurred, it is convenient to consider a pure tension field beam in which the web buckles when the shearing forces are initially applied. Such a web never exists in practice, since even very thin webs have enough buckling resistance to affect the stress distribution appreciably.

The beam shown in Fig. 11.48a has concentrated flange areas which are

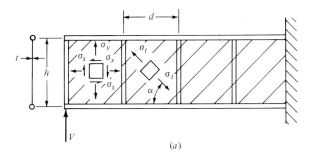

(a)

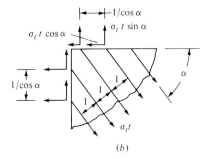

(b)

Figure 11.48

assumed to resist the entire beam bending moments. The beam web has thickness t and depth h between centroids of the flanges. The vertical stiffeners are spaced uniformly at a distance d along the span. The shear force V is constant for all cross sections. The shear flow at all points in the web is therefore equal to V/h, and the shear stress at all points τ_2 is $V/(th)$. If the web is shear-resistant, a web element at the neutral axis of the beam is stressed, as shown in Fig. 11.49. On the vertical and horizontal faces X and Y, the element resists only the shearing stresses τ and no normal stresses. The principal stresses σ_t and σ_c occur on planes at $45°$ to the horizontal, as shown in Fig. 11.49b. The magnitudes of the principal stresses are determined by the Mohr circle construction of Fig. 11.49c, from which $\sigma_t = \sigma_c = \tau$. If the beam web of Fig. 11.48a is assumed to be extremely flexible, it will not be capable of resisting the diagonal compressive stress. Then it will act as a group of parallel wires, inclined in the direction of the tension diagonal, or at an angle α of approximately $45°$, as shown. Such a group of wires cannot resist any of the beam bending moment, and it is customary to assume that every element in a tension field web resists the same stress as an element at the neutral axis. A web element for a pure tension field web is therefore stressed as shown in Fig. 11.50. The shearing stresses on the vertical and horizontal faces have the same values $\tau = V/(th)$ as for the shear-resistant web. These planes, X and Y, also have tensile stresses σ_x and σ_y, respectively, which are obtained from the Mohr circle construction in Fig. 11.50c. From the geometry of the circle, the lengths of lines QX and PY are $\tau/\sin\alpha$, and the lengths of lines PX and QY are $\tau/\cos\alpha$. The following stresses are obtained:

$$\sigma_x = \tau\cos\alpha \tag{11.77}$$

$$\sigma_y = \tau\tan\alpha \tag{11.78}$$

$$\sigma_z = \frac{\tau}{\sin\alpha\cos\alpha} = \frac{2\tau}{\sin 2\alpha} \tag{11.79}$$

The relationships expressed by Eqs. (11.77) to (11.79) may be obtained without the use of the Mohr circle construction by referring to Fig. 11.48b. The web of thickness t, which resists a maximum tensile stress σ_t, is assumed to be replaced by wires a unit distance apart which resist forces of σ_t. The vertical

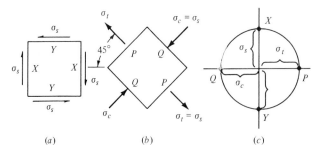

(a)   (b)   (c)

Figure 11.49

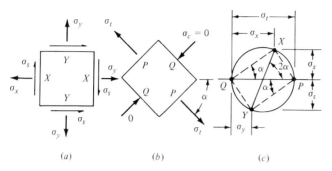

(a)   (b)   (c)

Figure 11.50

component of the wire tension is then $\sigma_t \sin \alpha$, and the horizontal component is $\sigma_t \cos \alpha$. Along a horizontal line through the wires, the spacing is $1/\sin \alpha$, corresponding to a horizontal web area of $t/\sin \alpha$. The web tension stress σ_y on the horizontal plane is obtained by dividing the vertical component of the wire tension $\sigma_t \sin \alpha$ by the web area $t/\sin \alpha$:

$$\sigma_y = \sigma_t \sin^2 \alpha \qquad (11.80)$$

The shearing stress τ on a horizontal plane is found by dividing the horizontal component of the wire tension $\sigma_t \cos \alpha$ by the web area $t/\sin \alpha$:

$$\tau = \sigma_t \sin \alpha \cos \alpha \qquad (11.81)$$

Equation (11.81) corresponds to Eq. (11.79). Similarly, a vertical line through the wires gives spacing of $1/\cos \alpha$ corresponding to a web area of $t/\cos \alpha$. The horizontal web stress σ_x is obtained by dividing the horizontal component of the wire tension by the area:

$$\sigma_x = \sigma_t \cos^2 \alpha \qquad (11.82)$$

Equations (11.77) and (11.78) may be obtained from Eqs. (11.80) to (11.82).

The tension field beam also differs from the shear-resistant beam in the manner in which stresses are transferred to the stiffeners, beam flanges, and riveted connections. The vertical stiffeners in a shear-resistant beam resist no compression load; they only divide the web into smaller unsupported rectangles and thus increase the web buckling stress as calculated from Eq. (11.43). In a tension field beam, however, the vertical web tension stresses σ_y tend to pull the beam flanges together, and this tendency must be resisted by compression forces in the stiffeners. Each stiffener must resist a compressive force P that is equal to the vertical tension force in the web for a length d equal to the stiffener spacing, as shown in Fig. 11.51a:

$$P = \sigma_y td = \frac{Vd}{h} \tan \alpha \qquad (11.83)$$

The vertical web tension stresses also tend to bend the beam flanges inward. The flanges act as continuous beams supported by the stiffeners. If the ends of the

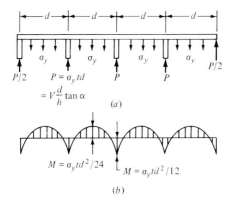

$M = \sigma_y td^2/24$ $M = \sigma_y td^2/12$

(b)   **Figure 11.51**

flanges are assumed to be fixed against rotation, the flange bending-moment diagram is as shown in Fig. 11.51b. At the stiffeners the bending moment is

$$M = \frac{\sigma_y td^2}{12} = \frac{Pd}{12} \tag{11.84}$$

Midway between the stiffeners the flange bending moment is

$$M = \frac{\sigma_y td^2}{24} = \frac{Pd}{24} \tag{11.85}$$

The direction of the bending moment is such that it produces tension on the outside of the flange at the stiffeners and on the inside of the flange between the stiffeners.

The horizontal components of the web stresses σ_x tend to pull the end stiffeners together with a force $\sigma_x th = V \cot \alpha$. This force is resisted equally by the two spar flanges, producing compression forces of $V(\cot \alpha)/2$, which must be superimposed on the forces M/h that result from beam bending.

The riveted connections for a shear-resistant web must be designed to resist a load $q = \tau_s t$ per unit length. In a tension field web connection, the horizontal riveted joints must resist shear flows q as well as tension forces of $q \tan \alpha$ per unit length in a perpendicular direction. Hence all horizontal riveted joints must resist forces of $q\sqrt{1 + \tan^2 \alpha} = q \sec \alpha$ per unit length. The vertical riveted joints at the ends of the beam or at web splices must resist shear flows of $q = \tau_s t$ and tensile forces of $\sigma_x t + q \cot \alpha$ per unit length. Thus, the joints must be designed for a load of $q\sqrt{1 + \cot^2 \alpha}$ or $q \csc \alpha$ per unit length. This force does not apply for connections between the web and intermediate stiffeners, since no appreciable load is transferred by this connection.

In earlier chapters, various shear-flow analyses are made in which the webs are assumed to resist pure shear. These analyses remain valid even though the webs are in tension field, since the tension stresses on the X and Y planes may be superimposed on the shearing stresses without affecting the shear-flow analysis.

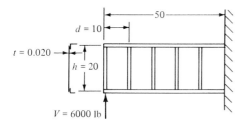

Figure 11.52

Example 11.8 The beam shown in Fig. 11.52 is assumed to have a pure tension field web. Draw free-body diagrams for the stiffeners and flanges, and plot the axial loads in the stiffeners and flanges. Assume $\alpha = 45°$.

SOLUTION The shearing stress on a horizontal or vertical plane of a web element is $\tau = V/(th) = 6000/(0.020 \times 20) = 15,000 \text{ lb/in}^2$. The running shear is $q = \tau_s t = 300 \text{ lb/in}$. The tension stresses σ_x and σ_y on these planes and the tension loads per inch also equal τ_s and q. The compression load on an intermediate stiffener, $P = Vd/h$, is 3000 lb. The stiffener at the left end has a compression load of $P/2$ and an additional compression force of 6000 lb applied at the lower end. Both beam flanges have compression loads of $V/2 = 3000$ lb at the left end. The beam flange loads vary linearly along the span. At the support, the flange loads from beam bending M/h are 6000 $\times \frac{60}{20} = 15,000$ lb. The compression flange resists a load $-M/h - V/2 =$

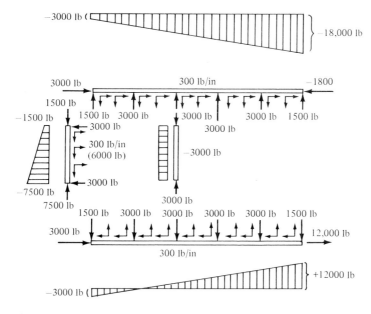

Figure 11.53

−18,000 lb, and the tension flange resists a load of $M/h - V/2 = 12,000$ lb. The free-body diagrams are shown in Fig. 11.53. All intermediate stiffeners resist the same loads.

11.18 ANGLE OF DIAGONAL TENSION IN WEB

This angle may be determined from the deflected geometry of the beam. framework, consisting of the flanges and stiffeners, has equal stiffness in resisting the horizontal tension σ_x and the vertical tension σ_y, the two tension stresses will be equal and α will be 45°. In practical beams, the flanges are much more rigid in resisting compression loads than are the stiffeners. The stiffeners deform in compression and permit the flanges to move together, while the stiffeners remain approximately the same distance apart. The horizontal web stress σ_x is therefore greater than σ_y, and the diagonal tension stress σ_y has an angle less than 45°. This angle may be determined from the deflected geometry of the beam.

The beam shown in Fig. 11.54 is initially horizontal, and it has a shearing deformation γ at all cross sections. Bending deflections are not considered here. The deformation γ is caused by axial elongations of the stiffeners and flanges and by the elongation of the web diagonal resulting from the diagonal tension stress. A section of the beam of length $h \cot \alpha$ and depth h is considered, and the value of α required to produce a minimum deformation γ is determined. All elongations are assumed positive as tension in the derivation, although the stiffeners and flanges will always be in compression and have negative elongations. The deformations of the length, $h \cot \alpha$, of the beam in Fig. 11.54 will be the same as those for the truss shown in Fig. 11.55, if the unit elongations in the horizontal, vertical, and diagonal directions are the same for the two structures.

The total elongation of a vertical stiffener is equal to the product of the unit elongation e_y and and the length h. This elongation causes a shearing deformation γ_1, as shown in Fig. 11.55a, which is obtained by dividing the total elongation by the radius $h \cot \alpha$:

$$\gamma_1 = \frac{he_y}{h \cot \alpha} = e_y \tan \alpha \tag{11.86}$$

The beam flanges have a unit elongation e_x, or a total elongation of $e_x h \cot$

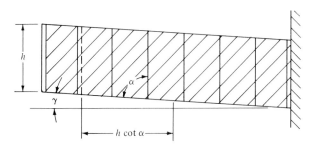

Figure 11.54

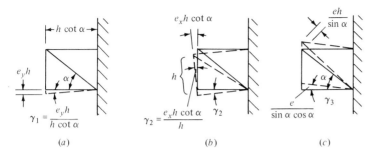

Figure 11.55

α in the horizontal length considered. The angular deformation γ_2 is obtained by dividing this deformation by the radius h, as shown in Fig. 11.55b:

$$\gamma_2 = e_x \cot \alpha \tag{11.87}$$

The diagonal strip of the web has a unit elongation e and a length $h/\sin \alpha$. The angular deformation γ_3 is obtained from the geometry of Fig. 11.55c:

$$\gamma_3 = \frac{e}{\sin \alpha \cos \alpha} \tag{11.88}$$

The total shearing deformation for the beam is the algebraic sum of the three components:

$$\gamma = -\gamma_1 - \gamma_2 + \gamma_3$$

Substituting from Eqs. (11.86) to (11.88) yields

$$\gamma = -e_y \tan \alpha - e_x \cot \alpha + \frac{e}{\sin \alpha \cos \alpha} \tag{11.89}$$

The angle of the web diagonal tension α will be such that the deformation γ is a minimum. Differentiating Eq. (11.89) and equating $d\gamma/d\alpha$ to zero yield

$$\tan^2 \alpha = \frac{e - e_x}{e - e_y} \tag{11.90}$$

where $e = \sigma_t/E$ is the unit strain along the web diagonal, e_x is the unit strain in the beam flanges resulting from the compression caused by the web tension σ_x, and e_y is the unit strain in the vertical stiffeners caused by the compression load P. All strains are positive for tension and negative for compression. The bending of the beam flanges and the slip in the riveted joints at the flanges have the same effect as an elongation e_y and may be included in the analysis.

The unit elongations used in Eq. (11.90) depend on the stresses, which in turn depend on the angle α. It is therefore necessary to solve this equation simultaneously with other equations obtained from the web stress conditions. For normal beam proportions, the flanges do not compress appreciably as a result of the tension field stress, and e_x may be assumed zero. The web diagonal strain e is σ_t/E, or $2\tau_s \csc 2\alpha/E$, from Eq. (11.79). The unit strain e_y is obtained as $-\tau_s td \tan \alpha/A_e E$, where A_e is the effective area of a vertical stiffener. Substituting these values into Eq. (11.90) gives

$$\cot^4 \alpha = \frac{td}{A_e} + 1 \tag{11.91}$$

The effective stiffener area is equal to the true stiffener area A if the stiffener consists of two members symmetrically attached on opposite sides of the web. Where a single stiffener is attached to only one side of the web, it is loaded in bending as well as compression. The compression load P has an eccentricity e measured from the center of the web to the centroid of the stiffener area. The combined bending and compression stress at a distance e from the neutral axis is

$$\sigma_c = \frac{P}{A} + \frac{Me}{I} = \frac{P}{A}\left[1 + \left(\frac{e}{\rho}\right)^2\right] = \frac{P}{A_e}$$

where ρ is the radius of gyration of the stiffener cross-sectional area and

$$A_e = \frac{A}{1 + (e/\rho)^2} \tag{11.92}$$

The differentiation of Eq. (11.89) may appear questionable, since the elongations e, e_x, and e_y are treated as constant with respect to α. Equation (11.91) also may be obtained by substituting values for the strains as functions of α into Eq. (11.89) before differentiation, which is a more rigorous mathematical procedure but yields the same angle α. Equation (11.90), however, is a general expression for the angle of the principal planes at a point in any structure with two-dimensional stress conditions when the strains e_x, e_y, and e are known. Lahde and Wagner[55] first applied this equation to the analysis of tension field webs. Langhaar[67] expressed the strain e in terms of a known distortion γ and equated $de/d\alpha$ to zero in order to find the angle α for the maximum or principal strain. The angle α thus obtained is equal to that yielded by Eq. (11.90). Langhaar also expressed the total strain energy as a function of α and equated the derivative of the strain energy to zero in order to find the angle α, which yielded a minimum of the total strain energy. This also gave the same result as that obtained from Eq. (11.90). If flange bending and other deformations are considered, the value of e_y is greater than that used in obtaining Eq. (11.91).

11.19 SEMITENSION FIELD BEAMS

In Sec. 11.18 we assume that the beam web is perfectly flexible and is not capable of resisting any diagonal compressive stress. In practical beams, the webs resist some diagonal compressive stress after buckling, and thus they act in an intermediate range between shear-resistant webs and pure tension field webs. Such beams are termed *semitension field beams*, *partial-tension field beams*, or *incom-*

pletely developed diagonal tension field beams. The pure tension field theory is conservative for the design of all parts of a practical beam, but may yield stiffener loads or flange bending moments which are as much as 5 times the true values. Hence a more accurate theory is necessary for design purposes.

The theory of pure tension field beams was first published by Wagner in 1929. Since then, many investigators have studied the problem of semitension field beams. Lahde and Wagner[55] published empirical data in 1936 which were based on strain measurements of buckled rectangular sheets. These data provided information for the practical design of beams, but the test points had considerable scatter because of the difficulty in making strain measurements of buckled sheet. Many aircraft manufacturers conducted tests and developed empirical design formulas, but usually, one particular type of beam has been tested, and the equations must be used with caution in designing beams of different materials or proportions different from those on which the tests were conducted. The most extensive test program has been conducted by the NACA under the direction of Paul Kuhn. Kuhn et al.[51] and Peterson[52] measured strains in the vertical stiffeners of a large number of beams and derived empirical equations from these measurements. The stiffener stresses supply information required for the stiffener design and for the design of the flanges to resist secondary bending. A theoretical analysis of the stresses in a buckled rectangular sheet has been made by Levy et al.[53,54] While some simplifying assumptions are made in the analysis, it provides valuable information regarding the stress distribution in the web. The analysis of Levy et al. shows that the stress conditions vary considerably at different points in the web and that any practical analysis in which the same stress conditions are assumed at all points will have some discrepancies with observed test conditions.

In the analysis of semitension field beams by Kuhn et al., it is assumed that part of the shear load kV is resisted by pure tension field action and that the remaining load $(1 - k)V$ is resisted by the beam acting as a shear-resistant beam. All points of the web are assumed to have the same stress distribution, except for the web adjacent to the vertical stiffeners. This portion of the web is riveted to the stiffeners and does not wrinkle. The stresses in the web may be found by multiplying the values shown for a shear-resistant web in Fig. 11.49 by $1 - k$ and those shown for a pure tension field web in Fig. 11.50 by k and then superimposing them. The values shown in Fig. 11.56c represent the total stresses on horizontal and vertical planes and are found by superimposing the conditions shown in Fig. 11.56a and b. The angle α is obtained from Eq. (11.90) with sufficient accuracy. There is a conservative error involved in using Eq. (11.90) for partial tension field webs, since this equation yields the angle of principal stress. The principal tension stress for the element of Fig. 11.56c has an angle which is between the 45° angle for the principal stress of the element of Fig. 11.56a and the angle α for the element of Fig. 11.56b.

The diagonal tension factor k is given by the empirical equation of Kuhn et al.:

$$k = \tanh\left(0.5 \log_{10} \frac{\tau}{\tau_{cr}}\right) \tag{11.93}$$

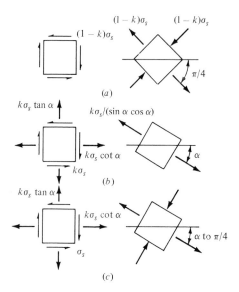

(a)

(b)

(c)

Figure 11.56

which is plotted in Fig. 11.57. The angle α is obtained from Eq. (11.90) after functions of k, α, and A_e are substituted for the strains. Values of $\tan \alpha$ are plotted as functions of k and td/A_e in Fig. 11.58. Since Eq. (11.90) must be solved by trial for $\tan \alpha$, it is much more convenient to use Fig. 11.58 than to solve the equation for each particular case.

The stress conditions for any web element are known after k and α are found and are as shown in Fig. 11.56c. The stiffener compression forces and the flange bending moments are proportional to the vertical component of the web tensile stress σ_y, which is shown in Fig. 11.56:

$$\sigma_y = k\tau \tan \alpha \tag{11.94}$$

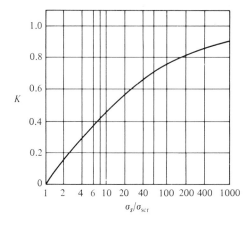

Figure 11.57

The stiffener compression load P is obtained as follows:

$$P = \sigma_y td = k\tau td \tan \alpha \tag{11.95}$$

The flange bending moments are obtained from the stress σ_y, as in Eqs. (11.84) and (11.85), for pure tension field webs, but the values of σ_y and P are smaller for semitension field webs:

$$M = \frac{\sigma_y td^2}{12} = \frac{Pd}{12} \qquad \text{at stiffeners} \tag{11.84}$$

$$M = \frac{\sigma_y td^2}{24} = \frac{Pd}{24} \qquad \text{between stiffeners} \tag{11.85}$$

The compression load on a vertical stiffener is resisted by the stiffener and by the effective web that is riveted to the stiffener. If sufficient rivets are provided that the web wrinkles do not extend through the riveted joint at the stiffener, as in the customary construction, the web must have the same vertical compression strain and approximately the same compression stress as the stiffener at the rivet line. An effective width of web equal to $0.5(1 - k)d$ is assumed by Kuhn et al. to act with the stiffener, and the stiffener compression stress then has the following value:

$$\sigma_c = \frac{P}{A_e + 0.5(1 - k)td} \tag{11.96}$$

Since the values of k are obtained empirically from measurements of σ_c, an approximate expression for the effective width of web will yield an accurate value of σ_c if the same effective width is assumed in calculating k from experimentally determined values of σ_c. It seems probable that the true effective widths of the web are less than those assumed and that the empirical values of k thus yield conservative values for σ_y and the flange bending moments.

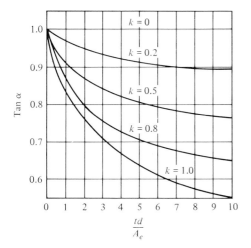

Figure 11.58

The riveted joints between the webs and the beam flanges must be designed for the resultant of the shearing stress τ and the tension stress σ_y or for a running load of $\sqrt{\tau^2 + (\sigma_y)^2}\ t$. Similarly, vertical web splices must be designed for a running load of $\sqrt{\tau^2 + (\sigma_x)^2}\ t$. The rivets connecting the vertical stiffeners to the beam flanges should be designed to transfer the load $P_u = A_e \sigma_c$, according to the above theory. In actual beams, frequently it is impractical to provide this strength. The theoretical analysis by Levy et al. and many tests indicate that the stiffener load decreases near the end of the stiffener and that as much as half of this load is transferred to the web near the end of the stiffener rather than to the beam flange. In practical beams, it is customary to provide a total strength in the rivets connecting the stiffener to the flange and the rivets connecting the end of the stiffener to the web to resist the load P_u. The stiffener-web rivets which are assumed to transfer part of this load P_u are spaced as close to the end of the stiffener as possible.

The allowable strength for beam webs has sometimes been obtained by equating the calculated diagonal tension stresses to the allowable tension stress for the web material, with empirical corrections for rivet holes and various stress concentration factors. More accurate web strength predictions can probably be obtained by equating the total web shearing stress τ to an allowable stress σ_{sw} obtained from tests of beams of common proportions. These allowable stresses are plotted in Fig. 11.59 as functions of τ/τ_{cr}. The curves of Fig. 11.59 were obtained by analyzing data from extensive tests of semitension field beams. The tests were conducted under the supervision of S. A. Gordon of the Glenn L. Martin Company.

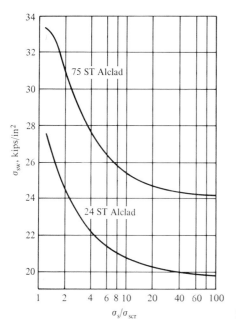

Figure 11.59

It is assumed that the web-flange rivet spacing is in the normal range of 3 to 5 times the rivet diameter. A smaller rivet spacing reduces the net area excessively, while a very large spacing may permit the web wrinkles to extend through the rivet lines. The allowable web stresses often are plotted as functions of the sheet thickness, with the heavier sheet gages resisting higher stresses. This practice is permissible because stiffener spacings usually are kept to a maximum of about 8 in, even for very heavy sheets, and the thicker webs therefore have lower ratios of τ/τ_{cr}. For geometrically similar webs, however, the allowable stress would be independent of the web thickness. Where beam flanges and web stiffeners are attached symmetrically to both sides of a web, the edge of the web is better supported and resists higher stresses than those given in Fig. 11.59 for flanges attached to only one side of the web.

Example 11.9 Determine the margin of safety for the web intermediate stiffeners, and riveted joints of the beam shown in Fig. 11.60. The web is, 2024S-T Alclad sheet, and the flanges and stiffeners are 2024S-T extrusions.

SOLUTION The buckling stress for the web in shear is obtained from Eq. (11.69) and Fig. 11.44. The web dimensions for computing buckling stresses are measured between rivet lines, as $a = 15$ and $b = 8$ in. By entering Fig. 11.44 with $a/b = 1.875$, the values $K = 10.3$ for clamped edges and $K = 5.9$ for simply supported edges are obtained. An average value, $K = 8.1$, is used because of the restraining effects of the flanges and stiffeners. It is given that $E = 9,700,000 \text{ lb/in}^2$. Substituting in Eq. (11.69) yields

$$\tau_{cr} = KE\left(\frac{t}{b}\right)^2 = 8.1 \times 9,700,000 \times \left(\frac{0.032}{8}\right)^2 = 1260 \text{ lb/in}^2$$

The shear stress is

$$\tau = \frac{V}{ht} = \frac{10,000}{16 \times 0.032} = 19,500 \text{ lb/in}^2$$

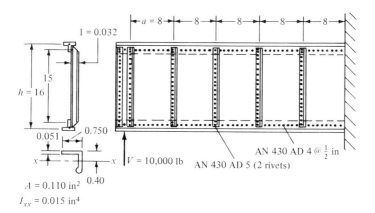

Figure 11.60

The web depth h is always the distance between centroids of the flange areas, rather than the distance between rivet lines, when used in shear stress or shear-flow calculations.

The effective stiffener area is computed from Eq. (11.92), since the stiffeners are attached to only one side of the web and are loaded eccentrically. The stiffener radius of gyration is $\rho = \sqrt{I/A} = \sqrt{0.015/0.110} = 0.37$ in. The stiffener eccentricity is $e = 0.40 + t/2 = 0.416$ in. The effective stiffener area is obtained from Eq. (11.92):

$$A_e = \frac{0.11}{1 + (0.416/0.37)^2} = 0.049 \text{ in}^2$$

The ratio of web area to effective stiffener area is

$$\frac{td}{A_e} = \frac{8 \times 0.032}{0.049} = 5.22$$

This ratio and τ/τ_{cr} determine the stress distribution in the beam:

$$\frac{\tau}{\tau_{cr}} = \frac{19,500}{1260} = 15.5$$

From Fig. 11.57, $k = 0.53$, and from Fig. 11.58, $\tan \alpha = 0.79$. From Fig. 11.59, the allowable web stress is $\tau_w = 20,400$ lb/in^2. The margin of safety for the web in shear is

$$\text{Web MS} = \frac{\tau_w}{\tau} - 1 = \frac{20,400}{19,500} - 1 = 0.04$$

The vertical component of the web tension σ_y is obtained as follows:

$$\sigma_y = \tau k \tan \alpha = 19,500 \times 0.53 \times 0.79 = 8160 \text{ lb/in}^2$$

The stiffener compression load is obtained from Eq. (11.93):

$$P = \sigma_y td = 8160 \times 0.032 \times 8 = 2090 \text{ lb}$$

The load per inch in the web-flange rivets is

$$q_r = \sqrt{\tau^2 + \sigma_y^2}\, t = \sqrt{19,500^2 + 8160^2} \times 0.032 = 660 \text{ lb/in}$$

The allowable load for one $\frac{1}{8}$-in A17S-T rivet is 375 lb shear and 477 lb bearing on a 0.032 gage sheet, as obtained from ANC-5. For the $\frac{1}{2}$-in spacing, the allowable rivet load is 750 lb/in, as determined from the shear strength:

$$\text{Rivet MS} = \frac{750}{660} - 1 = 0.14$$

The maximum compression stress in the stiffener, resulting from the eccentric compression load, is obtained from Eq. (11.96):

$$\sigma_c = \frac{2090}{0.049 + 0.5(1 - 0.53) \times 0.032 \times 8} = 19,200 \text{ lb/in}^2$$

This stress exists in the leg attached to the web and decreases to zero or a tension stress in the outstanding leg. The allowable stress is the compression crippling stress for a leg with $b = 0.70$, $t = 0.051$, or $b/t = 13.7$, as computed by the methods of Sec. 11.9. The allowable crippling stress is approximately 22,000 lb/in².

$$MS = \frac{22,900}{19,200} - 1 = 0.15$$

The allowable compression stress in the stiffener depends on many factors and is only roughly approximated here. The attached leg of a stiffener usually is made at least one gage thicker than the web in order to prevent a forced crippling from the web wrinkles. Since the outstanding leg is not highly stressed in compression, it supplies torsional rigidity, and the attached leg probably can be assumed to have one side clamped and the other side free in computing the crippling stress.

The two rivets connecting the stiffener to the flange are $\frac{5}{32}$-in A17S-T rivets with a single shear strength of 596 lb each. They must transfer the compression load in the stiffener:

$$P_u = \sigma_c \, A_e = 19,200 \times 0.049 = 940 \text{ lb}$$

The remaining part of the force P is resisted by compression in the effective sheet and is not transferred by the rivets. The margin of safety of the two stiffener-flange rivets is

$$MS = \frac{2 \times 596}{940} - 1 = 0.27$$

PROBLEMS

11.1 A long column has an initial curvature defined by the equation $y_0 = a \sin (\pi x/L)$. Derive an equation for the additional deflection y by integrating Eq. (11.6). Show that the center deflection δ is defined by

$$\delta + a = \frac{a}{1 - P/P_{cr}}$$

where P_{cr} is defined by Eq. (11.11). Compare values of $\delta + a$ from this equation to those obtained from Eq. (11.14) for $P/P_{cr} = 0.2, 0.4, 0.6, 0.8, 0.9, 0.95$, and 1.00.

11.2 From the expressions $R_a = 1 - KB$ and $R_a = 1/B^2$ derive Eqs. (11.24) and (11.31) by equating the slopes of the two curves at their point of tangency. Find the coordinates of the point of tangency.

11.3 From the expressions $R_a = 1 - KB^2$ and $R_a = 1/B^2$ derive Eqs. (11.23) and (11.29) by equating the slopes of the two curves at their point of tangency. Find the coordinates of the point of tangency.

11.4 Find the column loads which may be resisted by round steel tubes heat-treated to an ultimate tensile strength of 180,000 lb/in², with the ends welded before heat treatment. The dimensions are:

Tube size	L	
1×0.05	20	2
$1\frac{1}{2} \times 0.049$	20	
$1\frac{1}{2} \times 0.065$	40	1
$1\frac{1}{4} \times 0.05$	30	2

11.5 Repeat Prob. 11.4 for steel heat-treated to an ultimate tensile strength of 150,000 lb/in².

11.6 Repeat Prob. 11.4 for steel heat-treated to an ultimate tensile strength of 125,000 lb/in².

11.7 The skin of the upper side of an airplane wing is made of 2024-T6 Alclad. The stringer spacing is 5 in, and the rib spacing is 20 in. Assuming the edges to be simply supported, find the compression buckling stress for skin gages of (a) 0.020, (b) 0.032, (c) 0.040, and (d) 0.064 in.

11.8 Repeat Prob. 11.7, assuming the values of K to be the average of values for simply supported edges and clamped edges.

11.9 Calculate points on the curve for $m = 1$ of Fig. 11.20 for values of a/b of 0.25, 0.33, 0.5, 1, 2, 3, and 4. Calculate points on the curve for $m = 2$ for values of a/b of 0.50, 0.66, 1, 2, 4, 6, and 8. Note the similarity between the two curves, and devise a system of coordinates which would show all the curves of Fig. 11.20 as a single curve.

11.10 Calculate the compression buckling stress for a sheet with $a = 8$ in, $b = 4$ in, and $t = 0.156$ in. The tangent modulus column curve for the material is shown in Fig. 11.29.

    (a) Assume all four edges are simply supported.
    (b) Assume all four edges are clamped.
    (c) Assume the ends are simply supported and the sides are free.

11.11 Solve Prob. 11.10, using a dimensionless buckling curve for $n = 10$. Assume $E = 10,700,000$ and $\sigma_1 = 37,000$ lb/in².

11.12 Find the buckling stress for a column with (a) both ends fixed, (b) both ends free, (c) one end fixed and one end free.

11.13 Use the Rayleigh-Ritz method to find the buckling load for a column with both ends pinned.

11.14 The skin on a fuselage is supported by stringers which are spaced at 5 in and by rings spaced at 20 in. Assume $E = 10^7$ lb/in² and an average between simply supported and clamped-edge conditions. Find the shear buckling stresses for the flat sheet if (a) $t = 0.020$, (b) $t = 0.032$, (c) $t = 0.040$, and (d) $t = 0.064$ in.

11.15 Plot the axial loads in the flanges and stiffeners of a pure tension field beam similar to that shown in Fig. 11.48 with $h = 10$ in, $d = 10$ in, and $V = 10,000$ lb. Compute the flange bending moments and the load per inch on all rivets. Assume $\alpha = 45°$.

11.16 Solve Example 11.5 by the Needham method and the Gerard method. Compare your results.

11.17 Solve Example 11.6 by the Needham method and the Gerard method. Compare your results.

TWELVE

JOINTS AND FITTINGS

12.1 INTRODUCTION

A flight vehicle structure is manufactured from many parts. These parts are made from sheets, extruded sections, forgings, castings, tubes, or machined shapes, which must be joined to form subassemblies. The subassemblies must then be joined to form larger assemblies and then finally assembled into a complete flight vehicle. Many parts of the completed vehicle must be arranged so that they can be disassembled for shipping, inspection, repair, or replacement and are usually joined by bolts.

In order to facilitate assembly and disassembly, it is desirable for such connections to contain as few bolts as possible. For example, a semimonocoque metal wing usually resists bending stresses in numerous stringers and sheet elements distributed around the periphery of the wing cross section. The wing cannot be made as one continuous riveted assembly from tip to tip, but usually must be spliced at two or more cross sections. Often these splices are designed so that four bolts transfer all the loads across the splice. These bolts connect members called *fittings*, which are designed to resist the high concentrated loads and to transfer them to the spars, from which the loads are distributed to the sheet and stringers. The entire structure for transferring the distributed loads from the sheet and stringers outboard of the splice to a concentrated load at the fitting and then distributing this load to the sheet and stringers inboard of the splice is considerably heavier than the continuous structure which would be required if there were no splice.

Many uncertainties exist concerning the stress distribution in fittings. Manufacturing tolerances are such that bolts never fit the holes perfectly, and small variations in dimensions may affect the stress distribution. An additional margin of safety of 15 percent for military airplanes and 20 percent for civil airplanes is used in the design of fittings. A common procedure is to multiply the design loads by a fitting factor of 1.15 or 1.20 before the stresses are calculated. This fitting factor must be used in designing the entire fitting, including the riveted, bolted, or welded joint attaching the fitting to the structural members. The fitting factor need not be used in designing a continuous riveted joint, although the stress distribution in such a joint is also indeterminate.

The allowable stresses for rivets are rather conservative to account for such uncertainty.

12.2 BOLTED OR RIVETED JOINTS

Bolted or riveted joints must be investigated for four types of failure: bolt or rivet shear, as shown in Fig. 12.1; bearing, as shown in Fig. 12.2; tear-out, as shown in Fig. 12.3; and tension, as shown in Fig. 12.4. The true stress distribution is rather complex and is discussed later. It is customary to assume a simple uniform or

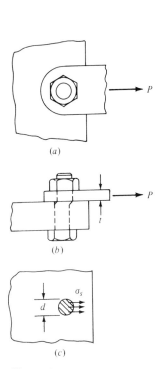

Figure 12.1

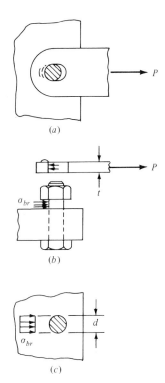

Figure 12.2

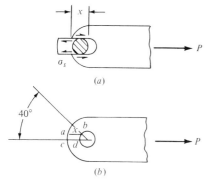

(a)

(b)

Figure 12.3

average stress distribution in all cases, and the allowable stresses which are used in design are also average stresses which have been obtained from tests of similar joints.

It is therefore possible to predict the strength of a joint with an accuracy of a few percent, although the true maximum stresses may be 3 or 4 times as much as the average stresses. The average stress for any of the four types of failure is

$$\sigma = \frac{P}{A} \tag{12.1}$$

where σ is the average stress, P is the load, and A is the area of the cross section on which failure may occur. The margin of safety (MS) is found from

$$MS = \frac{\sigma_a}{\sigma} - 1 \tag{12.2}$$

where σ_a is the allowable stress and the stress σ is obtained from the load P, which includes the safety factor of 1.5 and usually the fitting factor of 1.15 or 1.2 as well. If this fitting factor is included in the stress σ, the margin of safety should be zero or a small positive value. Some designers may not include the fitting factor in the stress σ, and thus they must show a minimum margin of safety of 0.15 or 0.20 from Eq. (12.2). In any analysis, it should be clearly stated whether the fitting factor is included in the margin of safety. The symbol σ_a always represents an allowable stress, and the symbol σ represents a calculated stress. A subscript is used to designate the type of stress; that is, τ_a and τ are shearing stresses, σ_{abr} and σ_{br} are bearing stresses, σ_{at} and σ_t are tensile stresses, σ_{ac} and σ_c are compression stresses, and σ_{ab} and σ_b are bending stresses.

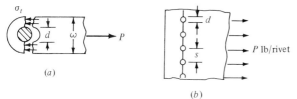

(a)

(b)

Figure 12.4

For investigating the shear strength of a bolt or rivet, the area to be used in Eq. (12.1) is the area of the bolt or rivet cross section, or $A = \pi d^2/4$, where d is the diameter of the bolt or rivet. See Fig. 12.1. The shearing stress is then obtained from Eq. (12.1):

$$\tau = \frac{4P}{\pi d^2} \tag{12.3}$$

In Figs. 12.1 through 12.4, the bolt is shown to be in single shear and one plate is assumed to be rigid in bending, so that the forces on the thin plate are in static equilibrium. Hence the bolt would resist a bending moment $Pt/2$ at the cross section subjected to shear. It is shown later that this bending moment on the bolt does not exist in most actual single-shear connections, and it is customary to disregard this bolt bending moment when the two plates are clamped together by the bolt. When a washer or a filler plate is used between the two stressed plates, the bolt bending must be considered.

The bearing failure of a riveted or bolted joint usually consists of an elongation of the hole in the plate, as shown in Fig. 12.2a. The allowable bearing stress usually depends on the permissible elongation of the hole. For riveted joints, the allowable bearing stress is determined by arbitrarily specifying a hole elongation equal to a certain percentage of the rivet diameter. The bearing failure is somewhat similar to the tear-out failure shown in Fig. 12.3, and the allowable bearing stress for rivets is reduced when the rivets are too close to the edge of the sheet. The bearing stress is assumed to be uniformly distributed over an area $A = td$, as shown in Fig. 12.2. By substituting this area into Eq. (12.1), the equation for the assumed average bearing stress is obtained:

$$\sigma_{br} = \frac{P}{td} \tag{12.4}$$

Bolt holes always must be slightly larger than the bolt diameter. If the joint is subjected to shock or vibrational loading, as in a landing-gear member, there is a much greater tendency for a bolt hole to elongate than when the joint resists only static loading. Similarly, when relative rotation of the two parts occurs, the bolt hole is more likely to become enlarged. In such cases, the bearing stress must be low in order to prevent frequent replacement of the bolt or the hole bushing. The licensing agencies therefore specify that a bearing factor of 2.0 or more be used in obtaining the bearing stress when a bolted joint is subject to relative rotation under design loads or to shock or vibration loads. This bearing factor is used in place of the fitting factor, *not* in addition to the fitting factor.

A tear-out failure of a bolt or rivet hole is shown in Fig. 12.3. The plate material fails in shear on the areas $A = 2xt$, and the tear-out stress is found from

$$\tau = \frac{P}{2xt} \tag{12.5}$$

The distance x is obtained as length ab in Fig. 12.3, but it is conservative to use length *cd*, which is easier to calculate. It is seldom necessary to calculate the tear-out stresses for riveted joints in a sheet of the type shown in Fig. 12.4. From practical considerations, it is desirable to keep the distance from the center of the rivets to the edge of the sheet equal to at least two diameters of the rivet, and there is no danger of tear-out with this edge distance.

A riveted or bolted joint must be investigated for a possible tension failure through the bolt or rivet holes, as shown in Fig. 12.4. The tension stress is assumed to be uniformly distributed over the area $A = (\omega - d)t$ for the bolted fitting shown in Fig. 12.4a.

$$\sigma_t = \frac{P}{(\omega - d)t} \tag{12.6}$$

For the riveted joint shown in Fig. 12.4b, the tension stress is

$$\sigma_t = \frac{P}{(s - d)t} \tag{12.7}$$

where P = load per rivet
  s = rivet spacing
  d = rivet diameter
  t = sheet thickness

Example 12.1 The fitting shown in Fig. 12.5 is made of a 1014 aluminum forging, for which $\sigma_{at} = 65,000$, $\tau_a = 39,000$, and $\sigma_{abr} = 98,000$ lb/in². The bolt and bushing are made of steel for which $\sigma_{at} = 125,000$, $\tau_a = 75,000$, and $\sigma_{abr} = 175,000$ lb/in². The fitting resists limit or applied loads of 15,000-lb compression and 12,000-lb tension. A fitting factor of 1.2 and a bearing factor of 2.0 are used. Find the margins of safety for the fitting for various types of failure.

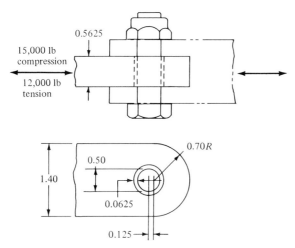

Figure 12.5

SOLUTION The design or ultimate fitting loads are obtained by multiplying the loads given by a safety factor of 1.5 and a fitting factor of 1.2:

Design fitting loads:

$$15,000 \times 1.5 \times 1.2 = 27,000 \text{ lb} \qquad \text{compression}$$

$$12,000 \times 1.5 \times 1.2 = 21,600 \text{ lb} \qquad \text{tension}$$

The bearing of the bolt on the bushing is investigated by using the bearing factor of 2.0 in place of the fitting factor of 1.2:

Design bearing loads:

$$15,000 \times 1.5 \times 2.0 = 45,000 \text{ lb} \qquad \text{compression}$$

$$12,000 \times 1.5 \times 2.0 = 36,000 \text{ lb} \qquad \text{tension}$$

The bolt is in double shear; therefore one-half of the 27,000-lb load must be resisted by each cross section of the bolt in shear. From Eqs. (12.3) and (12.2):

$$\tau = \frac{4 \times 13,500}{\pi(0.5)^2} = 68,600 \text{ lb/in}^2$$

and
$$MS = \frac{75,000}{68,600} - 1 = 0.09 \qquad \text{includes fitting factor}$$

The bearing stress also is calculated from the larger of the loads for tension and compression. From Eqs. (12.4) and (12.2), the bearing of the bolt on the bushing is investigated:

$$\sigma_{br} = \frac{45,000}{0.5625 \times 0.5} = 160,000 \text{ lb/in}^2$$

$$MS = \frac{175,000}{160,000} - 1 = 0.09 \qquad \text{includes bearing factor}$$

For bearing of the bushing on the forging, one need only use the fitting factor, because the bushing fits tightly in the hole:

$$\sigma_{br} = \frac{27,000}{0.5625 \times 0.625} = 76,800 \text{ lb/in}^2$$

$$MS = \frac{98,000}{76,800} - 1 = 0.29 \qquad \text{includes fitting factor}$$

The tear-out of the bolt hole is investigated first by assuming that the length x shown in Fig. 12.3 is equal to cd rather than ab:

$$cd = 0.70 + 0.125 - 0.3125 = 0.5125 \text{ in}$$

The tension load must be used in calculating the tear-out stress, since the compression load produces no stress on this cross section. From Eqs. (12.5)

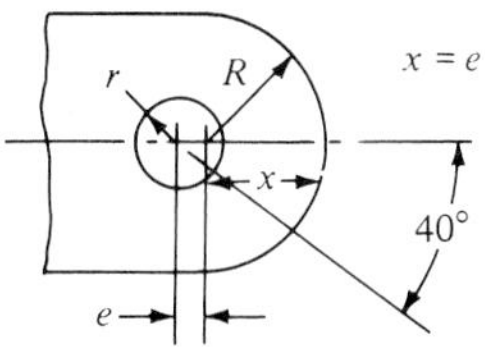

$$x = e + R \left[\sqrt{t - \frac{r^2}{R^2} \sin^2 40°} - \frac{r}{R} \cos 40° \right]$$

40°

e

Figure 12.6

and (12.2),

$$\tau = \frac{21{,}600}{2 \times 0.5125 \times 0.5625} = 37{,}400 \ \text{lb/in}^2$$

and $\qquad \text{MS} = \dfrac{39{,}000}{37{,}400} - 1 = 0.04 \qquad$ includes fitting factor

A more accurate value of the distance x may be calculated from the equation given in Fig. 12.6. The term in brackets may be plotted for various values of r/R in order to reduce the labor of the calculations, where it is necessary to repeat such calculations frequently. For $R = 0.7$, $r = 0.3125$, $e = 0.125$, and $x = 0.562$,

$$\tau = \frac{21{,}600}{2 \times 0.562 \times 0.5625} = 34{,}000 \ \text{lb/in}^2$$

and $\qquad \text{MS} = \dfrac{39{,}000}{34{,}200} - 1 = 0.14 \qquad$ includes fitting factor

The tension stress through the bolt hole is obtained from Eq. (12.6):

$$\tau = \frac{21{,}600}{(1.4 - 0.625)0.5625} = 49{,}600 \ \text{lb/in}^2$$

$$\text{MS} = \frac{65{,}000}{49{,}600} - 1 = 0.13 \qquad \text{includes fitting factor}$$

12.3 ACCURACY OF FITTING ANALYSIS

The ultimate strength of a fitting usually may be calculated accurately by the methods previously described. True stress distribution at stresses below the elastic limit often is much different from the assumed distribution. Before the ultimate strength of the fitting is reached, however, the material yields and the stresses are redistributed so that they usually approach the assumed stress distribution. Because of this plastic yielding of the material and because the allowable shear and bearing stresses are obtained from tests on specimens similar to those in the actual structure, it is possible to achieve accurate calculated strengths by means of inaccurate assumptions. While the conventional methods are satis-

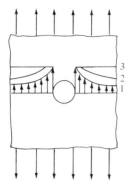

Figure 12.7

factory for making design calculations for fittings, the designer must keep in mind the true stress distribution and must avoid conditions of high local stress wherever possible.

One very common case of stress concentration, shown in Fig. 12.7, is that of a tension plate containing a circular hole. For small loads, the tensile stress at the side of the hole is 3 times the average tensile stress in the plate, as indicated by line 1. As the loads increase, the stress at the side of the hole exceeds the elastic limit, and local plastic yielding of the material occurs near the hole. The stresses near the hole remain almost constant at the yield point, while the stresses at a distance from the hole increase with the load, as indicated by line 2. Before failure occurs, yielding has progressed over the entire width of the plate, and the stress is constant over the net section, as shown by curve 3. Thus the customary assumption that failure occurs at a load equal to the product of the ultimate tensile stress and the net area is accurate for ductile materials. Brittle materials, which fail suddenly with no plastic elongation, should never be used for aircraft structural members.

Stress concentrations are much more serious in engine parts on which the loads are repeated millions of times than in airframe parts on which the maximum loads occur only a few times during the life of the airplane. In airframe design, usually it is safe to consider only average stresses and to neglect stress concentrations, although certain unfavorable conditions, such as radial cracking of sheet around holes when the holes are press-countersunk, may lead to service failures from stress concentrations.

The double-shear connection shown in Fig. 12.8 is assumed to resist one-half

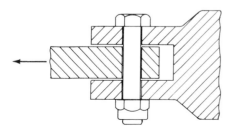

Figure 12.8

the load by shear on each bolt cross section. Manufacturing tolerances may permit the hole in the lower lug to be slightly to the left of the hole in the upper lug, as shown. For small loads, then, the entire load is resisted by shear on the upper cross section. As the load is increased, the parts deflect so that the lower end of the bolt is also bearing on the lug, but the upper lug continues to resist more than one-half the load. The fitting factor is intended to account for such eccentric loading conditions; in this case, the use of a fitting factor of 1.2 is equivalent to the assumption that one side of the fitting may resist 60 percent of the total ultimate load.

Most of the bolted and riveted joints in aircraft structures are single-shear joints. For the joints shown in Figs. 12.1 and 12.2, we assume that one member is rigid, and only the forces acting on the other member are considered. For this assumed loading, the bolt resists a bending moment of $Pt/2$ and the heavy member resists a larger bending moment. The usual single-shear joint has both members of comparable size. At first it might appear that each of the members shown in Fig. 12.9 could be treated in the same manner as the upper member of Fig. 12.2b. In fact, many textbooks show the forces as in Fig. 12.9, and this assumed stress distribution is customary and satisfactory for design. The forces shown in Fig. 12.9 cannot be in equilibrium, however, because there is an unbalanced moment Pt on the plates in Fig. 12.9a and a similar unbalanced moment on the pin in Fig. 12.9b. The correct stress distribution must be as shown in Fig. 12.10. For the forces P to balance, they must act on the same line, as shown in Fig. 12.10a. The stresses in the plate are no longer P/A, but must also include stresses from the bending moment $Pt/2$. If the plate width is b, the plate stress is $P/A \pm My/I$:

$$\sigma = \frac{P}{bt} \pm \frac{Pt}{2} \frac{6}{bt^2} = \frac{P}{bt} \pm \frac{3P}{bt} \tag{12.8}$$

At the inside faces of the plates, the tensile stress from Eq. (12.8) is $4P/A$, and at the outside faces the compressive stress is $2P/A$, as shown.

In order for the pin to be in equilibrium under the bearing stresses, it must bear on opposite corners of the hole, as shown in Fig. 12.10b. The most optimistic assumption of bearing stresses is the straight-line assumption shown in Fig. 12.10c, which yields maximum bearing stresses $4P/(tb)$ at the inside corner and $2P/(tb)$ at the outside corners. If the pin does not fit tight in the hole, the

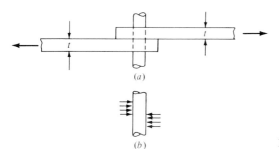

(a)

(b)

Figure 12.9

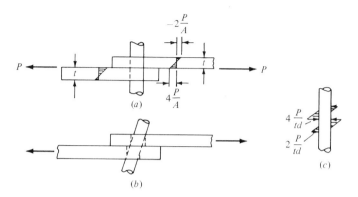

(a)

(b)

(c)

Figure 12.10

bearing stresses must be higher than those assumed. Thus, for the single-shear pin joint between plates of equal thickness, the maximum plate tension stresses and the bearing stresses are both 4 times the values assumed in Figs. 12.2 and 12.9.

The bending moment in the pin of Fig. 12.10 is zero at the cross section of maximum shear, and the maximum pin bending moment is $\frac{4}{27}Pt$ at a cross section a distance $t/3$ from the inside of the plates.

The ultimate strength of conventional riveted and bolted joints approaches that assumed in the original simple analysis, because of the clamping action of the rivet heads or bolt heads. For a riveted joint between two sheets in tension, the bending and tension stresses in the sheets exceed the elastic limit, and the sheets deform as shown in Fig. 12.11. The two forces P are almost in the center plane of the sheets, as shown in Fig. 12.11a, as the ultimate strength is approached. The moment Pt of the bearing forces on the rivet is balanced by the moment of clamping forces under the head of the rivet, as shown in Fig. 12.11b. These forces on the rivet head have a moment arm D which is slightly less than the diameter of the rivet head. The bending moment in the rivet shank varies from $Pt/2$ at each end of the shank to 0 at the plane of rivet shear. After plastic

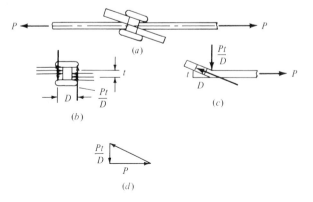

(a)

(b)

(c)

(d)

Figure 12.11

yielding has progressed in the sheet, the bending stresses shown in Fig. 12.10a are eliminated, and the sheet is in almost uniform tension at all points. The angular change in the sheet is arctan (t/D), as shown in Fig. 12.11c, and the force exerted by the rivet head on the sheet is just sufficient to keep the resultant tension in the center plane of the sheet. The force triangle at the bend in the sheet is represented by Fig. 12.11d. The angular change in the sheet is exaggerated in Fig. 12.11 For a rivet shank diameter of 4 times the sheet thickness and a rivet head diameter D of twice the shank diameter, the angle is arctan (t/D) or arctan $\frac{1}{8}$.

Where a tension joint has two lines of rivets, the deformation is as shown in Fig. 12.12. If the tension stresses in the sheet were uniform at all points, the sheet would deform as shown in Fig. 12.12a. Between the rivet lines, however, the sheets have only one-half the average tensile stress that they have at the ends and therefore may resist the bending deformation and assume the deformed shape shown in Fig. 12.12b. The forces on the rivets will remain approximately as shown in Fig. 12.11b, since the clamping forces on the rivet head must balance the moment of the bearing forces.

Any riveted or bolted single-shear joint will have stress conditions which vary between the extreme conditions of Fig. 12.10 and 12.11. At low loads, the sheet must resist bending stresses, as shown in Fig. 12.10; but as local yielding occurs, the stresses are redistributed so that they approach the conditions of Fig. 12.11. The ultimate strength is predicted accurately from an assumed average tension stress in the sheet and an average bearing stress on the bolt or rivet. Many types of "blind" rivets or of countersunk rivets do not provide a sufficient amount of clamping action by the rivet head, and strength calculations based on simple stress distributions must be verified by tests.

It is interesting to compare the action of aircraft rivets with the action of hot-driven steel rivets, such as those used in bridges, buildings, boilers, and other steel structures. The steel rivet is upset when red-hot and cools and contracts in place. The contraction makes the rivet slightly smaller in diameter than the hole and provides a residual tension stress in the rivet approximately equal to the yield stress of the rivet material. The rivet tension clamps the plates so tightly that small loads are resisted by friction between the plates, and the rivet shank bears on the hole only at higher loads. This tension does not exist in aircraft rivets, which are driven at room temperatures.

It is common aircraft practice to assume the same allowable bearing stresses

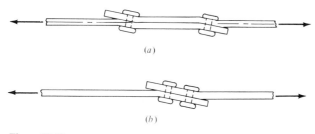

(a)

(b)

Figure 12.12

for single-shear joints and the double-shear joints of the type shown in Fig. 12.5. The common practice in bridge or structural steel design is to use higher allowable bearing stresses for joints in double shear. This practice is logical, since the eccentric distributions, as shown in Fig. 12.10, are eliminated in double-shear joints.

When several similar rivets or bolts act together in a joint, it is customary to assume that each rivet or bolt carries a proportionate share of the load. This assumption is very inacccurate when the joint is not highly stressed, but is more accurate as the loads approach the ultimate strength of the joint and local plastic yielding and rivet "slip" have occurred. In Fig. 12.13 the deformations of the various rivets in a double-shear joint are exaggerated in order to show the relative motion between the plates, although the actual plates would be in close contact and the actual deformation would consist of hole elongations as well as rivet shear deformations. It is assumed that the two outside plates have the same total area A as the inside plate and that the average stress in all plates is $p = P/A$. If each of the five rivets transfers one-fifth to the total load, as commonly assumed, stresses in the various plates between rivets will be $0.2p$, $0.4p$, $0.6p$, and $0.8p$, as shown in Fig. 12.13. Between rivets 1 and 2, the outside plates resist tensile stresses of $0.8p$, and the inside plate resists a tensile stress of $0.2p$; therefore, the outside plates must elongate 4 times as much as the inside plate. Thus, rivet 1 must be deformed much more than rivet 2 and must resist a higher shear. Between rivet 2 and rivet 3, the outside plates have 1.5 times the stress and deformation of the inside plate; therefore, rivet 2 must resist more load than rivet 3. Study of other deformations shows that the end rivets 1 and 5 are equally stressed and must resist much higher shears than the other rivets. Rivets 2 and 4 are equally stressed and resist higher shears than rivet 3.

In the case of a longer line of bolts or rivets than that shown in Fig. 12.13, the end bolts or rivets are still more highly stressed relative to the bolts and rivets near the center of the line. As the load is applied gradually, first the two end rivets must resist most of the load, until they slip or yield in shearing and bearing. Then the load is transferred to the next rivets in the line, until they also slip and transfer load to other rivets. The ultimate strength of the joint is accurately predicted as the sum of the strengths of the individual rivets, provided there is enough ductility to permit each rivet to slip considerably and yet still retain its maximum strength after slipping. It is desirable, however, to vary the plate areas in order to obtain approximately constant tension stresses in the plates and thus distribute small loads more equally to all the rivets or bolts. Bolted or riveted joints in brittle materials are undesirable, since the end bolts may fail before they

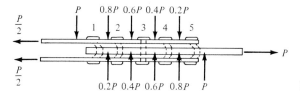

Figure 12.13

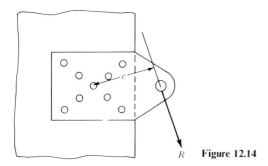

R **Figure 12.14**

deform enough to redistribute the load, and then each bolt in the line will fail in turn. Some spot welds do not have enough ductility to be satisfactory for this type of loading, and occasional rivets are used in most lines of spot welds so that a progressive failure will not extend past the rivet.

12.4 ECCENTRICALLY LOADED CONNECTIONS

In many connections, the resultant force does not act through the center of the bolt or rivet group. In such cases, it is usually convenient to superimpose the effects of an equal parallel force acting at the center of the rivet group and a moment about the center which is equal to the product of the force and its distance from the center. The rivet forces in the typical connection shown in Fig. 12.14 may be obtained by superimposing the forces for the concentric loading of Fig. 12.15a and for the moment Re, shown in Fig. 12.15b.

First we assume that all the rivets are critical in single shear and that all

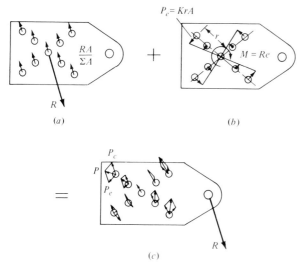

Figure 12.15

plates are rigid. For the concentric load shown in Fig. 12.15a, the shearing stresses on all rivets are assumed equal. The force P_c on any rivet resulting from the concentric load is

$$P_c = \frac{RA}{\Sigma A} \tag{12.9}$$

where A is the area of the rivet cross section and ΣA is the total cross-sectional area of all the rivets in the group. For a rivet group of n rivets of equal area, Eq. (12.9) reduces to the following form:

$$P_c = \frac{R}{n} \tag{12.10}$$

The resultant of the forces on the individual rivets passes through the centroid of the areas of the rivet cross sections; hence this point must be used as the center of moments for the rivet group.

When the rivets resist a moment, the shearing stresses are assumed to be proportional to the distance r from the centroid of the rivet areas. The force P_c on any rivet of area A resulting from this moment is

$$P_c = KrA \tag{12.11}$$

The constant K is obtained by equating the sum of the moments of the individual rivet forces to the external moment:

$$M = \Sigma P_c r = K\Sigma r^2 A \tag{12.12}$$

The constant K may be eliminated from Eq. (12.11) and (12.12) and the force P_c obtained:

$$P_c = \frac{MrA}{\Sigma r^2 A} \tag{12.13}$$

Equation (12.13) is similar in form to the common equations for bending or torsion.

The resultant force P on any rivet can be determined now from the component forces P_c and P_e, as shown in Fig. 12.15c. When an algebraic solution is desired, usually it is more convenient to obtain the horizontal and vertical components of the rivet forces. The distance r does not need to be calculated if the coordinates x and y are used. From Fig. 12.16 and Eq. (12.13), the following equations for the components P_{ex} and P_{ey} are obtained:

$$P_{ex} = \frac{-MyA}{\Sigma x^2 A + \Sigma y^2 A} \qquad P_{ey} = \frac{MxA}{\Sigma x^2 A + \Sigma y^2 A} \tag{12.14}$$

The method of analysis for an eccentric connection, like methods of analysis for several other types of fittings, must be considered only as a rough approximation. Where bearing stresses are critical in the design of the bolts or rivets, it is customary to substitute P_a, the allowable bearing load for each bolt or rivet, into

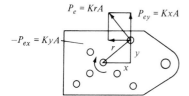

$P_e = KrA$ $P_{ey} = KxA$

$-P_{ex} = KyA$

y

r

x

Figure 12.16

Eqs. (12.13) and (12.14) in place of the shear area A. In some cases, it may be assumed that the loads on all bolts or rivets approach their ultimate strengths, rather than that the loads are proportional to r or that bolts or rivets near the center are not highly stressed. Frequently bolts or rivets are attached to members which are more rigid in one direction than another. If the supporting structure is rigid horizontally but flexible vertically, for example, it may be assumed that an applied moment is resisted by horizontal rivet or bolt forces, rather than by forces perpendicular to the radial line. Where both standard bolts and rivets are used in the same connection, it is necessary to design the connection so that either the rivets alone or the bolts alone can resist the total load. Rivets fill the holes completely, but bolts must be slightly smaller than the holes; consequently, bolts resist no loads until rivets slip enough to be permanently damaged. Close-tolerance, drive-fit bolts are occasionally used with rivets, and each may be assumed to resist a proportionate share of the load.

Example 12.2 Find the resultant force on each rivet of the connection shown in Fig. 12.17. Also find the margin of safety of the most highly stressed rivet. All rivets are $\frac{5}{32}$-in-OD aluminum alloy in single shear, and the sheet is 0.051 gage 2024-T3 aluminum Alclad.

SOLUTION The rivet loads are calculated in Table 12.1. The centroid is determined by inspection, and values of x, x^2, y, and y^2 are tabulated in columns 2 to 5, respectively. The rivet forces P_{cx} and P_{cy} are obtained by dividing the loads of 1800 and 300 lb by 6, since these loads are resisted equally by each

Table 12.1 Analysis of riveted connection

Rivet (1)	x (2)	y (3)	x^2 (4)	y^2 (5)	P_{cx} (6)	P_{ex} (7)	P_x (8)	P_{cy} (9)	P_{ey} (10)	P_y (11)	P (12)
1	-1	1.5	1	2.25	300	-120	180	50	-80	-30	183
2	1	1.5	1	2.25	300	-120	180	50	80	130	221
3	-1	0	1	0	300	0	300	50	-80	-30	302
4	1	0	1	0	300	0	300	50	80	130	327
5	-1	-1.5	1	2.25	300	120	420	50	-80	-30	422
6	1	-1.5	1	2.25	300	120	420	50	80	130	440
Σ			6	9.00							

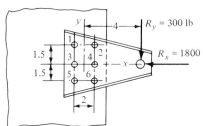

Figure 12.17

of the six rivets. The values of P_{ex} and P_{ey} are found from Eqs. (12.14). Since A is the same for all rivets, the values of A may be omitted from Eqs. (12.14). The moment M is 1200 in · lb. The values P_x and P_y are each obtained as the sum of the terms in the two preceding columns, with care being taken with regard to the algebraic signs. The resultant rivet forces P are found as the square root of the sum of the squares of the rectangular components P_x and P_y.

The allowable load for the rivet is obtained from MILHDBK-5 as 593 lb. Rivet 6 resists the greatest load, 440 lb. The margin of safety is obtained from these loads:

$$\text{MS} = \frac{593}{440} - 1 = 0.35$$

It has been assumed that the loads of 1800 and 300 lb were design fitting loads, or that they were obtained by multiplying the applied or limit loads by the safety factor of 1.5 and the fitting factor of 1.2 or 1.15.

12.5 WELDED JOINTS

Welding is used extensively for steel-tube truss structures, such as engine mounts and fuselages, and for steel landing gears and fittings. The most common type of welding consists of heating the parts to be joined by means of an oxyacetylene torch and then using them together with a suitable welding rod. The grain structure of the material at the weld becomes similar to that of cast metal, and it is more brittle and less able to resist shock and vibration loading than is the original material. Aircraft tube walls are thin and more difficult to weld than other machine and structural members. All aircraft welding was previously torch welding, but electric arc welding has been developed so that it is also satisfactory for the thin aircraft members. In arc welding, the welding rod forms an electrode from which current passes in an arc to the parts being joined. The electric arc simultaneously heats the parts and deposits the weld metal from the electrode. The heating is much more localized than in torch welding, and the strength of heat-treated parts is not impaired as much by arc welding as by torch welding. Design specifications normally require that the same allowable stresses be used for arc welding and for torch welding.

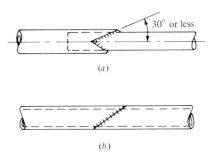

(a)

(b)   **Figure 12.18**

The strength of welded joints depends greatly on the skill of the welder. Often the stress conditions are uncertain, and it is customary to design welded joints with liberal margins of safety. It is preferable to design joints so that the weld is in shear or compression rather than tension, but frequently it is necessary to have welds in tension. Steel tubes in tension are usually spliced by "fish mouth" joints, as shown in Fig. 12.18a, which are designed so that most of the weld is in shear and the local heating of the tube at the weld is not confined to one cross section. Where a butt weld must be used, as shown in Fig. 12.18b, the weld is not perpendicular to the centerline of the tube.

Fuselage truss members often are welded as shown in Fig. 12.19a. Only the horizontal member is highly stressed, and usually the size of the other members is determined as a minimum tube size, because they resist small loads. When these members are highly stressed, it is necessary to insert gusset plates, as shown in Fig. 12.19b. Steel tubes often have walls as thin as 0.035 in and the welder must control the temperature to keep from overheating the thin walls and burning holes in them. It is extremely difficult to weld a thin member to a heavy one,

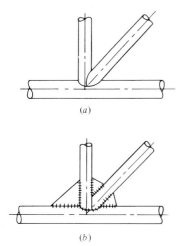

(a)

(b)   **Figure 12.19**

because more heat is required for the heavy member. The thickness ratio of parts being welded should always be less than 3 : 1 and preferably less than 2 : 1.

The allowable load on the weld metal in welded seams is specified in MIL-HDBK-5 by the following equations:

$$P = 32,000Lt \qquad \text{(low-carbon steel)}$$

$$(12.15)$$

$$P = 0.48Lt\sigma_{tu} \qquad \text{(chrome-molybdenum steel)}$$

where P = allowable load, lb

L = length of welded seam, in

t = thickness of thinnest material joined by weld in lap welds between two steel plates or between plates and tubes, in

t = average thickness of weld metal in tube assemblies (cannot be assumed greater than 1.25 times thickness of welded stock), in

σ_{tu} = 90,000 lb/in² for material not heat-treated after welding

σ_{tu} = ultimate tensile stress of material heat-treated after welding, (heat-treatable welding rod must be used), but not to exceed 150,000 lb/in²

The local heating during welding also reduces the allowable tension or bending stress in the material near the weld. For normalized tubing with no heat treatment after welding, the allowable tensile stress is 90,000 lb/in² near the weld for tapered welds making an angle of 30° or less with the axis of the tube and 80,000 lb/in² for other welds. For tubing which is heat-treated after welding, the allowable tensile stress is σ_{tu}.

Example 12.3 The $1\frac{1}{2}$- by 0.065-in chrome-molybdenum steel tube shown in Fig. 12.20 resists a limit or applied tension load of 15,000 lb. Find the margin of safety of the weld and of the tube near the weld if $L_1 = 2.5$, $L_2 = 3$, and $t_1 = 0.20$ in; the tube area is $A = 0.293$ in².
(a) Assume that the ultimate tensile stress σ_{tu} is 100,000 lb/in² before welding and that there is no subsequent heat treatment.
(b) Assume that the tube assembly is heat-treated to a σ_{tu} of 180,000 lb/in² after welding and that the limit load is 22,000 lb.

SOLUTION (a) The weld on the curved end of the tube is neglected, since loads transmitted to this portion of the tube tend only to straighten and flatten the end of the tube and do not increase the tensile strength of the weld appreciably. Because the load P is applied at the center of the tube, one-half of this load is resisted by the weld on each side of the tube. Thus the two welds of length L_1 must resist the load of $P/2$. The ultimate or design load is obtained by multiplying the applied load by the safety factor or 1.5:

$$P = 15,000 \times 1.5 = 22,500 \text{ lb}$$

The allowable load P_a is obtained from Eq. (12.15). The tube thickness $t = 0.065$ is critical since the forging thickness t_1 is more than twice the tube wall thickness. The length L_1 is welded to the tube on both sides of the

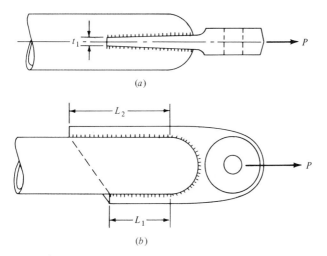

(a)

(b)

Figure 12.20

forging; therefore a length $L = 5$ in must resist half of the load. The allowable load is

$$\frac{P_a}{2} = 0.48Lt\sigma_{tu} = 0.48 \times 5 \times 0.065 \times 90,000$$

or

$$P_a = 28,000 \text{ lb}$$

The fitting factor of 1.20 must be included in the calculation of the margin of safety:

$$\text{MS} = \frac{P_a}{1.20P} - 1 = \frac{28,000}{1.2 \times 22,500} - 1 = 0.04$$

This margin appears small for a weld, but was calculated conservatively. The ultimate tensile stress in the tube near the weld is 90,000 lb/in^2 since for a slotted tube, the weld makes an angle of 0° with the tube axis. The allowable tension in the tube near the weld is

$$P_a = 90,000 \times 0.293 = 26,400 \text{ lb}$$

It is not necessary to use a fitting factor here, since the tube itself, rather than the fitting, is being investigated:

$$\text{MS} = \frac{26,400}{22,500} - 1 = 0.17$$

(b) The allowable load is computed in the same manner as for part (a), but now $\sigma_{tu} = 150,000$ lb/in^2:

$$\frac{P_a}{2} = 0.48 \times 5 \times 0.065 \times 150,000$$

or

$$P_a = 46,400 \text{ lb}$$

The design load is

$$P = 22,000 \times 1.5 = 33,000 \text{ lb}$$

The fitting factor is included in calculating the margin of safety:

$$MS = \frac{46,400}{33,000} 1.2 - 1 = 0.18$$

The allowable tensile stress in the tube near the weld is σ_{tu}, and

$$P_a = \sigma_{tu} A = 180,000 \times 0.293 = 52,700 \text{ lb}$$

$$MS = \frac{52,700}{33,000} - 1 = 0.60$$

PROBLEMS

12.1 An end fitting similar to those shown in Figs. 12.5 and 12.6 is made of steel with an ultimate tensile strength σ_{tu} of 180,000 lb/in^2 and has no bushing. It has a $\frac{1}{2}$-in steel bolt in double shear, a thickness of 0.5 in, and dimensions $R = 0.5$ and $e = 0.05$ in, as shown in Fig. 12.6. Find the maximum limit loads in tension and in compression if the fitting factor is 1.2 and the bearing factor is 1.0. Obtain allowable stresses from MIL-HDBK-5.

12.2 Design a fitting to resist a limit tension load of 15,000 and a limit compression load of 20,000 lb. Assume the materials and unit stresses to be the same as those used in Prob. 12.1.

12.3 Design an end fitting of steel with an ultimate tensile strength of 125,000 lb/in^2. The applied or limit loads are 15,000-lb tension and 20,000-lb compression. Use a fitting factor of 1.2 and a bearing factor of 2.0.

12.4 Find the margin of safety for the joint shown in Fig. 12.17 if $R_x = 3000$ lb, $R_y = 200$ lb, and the rivets are made of $\frac{3}{8}$-in-diameter 2017-T$_3$ in single shear. The plate is 0.072 gage 2024-T$_3$ clad aluminum.

12.5 Assume the tube of Fig. 12.20 to be 2 by 0.083 in with $t_1 = 0.2$, $L_1 = 3.0$, and $L_2 = 4.0$ in. Find the allowable load P if (a) the tube has an allowable stress σ_{tu} of 95,000 lb/in^2 and (b) the assembly is heat-treated after welding to a tensile strength σ_{tu} of 150,000 lb/in^2.

MOMENTS OF INERTIA, MOHR'S CIRCLE

A.1 CENTROIDS

The force of gravity acting on any body is the resultant of a group of parallel forces acting on all elements of the body. The magnitude of the resultant of several parallel forces is equal to the algebraic sum of the forces, and the position of the resultant is such that it has a moment about any axis equal to the sum of the moments of the component forces. The resultant gravity force on a body is its weight W, which must be equal to the sum of the weights w_i of all elements of the body. If the forces of gravity act parallel to the z axis as shown in Fig. A.1, the moments of all forces about the x and y axes must be equal to the moment of the resultant:

$$W\bar{x} = x_1w_1 + x_2w_2 + \cdots = \Sigma xw \quad \text{or} \quad \int x \, dW \qquad (A.1)$$

$$W\bar{y} = y_1w_1 + y_2w_2 + \cdots = \Sigma yw \quad \text{or} \quad \int y \, dW \qquad (A.2)$$

If the body and the axes are rotated so that the forces are parallel to one of the other axes, a third moment equation can be used:

$$W\bar{z} = z_1w_1 + z_2w_2 + \cdots = \Sigma zw \quad \text{or} \quad \int z \, dW \qquad (A.3)$$

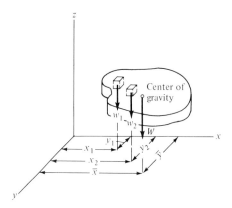

Figure A.1

The three coordinates $\bar{x}$, $\bar{y}$, and $\bar{z}$ of the center of gravity may be obtained from Eqs. (A.1) to (A.3):

$$\bar{x} = \frac{\Sigma xw}{W} \quad \text{or} \quad \frac{\int x \, dW}{W} \tag{A.4}$$

$$\bar{y} = \frac{\Sigma yw}{W} \quad \text{or} \quad \frac{\int y \, dW}{W} \tag{A.5}$$

$$\bar{z} = \frac{\Sigma zw}{W} \quad \text{or} \quad \frac{\int z \, dW}{W} \tag{A.6}$$

The summations or integrals for Eqs. (A.4) to (A.6) must include all elements of the body. In many engineering problems, the weights and coordinates of the various items are known, and the center of gravity is obtained by a summation procedure, rather than by an integration procedure.

In the case of a plate of uniform thickness and density which lies in the xy plane, as shown in Fig. A.2, the coordinates of the center of gravity are

$$\bar{x} = \frac{\int x \, dW}{W} = \frac{w \int x \, dA}{wA} = \frac{\int x \, dA}{A} \tag{A.7}$$

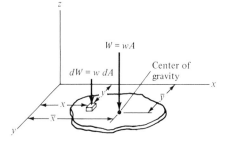

Figure A.2

$$\bar{y} = \frac{\int y \, dW}{W} = \frac{w\int y \, dA}{wA} = \frac{\int y \, dA}{A} \qquad \text{(A.8)}$$

where A is the area of the plate and w is the weight per unit area. It is seen that the coordinates $\bar{x}$ and $\bar{y}$ will be the same regardless of the thickness or weight of the plate. In many engineering problems, the properties of areas are important, and the point in the area having coordinates $\bar{x}$ and $\bar{y}$ as defined by Eqs. (A.7) and (A.8) is called the *centroid* of the area.

A.2 MOMENT OF INERTIA

In considering inertia forces on rotating masses, it was found that the inertia forces on the elements of mass had a moment about the axis of rotation of

$$L_z = \alpha \int r^2 \, dM$$

as shown in Fig. A.3. The term under the integral sign is defined as the moment of inertia of the mass about the z axis:

$$I_z = \int r^2 \, dM \qquad \text{(A.9)}$$

Since the x and y coordinates of the elements are easier to tabulate than the radius, frequently it is convenient to use the relation

$$r^2 = x^2 + y^2$$

or

$$I_z = \int x^2 \, dM + \int y^2 \, dM \qquad \text{(A.10)}$$

An area has no mass, and consequently no inertia, but it is customary to designate the following properties of an area as the moments of inertia of the area since they are similar to the moments of inertia of masses:

$$I_x = \int y^2 \, dA \qquad \text{(A.11)}$$

$$I_y = \int x^2 \, dA \qquad \text{(A.12)}$$

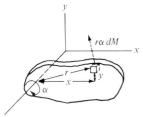

Figure A.3

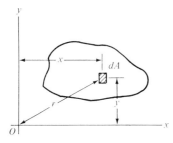

Figure A.4

The coordinates are shown in Fig. A.4. The *polar moment of inertia* of an area is defined as

$$I_p = \int r^2 \, dA \tag{A.13}$$

From the relationship used in Eq. (A.10),

$$r^2 = x^2 + y^2$$

$$I_p = \int x^2 \, dA + \int y^2 \, dA = I_y + I_x \tag{A.14}$$

It is frequently necessary to find the moment of inertia of an area about an axis when the moment of inertia about a parallel axis is known. The moment of inertia about the y axis shown in Fig. A.5 is defined as follows:

$$I_y = \int x^2 \, dA \tag{A.15}$$

Substituting the relation $x = d + x'$ in Eq. (A.15) yields

$$I_y = \int (d + x')^2 \, dA$$

$$= d^2 \int dA + 2d \int x' \, dA + \int x'^2 \, dA$$

or

$$I_y = Ad^2 + 2\bar{x}' \, Ad + I'_y \tag{A.16}$$

where $\bar{x}'$ represents the distance of the centroid of the area from the y' axis, as defined in Eq. (A.7), I'_y represents the moment of inertia of the area about the y' axis, and A represents the total area. Equation (A.16) is simplified when the y' axis is through the centroid of the area, as shown in Fig. A.6:

$$I_y = Ad^2 + I_c \tag{A.17}$$

The term I_c represents the moment of inertia of the area about a centroidal axis.

The moment of inertia of a mass may be transferred to a parallel axis by a

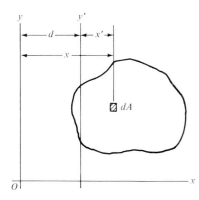

Figure A.5

similar procedure to that used for the moment of inertia of an area. For the mass shown in Fig. A.7, the following relations apply, where the centroidal axis C lies in the xz plane:

$$I_z = \int r^2 \, dM = \int (x^2 + y^2) \, dM$$

$$= \int [(d + x')^2 + y^2] \, dM$$

$$= \int (d^2 + 2dx' + x'^2 + y^2) \, dM$$

and substituting $r_c^2 = x'^2 + y^2$ gives

$$I_z = d^2 \int dM + 2d \int x' \, dM + \int r_c^2 \, dM$$

Since x' is measured from the centroidal axis, the second integral is zero. The last integral represents the moment of inertia about the centroidal axis:

$$I_z = Md^2 + I_c \tag{A.18}$$

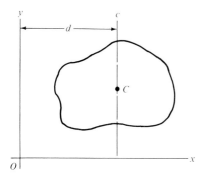

Figure A.6

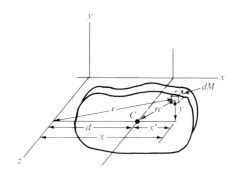

Figure A.7

Equations (A.17) and (A.18) can be used to find either the moment of inertia about any axis when the moment of inertia about a parallel axis through the centroid is known or the moment of inertia about the centroidal axis when the moment of inertia about any other parallel axis is known. In transferring moments of inertia between two axes, neither of which is through the centroid, it is necessary first to find the moment of inertia about the centroidal axis, then to transfer this to the desired axis, by using Eq. (A.17) or (A.18) twice. It is seen that the moment of inertia is always a positive quantity and that the moment of inertia about a centroidal axis is always smaller than that about any other parallel axis. If Eq. (A.16) is used, it is necessary to use the proper sign for the term $\bar{x}'$. All terms in Eqs. (A.17) and (A.18) are always positive.

The radius of gyration ρ of a body is the distance from the inertia axis over which the entire mass would be concentrated in order to give the same moment of inertia. Equating the moment of inertia of the concentrated mass to that for the body yields

$$\rho^2 M = I$$

or

$$\rho = \sqrt{\frac{I}{M}} \tag{A.19}$$

It is seen that the point where the mass is assumed to be concentrated is not the same as the center of gravity, except for the case where I_c in Eq. (A.18) is zero. The point at which the mass is assumed to be concentrated is also different for each inertia axis chosen.

The *radius of gyration* of an area is defined as the distance from the inertia axis to the point where the area would be concentrated in order to produce the same moment of inertia:

$$\rho^2 A = I$$

or

$$\rho = \sqrt{\frac{I}{A}} \tag{A.20}$$

The moment of inertia of an area is obtained as the product of an area and the square of a distance and usually is expressed in units of inches to the fourth power. The moments of inertia for the common areas shown in Fig. A.8 should

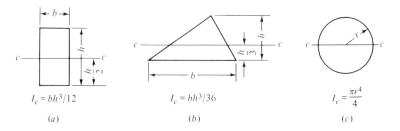

$I_c = bh^3/12$

(a)

$I_c = bh^3/36$

(b)

$I_c = \dfrac{\pi r^4}{4}$

(c)

Figure A.8

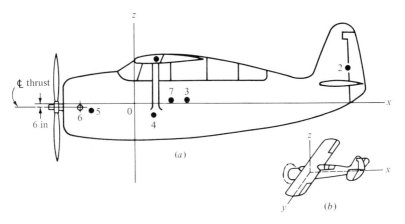

Figure A.9

be memorized, since they are used frequently. Moments of inertia for other areas may be found by integration or from engineering handbooks.

Example A.1 Find the center of gravity of the airplane shown in Fig. A.9a. The various items of weight and the coordinates of their individual centers of gravity are shown in Table A.1. It is customary to take reference axes in the directions shown in Fig. A.9b with the x axis parallel to the thrust line and the z axis vertical. While the z axis is at the wing leading edge in this example, it may be taken through the propeller or through some other convenient reference point.

SOLUTION The y coordinate of the center of gravity is in the plane of symmetry of the airplane. The coordinates $\bar{x}$ and $\bar{z}$ are obtained from Eqs. (A.4) and (A.6). The terms W, Σxw, and Σzw are obtained by totaling columns 3, 5, and 7 of Table A.1.

$$\bar{x} = \frac{\Sigma wx}{W} = \frac{50,723}{4243} = 12.16$$

$$\bar{z} = \frac{\Sigma wz}{W} = \frac{26,109}{4243} = 6.2$$

Table A.1

No.	Item	Weight w	x	wx	z	wz
(1)	(2)	(3)	(4)	(5)	(6)	(7)
1	Wing group	697	22.6	+15,781	40.9	+28,574
2	Tail group	156	198.0	30,904	33.1	5,171
3	Fuselage group	792	49.8	39,430	3.9	3,092
4	Landing gear (up)	380	19.2	7,297	−11.7	−4,429
5	Engine section group	160	−38.6	−6,179	−7.1	−1,138
6	Power plant	1,302	−48.8	−63,674	−6.0	−7,782
7	Fixed equipment	756	35.9	27,164	3.5	2,621
	Total weight empty	4,243		50,723		26,109

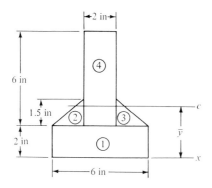

Figure A.10

Example A.2 Find the centroid and the moment of inertia about a horizontal axis through the centroid of the area shown in Fig. A.10. Find the radii of gyration about axis xx and about axis cc.

SOLUTION The area is divided into rectangles and triangles, as shown. The areas of the individual parts are tabulated in column 2 of Table A.2. The y coordinates of the centroids of the elements are tabulated in column 3, and the moments of the areas Ay are tabulated in column 4. The centroid of the total area is now obtained by dividing the summation of the terms in column 4 by the summation of the terms in column 2:

$$\bar{y} = \frac{\Sigma Ay}{A} = \frac{79.5}{27} = 2.94 \text{ in}$$

The moment of inertia of the total area about the x axis will be obtained as the sum of the moments of inertia of the elements about this axis. In finding the moment of inertia of any element about the x axis, Eq. (A.17) may be written as

$$I_x = Ay^2 + I_0$$

where I_x is the moment of inertia of the element of area A about the x axis, y is the distance from the centroid of the element to the x axis, and I_0 is the moment of inertia of the element about its own centroid. The terms Ay^2 for

Table A.2

Element (1)	A (2)	y (3)	Ay (4)	Ay^2 (5)	I_0 (6)
1	12	1	12	12	4.0
2	1.5	2.5	3.75	9.4	0.2
3	1.5	2.5	3.75	9.4	0.2
4	12	5	60	300	36.0
Total	27.0		79.5	330.8	40.4

all the elements are obtained in column 5 as the product of terms in columns 3 and 4. The values of I_0 are found from the equations shown in Fig. A.8. The moment of inertia of the entire area about the x axis is equal to the sum of all terms in columns 5 and 6:

$$I_x = 330.8 + 40.4 = 371.2 \text{ in}^4$$

This moment of inertia may be transferred to the centroid of the entire area by using Eq. (A.17) as follows:

$$I_c = I_x - (\Sigma A)\bar{y}^2$$

where ΣA represents the total area and $\bar{y}$ represents the distance from the x axis to the centroid of the total area. Thus

$$I_c = 371.2 - 27.0(2.94^2) = 138.0 \text{ in}^4$$

The radii of gyration may now be obtained as defined in Eq. (A.20):

$$\rho_x = \sqrt{\frac{I_x}{A}} = \sqrt{\frac{371.2}{27.0}} = 3.71 \text{ in}$$

$$\rho_c = \sqrt{\frac{I_c}{A}} = \sqrt{\frac{138.0}{27.0}} = 2.26 \text{ in}$$

Example A.3 In a metal stressed-skin airplane wing, the sheet-metal covering acts with the supporting spanwise spars and stringers to form a beam which resists the wing bending. Figure A.11a shows a cross section of a typical wing which has a vertical web and extruded angle sections riveted to the spar web and to the skin. The stringers are extruded Z sections which are riveted to the skin. The upper surface of the wing is in compression, and the sheet-metal skin buckles between the stringers and is ineffective in carrying load. The

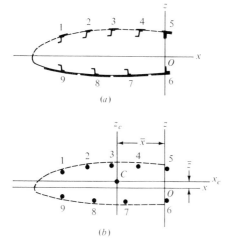

(a)

(b)   **Figure A.11**

skin is riveted to the stringers at frequent intervals, and a narrow strip of skin adjacent to each stringer is prevented from buckling and acts with the stringer in carrying compressive load. The effective width of skin acting with each stringer is usually about 30 times the skin thickness. On the underside of the wing, the entire width of the skin is effective in resisting tension. It is usually sufficiently accurate to assume the area of each stringer and its effective skin to be concentrated at the centroid of its area in computing the moment of inertia of the area. The wing cross section would then be represented by the nine elements of area shown in Fig. A.11b. The moment of inertia of each element about its own centroid is neglected. In this particular wing, the skin and stringers to the right of the spar are very light and are assumed to be nonstructural.

The moment of inertia of the area shown in Fig. A.11b is obtained about horizontal and vertical axes through the centroid of the total area. The areas and coordinates of the elements are given in columns 2, 3, and 6 of Table A.3.

SOLUTION This example is solved by the method used for Example A.2, except that the column for I_0 is omitted. Table A.3 shows the calculations for the moments of inertia about both the horizontal and vertical axes.

Then
$$\bar{x} = \frac{-82.4}{7.135} = -11.56$$

$$I_{zc} = 2310 - 7.135(11.56^2) = 1358 \text{ in}^4$$

$$\bar{z} = \frac{4.72}{7.135} = 0.66$$

$$I_{xc} = 331.6 - 7.135(0.66^2) = 328 \text{ in}^4$$

Table A.3

Element (1)	A (2)	x (3)	Ax (4)	Ax^2 (5)	z (6)	Az (7)	Az^2 (8)
1	0.358	-34.5	-12.34	426	$+8.6$	3.08	26.5
2	0.204	-28.1	-5.73	161	$+9.6$	1.96	18.8
3	0.395	-19.9	-7.85	156	$+10.0$	3.95	39.5
4	0.204	-10.1	-2.06	21	$+9.6$	1.96	18.8
5	1.615	$+0.5$	$+.81$	0	8.8	14.21	125.2
6	1.931	$+0.5$	$+.97$	1	-5.7	-11.02	62.8
7	0.752	-10.1	-7.60	77	-5.2	-3.91	20.4
8	0.784	-22.4	-17.65	394	-4.3	-3.37	14.5
9	0.892	-34.7	-30.92	1074	-2.4	-2.14	5.1
Total	7.135		-82.40	2310		4.72	331.6

A.3 MOMENTS OF INERTIA ABOUT INCLINED AXES

The moment of inertia of an area about any inclined axis may be obtained from the properties of the area with respect to the horizontal and vertical axes. For the area shown in Fig. A.12, a relationship between the moments of inertia about the inclined axes x' and y' and the axes x and y may be found. The moment of inertia about the x' axis is

$$I_{x'} = \int y'^2 \, dA \tag{A.21}$$

The coordinate y' of any point is

$$y' = y \cos \phi - x \sin \phi \tag{A.22}$$

If this value of y' is substituted in Eq. (A.21), the following value is obtained:

$$I_{x'} = (\cos^2 \phi) \int y^2 \, dA - (2 \sin \phi \cos \phi) \int xy \, dA + (\sin^2 \phi) \int x^2 \, dA \tag{A.23}$$

The integrals must extend over the entire area. The angle ϕ is the same regardless of the element of area considered and is therefore a constant with respect to the integrals. The first and last integrals of Eq. (A.23) represent the moments of inertia of the area about the x and y axes. The second integral represents a term which is called the *product of inertia* I_{xy} :

$$I_{xy} = \int xy \, dA \tag{A.24}$$

Equation (A.23) may now be written as

$$I_{x'} = I_x \cos^2 \phi - I_{xy} \sin 2\phi + I_y \sin^2 \phi \tag{A.25}$$

A similar expression may be derived for the moment of inertia about the y' axis:

$$I_{y'} = I_x \sin^2 \phi + I_{xy} \sin 2\phi + I_y \cos^2 \phi \tag{A.26}$$

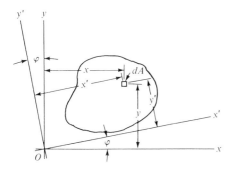

Figure A.12

If Eqs. (A.25) and (A.26) are added, the following relationship is obtained:

$$I_x + I_y = I_{x'} + I_{y'}$$

From Eq. (A.14), the sum of the moments of inertia about any two perpendicular axes is seen to be equal to the polar moment of inertia, which is the same regardless of the angle ϕ of the axes.

The product of inertia about the x' and y' axes is defined as follows:

$$I_{x'y'} = \int x'y' \, dA \tag{A.27}$$

By substituting the relations

$$x' = x \cos \phi + y \sin \phi$$

and

$$y' = y \cos \phi - x \sin \phi$$

into Eq. (A.27), the following value of $I_{x'y'}$ is obtained:

$$I_{x'y'} = (\cos^2 \phi) \int xy \, dA - (\sin^2 \phi) \int xy \, dA$$

$$+ (\sin \phi \cos \phi) \int y^2 \, dA - (\sin \phi \cos \phi) \int x^2 \, dA$$

or
$$I_{x'y'} = I_{xy}(\cos^2 \phi - \sin^2 \phi) + (I_x - I_y) \sin \phi \cos \phi \tag{A.28}$$

A.4 PRINCIPAL AXES

The moment of inertia of any area about an inclined axis is a function of the angle ϕ, as given in Eqs. (A.25) and (A.26). The angle ϕ at which the moment of inertia $I_{x'}$ is a maximum or minimum is obtained from the derivative of Eq. (A.25) with respect to ϕ:

$$\frac{dI_{x'}}{d\phi} = -2I_x \cos \phi \sin \phi - 2I_{xy} \cos 2\phi + 2I_y \sin \phi \cos \phi$$

This derivative is zero when $I_{x'}$ is a maximum or minimum. Equating the derivative to zero and simplifying yield

$$(I_y - I_x) \sin 2\phi = 2I_{xy} \cos 2\phi$$

or
$$\tan 2\phi = \frac{2I_{xy}}{I_y - I_x} \tag{A.29}$$

Since there are two angles under $360°$ which have the same tangent, Eq. (A.29) defines two values of the angle 2ϕ, which will be at $180°$ intervals. The two corresponding values of the angle ϕ will be at $90°$ intervals. It can be shown that the value of $I_{x'}$ will be a maximum about one of these axes and a minimum about

the other. These two perpendicular axes about which the moment of inertia is a maximum or minimum are called the *principal axes.*

The product of inertia about the inclined axes may be expressed in terms of the angle 2ϕ, by making use of the trigonometric relations

$$\sin 2\phi = 2 \sin \phi \cos \phi \qquad (A.30)$$

and

$$\cos 2\phi = \cos^2 \phi - \sin^2 \phi \qquad (A.31)$$

Substituting these values in Eq. (A.28) yields

$$I_{x'y'} = I_{xy} \cos 2\phi + \frac{I_x - I_y}{2} \sin 2\phi \qquad (A.32)$$

An important relation is obtained for the angle at which $I_{x'y'}$ is zero. Substituting $I_{x'y'} = 0$ in Eq. (A.32) produces

$$\tan 2\phi = \frac{2I_{xy}}{I_y - I_x}$$

This is identical to the expression defining the principal axes in Eq. (A.29). The product of inertia about the principal axes is therefore zero.

The moments of inertia about the principal axes may be obtained by substituting the value of ϕ obtained from Eq. (A.29) into Eq. (A.25):

$$I_p = \frac{I_x + I_y}{2} + \sqrt{I_{xy}^2 + \left(\frac{I_x - I_y}{2}\right)^2} \qquad (A.33)$$

and

$$I_q = \frac{I_x + I_y}{2} - \sqrt{I_{xy}^2 + \left(\frac{I_x - I_y}{2}\right)^2} \qquad (A.34)$$

where I_p represents the maximum value of $I_{x'}$ and I_q represents the minimum value of $I_{x'}$. These values are moments of inertia about perpendicular axes defined by Eq. (A.29).

A.5 PRODUCT OF INERTIA

The product of inertia of an area is evaluated by methods similar to those used in evaluating the moment of inertia. Products of inertia for various elements of the area usually are evaluated separately and then added to find the product of inertia for the entire area. When both x and y are positive or negative, the product of inertia is positive; but when one coordinate is positive and the other negative, the product of inertia is negative. In the case of an area which is symmetrical with respect to the x axis, as shown in Fig. A.13, each element of area dA in the first quadrant will have a corresponding area in the fourth quadrant with the same x coordinate but with the y coordinate changed in sign. The sum of the products of inertia for the two elements will be zero, and the integral

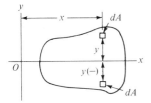

Figure A.13

of these terms for the entire area will be zero:

$$I_{xy} = \int xy \, dA = 0$$

The same relation is true if the area is symmetrical with respect to the y axis. Therefore, when either axis is an axis of symmetry, the product of inertia is zero and the axes are principal axes.

When the product of inertia of an area about one set of coordinate axes is known, the product of inertia about a set of parallel axes can be found. For the area shown in Fig. A.14, the product of inertia about the x and y axes is defined as

$$I_{xy} = \int xy \, dA$$

By substituting the values

$$x = h + u$$
$$y = k + v$$

the transfer theorem is obtained:

$$I_{xy} = \int (h + u)(k + v) \, dA$$

$$= hk \int dA + h \int v \, dA + k \int u \, dA + \int uv \, dA$$

$$I_{xy} = hkA + h\bar{v}A + k\bar{u}A + I_{uv} \tag{A.35}$$

where $\bar{u}$ and $\bar{v}$ are the coordinates of the centroid of the area and I_{uv} is the

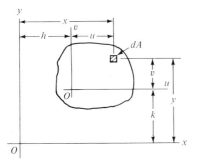

Figure A.14

product of inertia of the area with respect to the u and v axes. If the u and v axes are through the centroid of the area, Eq. (A.35) becomes

$$I_{xy} = hkA + I_{uv} \tag{A.36}$$

With u and v axes also principal axes of the area, $I_{uv} = 0$ and Eq. (A.36) becomes

$$I_{xy} = hkA \tag{A.37}$$

For an area composed of several symmetrical elements, the product of inertia may be obtained as the sum of the values found by using Eq. (A.37) for each element.

Example A.4 Find the product of inertia for the area shown in Fig. A.15.

SOLUTION The total area is divided into the three rectangular elements A, B, and C. Rectangle A is symmetrical about both the x and y axes; hence the product of inertia is zero. Rectangle B is symmetrical about axes through its centroid; therefore the product of inertia may be found from Eq. (A.37):

$$I_{xy} = hkA = -3 \times 5 \times 8 = -120 \text{ in}^4$$

For rectangle C,

$$I_{xy} = hkA = 3 \times -5 \times 8 = -120 \text{ in}^4$$

For the total area,

$$I_{xy} = 0 - 120 - 120 = -240 \text{ in}^4$$

Example A.5 Find the product of inertia about horizontal and vertical axes through the centroid of the area shown in Fig. A.16.

SOLUTION The total area is divided into the two rectangular elements A and B as shown. The x and y reference axes are chosen through the centroids of

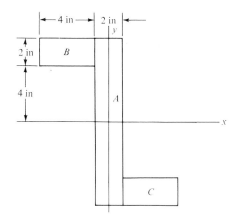

Figure A.15

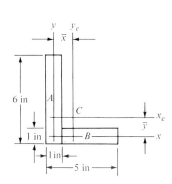

Figure A.16

these rectangles. Since rectangle A is symmetrical about the y axis and rectangle B is symmetrical about the x axis, $I_{xy} = 0$. The centroid of the area is obtained as follows:

$$\bar{x} = \frac{4 \times 2.5}{10} = 1.0$$

$$\bar{y} = \frac{6 \times 2.5}{10} = 1.5$$

The product of inertia about the centroidal axes may now be derived from Eq. (A.36):

$$I_{xy} = \bar{x}\bar{y}A + I_{x_c y_c}$$

$$0 = 1.0 \times 1.5 \times 10 + I_{x_c y_c}$$

or

$$I_{x_c y_c} = -15 \text{ in}^4$$

Example A.6 Find the product of inertia about horizontal and vertical axes through the centroid of the area shown in Fig. A.11. The areas and coordinates of the elements are given in Table A.3 and are repeated in columns 2, 3, and 4 of Table A.4.

SOLUTION The product of inertia about the x and z axes is obtained as the summation of the terms Axz, in column 5 of Table A.4. The centroidal axes were found in Example A.3 to have coordinates $\bar{x} = -11.56$, $\bar{z} = 0.66$. From Eq. (A.36),

$$I_{xz} = \bar{x}\bar{z}A + I_{x_c z_c}$$

$$-68.2 = -11.56 \times 0.66 \times 7.135 + I_{x_c z_c}$$

or

$$I_{x_c z_c} = -13.8$$

Table A.4

Element (1)	A (2)	x (3)	z (4)	Axz (5)
1	0.358	−34.5	+8.6	−106.2
2	0.204	−28.1	+9.6	−55.0
3	0.395	−19.9	+10.0	−78.5
4	0.204	−10.1	+9.6	−19.8
5	1.615	+0.5	+8.8	7.1
6	1.931	+0.5	−5.7	−5.5
7	0.752	−10.1	−5.2	39.5
8	0.784	−22.4	−4.3	75.9
9	0.892	−34.7	−2.4	74.3
Total	7.135			−68.2

A.6 MOHR'S CIRCLE FOR MOMENTS OF INERTIA

The equations for the moments and products of inertia about inclined axes are difficult to remember. It is often convenient to use a semigraphic solution, which is easier to remember and which is an aid in visualizing the relationship between the moments of inertia about various axes. If the values of $I_{xy'}$ from Eq. (A.32) are plotted against values of $I_{x'}$ obtained from Eq. (A.25) for corresponding values of ϕ, the points all fall on the circle shown in Fig. A.17. The maximum and minimum values of the moments of inertia are represented by points P and Q.

In order to prove that the circle shown in Fig. A.17 represents the values given by Eqs. (A.25) and (A.32), the values of $I_{x'}$ and $I_{xy'}$ are expressed in terms of the moments of inertia about the principal axes I_p and I_q and the angle θ from the x' axis to the principal axis. If the x and y axes are principal axes, a substitution of the values $I_{xy} = 0$, $I_x = I_p$, $I_y = I_q$, and $\phi = \theta$ into Eqs. (A.25) and (A.32) yields

$$I_{x'} = I_p \cos^2 \theta + I_q \sin^2 \theta \qquad (A.38)$$

$$I_{xy'} = \frac{I_p - I_q}{2} \sin 2\theta \qquad (A.39)$$

The following trigonometric relations for double angles are used:

$$\sin^2 \theta = \tfrac{1}{2} - \tfrac{1}{2} \cos 2\theta$$

$$\cos^2 \theta = \tfrac{1}{2} + \tfrac{1}{2} \cos 2\theta$$

Substituting these values in Eq. (A.38) yields

$$I_{x'} = \frac{I_p + I_q}{2} + \frac{I_p - I_q}{2} \cos 2\theta \qquad (A.40)$$

Equations (A.39) and (A.40) correspond to the coordinates $I_{xy'}$ and $I_{x'}$ which are computed from the geometry of the circle shown in Fig. A.17. The angle of inclination of the x' axis from the principal axis is one-half the angle 2θ measured between the corresponding points on the circle. If $I_{xy'}$ is measured as positive

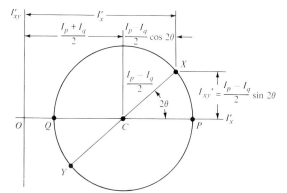

Figure A.17

upward on the circle, a counterclockwise rotation of the x' axis corresponds to a counterclockwise rotation around the circle. Points at opposite ends of the diameter of the circle correspond to perpendicular inertia axes. The products of inertia about perpendicular axes are always equal numerically but opposite in sign, since rotation of the axes through $90°$ interchanges the numerical values of the coordinates x' and y' for any element of area and changes the sign of one coordinate.

Example A.7 Find the moments of inertia about principal axes through the centroid of the area shown in Fig. A.18. Find the moments and product of inertia about axes x_1 and y_1 and axes x_2 and y_2.

SOLUTION The moments of inertia about the x and y axes are

$$I_x = \frac{2 \times 12^3}{12} + 2\left(\frac{4 \times 2^3}{12} + 8 \times 5^2\right) = 693.3 \text{ in}^4$$

$$I_y = \frac{12 \times 2^3}{12} + 2\left(\frac{2 \times 4^3}{12} + 8 \times 3^2\right) = 173.3 \text{ in}^4$$

From Example A.4,

$$I_{xy} = -240 \text{ in}^4$$

Mohr's circle for the moments and products of inertia about all inclined axes may now be plotted from these three values. The products of inertia $I_{xy'}$ are plotted against the moments of inertia I'_x, as shown in Fig. A.19. Point x in Fig. A.19 has coordinates 693.3 and -240.0, as shown. If the x' axis is rotated through $90°$, it will coincide with the y axis, and the coordinates of point Y will be $I_{x'} = 173.3$, $I_{xy'} = 240$. The positive sign for $I_{xy'}$ results from the fact that after the x' axis is rotated through $90°$ from the x axis, the coordinate x' is positive up and the coordinate y' is positive to the left. The

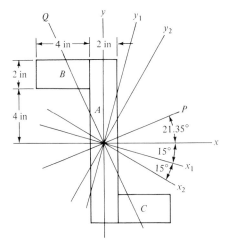

Figure A.18

value of $I_{xy'}$ is numerically equal but opposite in sign to the value of I_{xy}. Points X and Y are at opposite ends of a diameter of the circle, since a rotation of the axes of 90° corresponds to an angle of 180° on the circle.

The center of the circle is a distance $\frac{1}{2}(693.3 + 173.3)$, or 433.3, from the origin. Point X is a distance 260 horizontally and 240 vertically from the center of the circle; therefore

$$\text{Radius} = \sqrt{240^2 + 260^2} = 353.8$$

$$2\theta = \arctan \tfrac{240}{260} = 42.7°$$

or $$\theta = 21.35°$$

The principal axes are represented by points P and Q on the circle. The moments of inertia have maximum and minimum values, and the product of inertia is zero for these axes. The principal moments of inertia are equal to the distance from the origin to the center of the circle plus or minus the radius of the circle:

$$I_p = 433.3 + 353.8 = 787.1 \text{ in}^4$$
$$I_q = 433.3 - 353.8 = 79.5 \text{ in}^4$$

The P axis is counterclockwise from the x axis, at an angle $\theta = 21.35°$. Similarly, since point Q on the circle is counterclockwise from point Y, the Q axis is counterclockwise from the y axis.

The moments and product of inertia about the x_1 and y_1 axes are obtained from the coordinates of points on the circle. Since the x_1 and y_1 axes are 15° clockwise from the x and y axes, the points X_1 and Y_1 on the circle will be 30° clockwise from points X and Y. The coordinates of points X_1 and Y_1 may be obtained from the geometry of the circle as follows:

$$I_{x_1} = 433.3 + 353.8 \cos 72.7° = 538 \text{ in}^4$$
$$I_{y_1} = 433.3 - 353.8 \cos 72.7° = 328 \text{ in}^4$$
$$I_{x_1y_1} = -353.8 \sin 72.7° = -338 \text{ in}^4$$

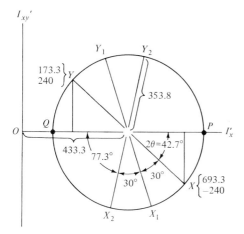

Figure A.19

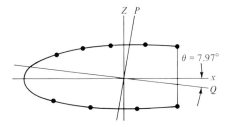

Figure A.20

The moments and product of inertia about the x_2 and y_2 axes are also derived from the geometry of the circle:

$$I_{x2} = 433.3 - 353.8 \cos 77.3° = 355 \text{ in}^4$$
$$I_{y2} = 433.3 + 353.8 \cos 77.3° = 511 \text{ in}^4$$
$$I_{x2 y2} = -353.8 \sin 77.3° = -345 \text{ in}^4$$

Example A.8 Find the principal axes through the centroid and the moments of inertia about these axes for the area shown in Fig. A.20. The x and z axes are through the centroid, and the moments and product of inertia have the values $I_x = 320$, $I_z = 1160$, and $I_{xz} = -120$.

SOLUTION The coordinates of point X on the circle of Fig. A.21 are 320 and -120. The coordinates of point Z are 1160 and $+120$. These points are at opposite ends of the diameter and thus determine the circle. It is important to show I_{xz} with the correct sign at point X so that the direction of the principal axes may be obtained correctly. The distance of the center of the circle from the origin of coordinates is $\frac{1}{2}(320 + 1160) = 740$. From the geometry of the circle,

$$\text{Radius} = \sqrt{420^2 + 120^2} = 437$$

$$\tan 2\theta = \tfrac{120}{420} \qquad 2\theta = 15.94°$$

or
$$\theta = 7.97°$$

The principal moments of inertia are represented by points P and Q on the circle, at which the moments of inertia have maximum and minimum values

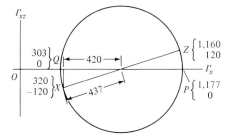

Figure A.21

and the product of inertia is zero:

$$I_p = 740 + 437 = 1177 \text{ in}^4$$
$$I_q = 740 - 437 = 303 \text{ in}^4$$

Since point P is $15.94°$ clockwise from point Z on the circle, the P axis will be half this angle, or $7.97°$ clockwise from the z axis, as shown in Fig. A.20.

A.7 MOHR'S CIRCLE FOR COMBINED STRESSES

The relationship between normal stresses and shearing stresses on planes at various angles of inclination is similar to the relationship between moments and products of inertia about inclined axes. Most structural members are subjected simultaneously to normal and shearing stresses, and it is necessary to consider the combined effect of the stresses in order to design the members. The landing-gear strut shown in Fig. A.22, for example, is subjected to bending stresses which produce tension in the direction of the strut, internal oil pressure which produces a circumferential tension, and torsion which produces shearing stresses on the horizontal and vertical planes. The maximum tensile stress does not occur on either the horizontal or the vertical plane, but on a plane inclined at some angle to them. It can be shown that there are always two perpendicular planes on which the shearing stresses are zero. These planes are called *principal planes*, and the stresses on these planes are called *principal stresses*.

Any condition of two-dimensional stresses can be represented as shown in Fig. A.23, in which the principal stresses σ_p and σ_q act on the perpendicular principal planes. The orientation of these planes depends on the condition of stress and will be found from known stress conditions. The normal and shearing stresses σ_n and τ can be found on a plane at an angle θ to the principal planes, from the equations of statics. The stresses are in pounds per square inch and must always be multiplied by the area in square inches in order to obtain the force. In Fig. A.24, a small triangular element is shown, with principal planes forming two sides of the element and the inclined plane a third side. If the inclined plane has

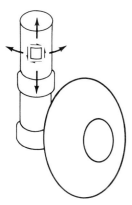

Figure A.22

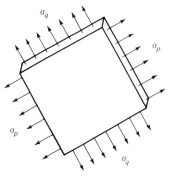

Figure A.23

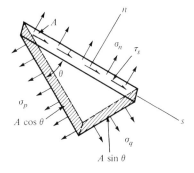

Figure A.24

an area A, the sections of the principal planes have areas $A \cos \theta$ and $A \sin \theta$, as shown. From a summation of forces along the n and s axes, which are perpendicular and parallel to the inclined plane, the following equations are obtained:

$$\sum = \sigma_n A - \sigma_p A \cos^2 \theta - \sigma_q A \sin^2 \theta = 0 \tag{A.41}$$

$$\sum = \tau A - \sigma_p A \cos \theta \sin \theta + \sigma_q A \sin \theta \cos \theta = 0 \tag{A.42}$$

Using the trigonometric relations for functions of double angles yields

$$\cos^2 \theta = \tfrac{1}{2} + \tfrac{1}{2} \cos 2\theta$$

$$\sin^2 \theta = \tfrac{1}{2} - \tfrac{1}{2} \cos 2\theta$$

$$\sin \theta \cos \theta = \tfrac{1}{2} \sin 2\theta$$

Dividing Eqs. (A.41) and (A.42) by A gives

$$\sigma_n = \frac{\sigma_p + \sigma_q}{2} + \frac{\sigma_p + \sigma_q}{2} \cos 2\theta \tag{A.43}$$

$$\tau = \frac{\sigma_p - \sigma_q}{2} \sin 2\theta \tag{A.44}$$

If values of τ and σ_n are plotted for different values of the angle θ, as shown in Fig. A.25, all the points will lie on the circle. This construction, first used by Mohr, is similar to that used for finding moments of inertia about inclined axes.

Normal stresses are considered positive when tensile and negative when compressive. Compressive stresses are therefore shown to the left of the origin on Mohr's circle. In the following examples, shearing stresses are considered positive when they tend to rotate the element clockwise and negative when they tend to rotate the element counterclockwise. Thus, on any two perpendicular planes, the shearing stress on one plane would tend to rotate the element clockwise and would be measured upward on the circle, and the shearing stress on the other

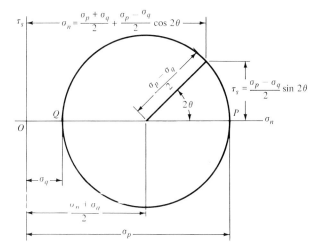

Figure A.25

plane would be equal numerically but opposite in sign and would be measured downward. If this sign convention for shearing stresses is followed, a clockwise rotation of the planes of stress corresponds to a clockwise rotation on the circle. In some books, the opposite sign convention for shearing stresses is used, and a clockwise rotation of the planes corresponds to a counterclockwise rotation on the circle.

Example A.9 The small element shown in Fig. A.26 represents the conditions of two-dimensional stress at a point in a structure. Find the normal and shearing stresses on planes inclined at an angle θ with the vertical plane for values of θ at 30° intervals. Find the principal planes and principal stresses. Find the planes of maximum shear and the stresses on these planes.

SOLUTION The values of the normal stress σ_n and the shearing stress τ on the horizontal and vertical planes are plotted as shown in Fig. A.27. The vertical plane has a normal stress of 10,000 and a shearing stress of 4500 lb/in², and these coordinates are shown for point A on Mohr's circle. The stresses on the horizontal plane are represented by point B on the circle, with coordinates of -2000 and $+4500$. The circle is now drawn with line AB as a diameter. The

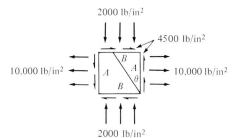

Figure A.26

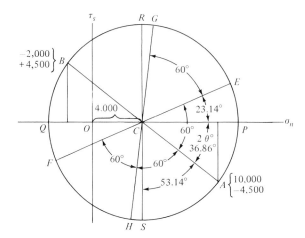

distance OC is $\frac{1}{2}(-2000 + 10,000) = 4000$. Points A and B have a horizontal distance of 6000 and a vertical distance of 4500 from the center of the circle:

$$\text{Radius} = \sqrt{6000^2 + 4500^2} = 7500$$

$$\tan 2\theta = \frac{4500}{6000} = 0.75$$

or
$$2\theta = 36.86° \quad \text{and} \quad \theta = 18.43°$$

The principal stresses are represented on the circle by points P and Q. The coordinates of these points are obtained by adding and subtracting the radius from distance OC:

$$\sigma_p = 4000 + 7500 = 11,500 \text{ lb/in}^2$$
$$\sigma_q = 4000 - 7500 = -3500 \text{ lb/in}^2$$

The point P on the circle is at an angle 2θ counterclockwise around the circle from point A. The principal plane P is therefore at the angle $\theta = 18.43°$ counterclockwise from the vertical plane A, as shown in Fig. A.28a. Similarly, point Q is counterclockwise from the horizontal plane B, or perpendicular to plane P.

The planes of maximum shearing stress are always at 45° to the principal planes, regardless of the stress conditions. Points R and S at extremities of the vertical diameter of Mohr's circle represent the maximum shearing stresses. On the circle, these points are always 90° from the points at the extremities of the horizontal diameter, which represent the princip. stresses. The normal stresses on the two planes of maximum shear are always equal, since they are both equal to the distance OC on the circle diagram. The maximum shearing stresses are shown on an element in Fig. A.28b. Plane S is

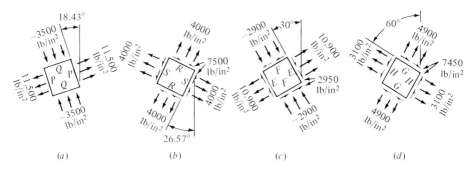

26.57° clockwise from the vertical, since point S is twice this angle clockwise from point A on the circle. Plane S is also 45° clockwise from plane P, and plane R is 45° counterclockwise from plane P. The shearing stress on plane R is positive, tending to rotate the element clockwise. The shearing stress on plane S is negative, tending to rotate the element counterclockwise.

Planes E and F are 30° counterclockwise from planes A and B. Points E and F on the circle must be 60° counterclockwise from points A and B, respectively. The stresses on plane E are obtained by calculating the coordinates of point E on the circle:

$$\tau = 7500 \sin 23.14° = 2950 \text{ lb/in}^2$$
$$\sigma_n = 4000 + 7500 \cos 23.14° = 10{,}900 \text{ lb/in}^2$$

The stresses on planes E and F are shown in Fig. A.28c in the correct directions.

The stresses on plane G, which is 60° counterclockwise from the vertical plane A, are

$$\tau = 7500 \sin 83.14° = 7450 \text{ lb/in}^2$$
$$\sigma_n = 4000 + 7500 \cos 83.14° = 4900 \text{ lb/in}^2$$

The stresses on planes H, which are perpendicular to plane G, are

$$\tau = -7500 \sin 83.14° = -7450 \text{ lb/in}^2$$
$$\sigma_n = 4000 - 7500 \cos 83.14° = 3100 \text{ lb/in}^2$$

These stresses are shown in Fig. A.28d.

B

MATRIX ALGEBRA

B.1 MATRIX DEFINITIONS

A *matrix* is defined as a rectangular array of elements (symbolic or numerical) arranged in rows and columns as follows:

$$[K] = \begin{bmatrix} k_{11} & k_{12} & \cdots & k_{1j} \\ k_{21} & k_{22} & \cdots & k_{2j} \\ \hdotsfor{4} \\ k_{i1} & k_{i2} & \cdots & k_{ij} \end{bmatrix} \qquad (i = 1, 2, 3, \ldots; j = 1, 2, 3, \ldots) \qquad \text{(B.1)}$$

where an element k_{ij} has the subscripts i and j to indicate the element location in the ith row and jth column, respectively. For instance, element k_{21} is located in row 2 and column 1.

For i not equal to j, as in Eq. (B.1), the matrix is called a *rectangular matrix* of order $i \times j$. For $i = j = n$, the matrix is defined as a *square matrix* of order $n \times n$:

$$[K] = \begin{bmatrix} k_{11} & k_{12} & \cdots & k_{1n} \\ k_{21} & k_{22} & \cdots & k_{2n} \\ \hdotsfor{4} \\ k_{n1} & k_{n2} & \cdots & k_{nn} \end{bmatrix} \qquad \text{(B.2)}$$

Elements $k_{11}, k_{22}, \ldots, k_{nn}$ form the *main diagonal* of a square matrix. Elements k_{ij} ($i \neq j$) are referred to as the *off-diagonal elements*.

For $i = 1$ and $j > 1$, the matrix is said to be a *row matrix* of order $1 \times j$ and is written as

$$[K] = [k_{11} \quad k_{12} \cdots k_{1j}] \qquad (j = 1, 2, 3, \ldots) \qquad \text{(B.3)}$$

For $i > 1$ and $j = 1$, the matrix is defined as a *column matrix* of order $i \times 1$

and is written as

$$\{K\} = \begin{bmatrix} k_{11} \\ k_{21} \\ \vdots \\ k_{i1} \end{bmatrix} \qquad (i = 1, 2, 3, \ldots) \tag{B.4}$$

Diagonal, Identity and Null Matrices

A *diagonal matrix* $[K]$ is one whose off-diagonal elements $k_{ij} = 0$ $(i \neq j)$ and main-diagonal elements $k_{ij} \neq 0$ $(i = j)$:

$$[K] = \begin{bmatrix} k_{11} & 0 & \ldots & 0 \\ 0 & k_{22} & \ldots & 0 \\ \hdotsfor{4} \\ 0 & 0 & \ldots & k_{11} \end{bmatrix} \tag{B.5}$$

An *identity*, or *unit*, *matrix* $[K]$ is a diagonal matrix whose main-diagonal elements $k_{ij} = 1$ $(i = j)$:

$$[K] = [I] = \begin{bmatrix} 1 & 0 & \ldots & 0 \\ 0 & 1 & \ldots & 0 \\ \hdotsfor{4} \\ 0 & 0 & \ldots & 1 \end{bmatrix} \tag{B.6}$$

A *null matrix* $[K]$ is one whose every element $k_{ij} = 0$:

$$[K] = \begin{bmatrix} 0 & 0 & \ldots & 0 \\ 0 & 0 & \ldots & 0 \\ \hdotsfor{4} \\ 0 & 0 & \ldots & 0 \end{bmatrix} \tag{B.7}$$

Symmetric and Transposed Matrices

A square matrix $[K]$ is said to be *symmetric* if the following holds true:

$$k_{ij} = k_{ji} \qquad (i \neq j) \tag{B.8}$$

A numerical example of a 3×3 symmetric matrix is

$$[K] = \begin{bmatrix} 15 & -2 & 7 \\ -2 & 9 & 21 \\ 7 & 21 & 16 \end{bmatrix}$$

Notice that the main diagonal (dashed line) is the line of symmetry.

The *transpose matrix* $[K]^T$ of a matrix $[K]$ is obtained by interchanging rows and corresponding columns. For example, if a matrix $[K]$ is given by

$$[K] = \begin{bmatrix} k_{11} & k_{12} & \ldots & k_{1j} \\ k_{21} & k_{22} & \ldots & k_{2j} \end{bmatrix} \tag{B.9}$$

then its transpose $[K]^T$ can be easily written as

$$[K]^T = \begin{bmatrix} k_{11} & k_{21} \\ k_{12} & k_{22} \\ \cdots\cdots\cdots \\ k_{1j} & k_{2j} \end{bmatrix} \tag{B.10}$$

Determinants

Determinants are defined for square matrices only. Thus, the *determinant* $|K|$ of a square matrix $[K]$ is defined as the quantity which results upon performing the following arithmetic operation on $[K]$:

$$|K| = \sum_{i=1}^{m} k_{ij} c_{ij} \qquad (j = \text{any integer between 1 and } m) \tag{B.11}$$

or

$$|K| = \sum_{j=1}^{m} k_{ij} c_{ij} \qquad (i = \text{any integer between 1 and } m) \tag{B.12}$$

Elements k_{ij} in Eqs. (B.11) and (B.12) appear in the jth column and the ith row, respectively, of matrix $[K]$. Likewise, c_{ij} are the *cofactors* which correspond to the jth column of matrix $[K]$ in Eq. (B.11) or to the ith row of matrix $[K]$ in Eq. (B.12).

The cofactors c_{ij} of a matrix $[K]$ are determined by

$$c_{ij} = (-1)^{i+j} M_{ij} \tag{B.13}$$

where M_{ij} are the minors of matrix $[K]$ and are defined as the determinant of the matrix which results after deleting the ith row and jth column of matrix $[K]$.

Let us illustrate the calculation of a determinant by considering the following 3×3 matrix:

$$[K] = \begin{bmatrix} 2 & 1 & 6 \\ 3 & 5 & 4 \\ 2 & 2 & 1 \end{bmatrix} \tag{a}$$

Using Eq. (B.11) and choosing $j = 3$ arbitrarily, we get

$$|K| = \sum_{i=1}^{m} k_{i3} c_{i3} = k_{13} c_{13} + k_{23} c_{23} + k_{33} c_{33}$$

where, from Eq. (a),

$$k_{13} = 6, \qquad k_{23} = 4, \qquad k_{33} = 1$$

For $j = 3$, Eq. (B.13) becomes

$$c_{i3} = (-1)^{i+3} M_{i3} \qquad i = 1, 2, 3$$

Hence

$$c_{13} = M_{13} \qquad c_{23} = -M_{23} \qquad c_{33} = M_{33}$$

As defined previously, the minor M_{13}, of the given matrix in Eq. (a) is the determinant of the matrix resulting from deleting the first row and the third column, corresponding to subscripts 1 and 3 on M_{13}. Or,

$$M_{13} = \begin{vmatrix} 3 & 5 \\ 2 & 2 \end{vmatrix} = 3 \times 2 - 5 \times 2 = -4$$

Similarly,

$$M_{23} = \begin{vmatrix} 2 & 1 \\ 2 & 2 \end{vmatrix} = 2 \times 2 - 1 \times 2 = 2$$

$$M_{33} = \begin{vmatrix} 2 & 1 \\ 3 & 5 \end{vmatrix} = 2 \times 5 - 1 \times 3 = 7$$

Therefore,

$$c_{13} = -4 \qquad c_{23} = -2 \qquad c_{33} = 7$$

and the determinant is

$$|K| = 6(-4) + 4(-2) + 1(7) = -25$$

Properties of Determinants

1. Interchanging any two rows or columns of a determinant changes the sign of the determinant:

$$\begin{vmatrix} 2 & 5 \\ 1 & 10 \end{vmatrix} = -\begin{vmatrix} 1 & 10 \\ 2 & 5 \end{vmatrix} = -\begin{vmatrix} 5 & 2 \\ 10 & 1 \end{vmatrix} = 15$$

2. If two rows or two columns of a determinant are identical, then the value of the determinant is zero.
3. If a determinant has dependent rows or columns, the value of the determinant will be zero:

$$\begin{vmatrix} 2 & 4 & 6 \\ 4 & 1 & 5 \\ 6 & 4 & 10 \end{vmatrix} = 0$$

Column 3 is dependent on columns 1 and 2:

$$\text{Column 3} = \text{column 1} + \text{column 2}$$

4. If rows and corresponding columns of a determinant are interchanged, the value of the determinant is not changed:

$$\begin{vmatrix} 5 & 10 \\ 1 & 6 \end{vmatrix} = \begin{vmatrix} 5 & 1 \\ 10 & 6 \end{vmatrix} = 20$$

5. Multiplying the determinant by a scalar quantity a, is equivalent to multiplying the elements of one row or one column by that quantity:

$$a \begin{vmatrix} k_{11} & k_{12} \\ k_{21} & k_{22} \end{vmatrix} = \begin{vmatrix} ak_{11} & k_{12} \\ ak_{21} & k_{22} \end{vmatrix} = \begin{vmatrix} ak_{11} & ak_{12} \\ k_{21} & k_{22} \end{vmatrix}$$

6. If a row or a column of a determinant is changed by adding to or subtracting from its elements the corresponding elements (or corresponding elements multiplied by a common factor) of any other row or column, then the value of the determinant is not changed:

$$\begin{vmatrix} 2 & 1 \\ 4 & 6 \end{vmatrix} = \begin{vmatrix} 6 & 7 \\ 4 & 6 \end{vmatrix} = \begin{vmatrix} 3 & 1 \\ 10 & 6 \end{vmatrix} = \begin{vmatrix} 2 & 2 \\ 4 & 8 \end{vmatrix} = \begin{vmatrix} 4 & 4 \\ 4 & 6 \end{vmatrix} = 8$$

row 2	column 2	half of	half of
added	added to	column 1	row 2
to row 1	column 1	added to	added to
		column 2	row 1

Singular Matrices

The matrix $[K]$ is said to be *singular* if the value of its determinant is zero.

If
$$[K] = \begin{bmatrix} 4 & 2 \\ 8 & 4 \end{bmatrix}$$

Then
$$|K| = \begin{bmatrix} 4 & 2 \\ 8 & 4 \end{bmatrix} = 16 - 16 = 0$$

Hence, matrix $[K]$ is singular.

B.2   MATRIX ALGEBRA

Matrix Addition

Two matrices $[B]$ and $[C]$ of order $m \times n$ can be added by adding each element b_{ij} of matrix $[B]$ to the corresponding element c_{ij} of matrix $[C]$:

$$\begin{bmatrix} b_{11} & b_{12} \\ b_{21} & b_{22} \end{bmatrix} + \begin{bmatrix} c_{11} & c_{12} \\ c_{21} & c_{22} \end{bmatrix} = \begin{bmatrix} b_{11} + c_{11} & b_{12} + c_{12} \\ b_{21} + c_{21} & b_{22} + c_{22} \end{bmatrix}$$

or, in general, $[A]^{m \times n} = [B]^{m \times n} + [C]^{m \times n}$ (B.14)

where $a_{ij} = b_{ij} + c_{ij}$ $i = 1, 2, \ldots, m; j = 1, 2, \ldots, n)$ (B.15)

Matrix Subtraction

Matrix subtraction is similar to matrix addition except elements b_{ij} and c_{ij} are subtracted instead of being added:

$$[S]^{m \times n} = [B]^{m \times n} - [C]^{m \times n}$$ (B.16)

where $\qquad s_{ij} = b_{ij} - c_{ij} \qquad (i = 1, 2, \ldots, m; j = 1, 2, \ldots, n)$ $\qquad$ (B.17)

Matrix Multiplication

The multiplication of two matrices $[B]^{m \times n}$ and $[C]^{n \times r}$ is performed by multiplying each element of a row i of matrix $[B]$ by the corresponding element in a column j of matrix $[C]$ and summing the products:

$$[P]^{m \times r} = [B]^{m \times n}[C]^{n \times r} \qquad (B.18)$$

where $\qquad p_{ij} = \sum_{s=1}^{n} b_{is} c_{sj} \qquad (i = 1, 2, \ldots, m; j = 1, 2, \ldots, r)$ $\qquad$ (B.19)

In expanded form, Eq. (B.19) becomes

$$p_{ij} = b_{i1}c_{1j} + b_{i2}c_{2j} + \cdots + b_{in}c_{nj} \qquad (B.20)$$

The two matrices $[B]$ and $[C]$ can be multiplied only if the number of columns in $[B]$ is equal to the number of rows in $[C]$.

Let us consider the following example:

$$\lfloor B \rfloor = [2 \quad 1 \quad 4]^{1 \times 3}$$

$$[C] = \begin{bmatrix} 2 & 1 \\ 4 & 2 \\ 3 & 5 \end{bmatrix}^{3 \times 2}$$

$$[P]^{1 \times 2} = \lfloor B \rfloor^{1 \times 3}[C]^{3 \times 2}$$

$$= [2 \quad 1 \quad 4] \begin{bmatrix} 2 & 1 \\ 4 & 2 \\ 3 & 5 \end{bmatrix} = [20 \quad 24]$$

or $\qquad p_{ij} = \sum_{k=1}^{3} b_{ik} c_{kj} = b_{i1}c_{1j} + b_{i2}c_{2j} + b_{i3}c_{3j}$

For $i = 1$ and $j = 1$,

$$p_{11} = b_{11}c_{11} + b_{12}c_{21} + b_{13}c_{31}$$
$$= 2 \times 2 + 1 \times 4 + 4 \times 3 = 20$$

For $i = 1$ and $j = 2$,

$$p_{12} = b_{11}c_{12} + b_{12}c_{22} + b_{13}c_{32}$$
$$= 2 \times 1 + 1 \times 2 + 4 \times 5 = 24$$

Properties of Matrix Multiplication

1. Distributive law:

$$[A] \times ([B] + [C]) = [A][B] + [A][C]$$

2. Commutative law:

$$[A][B] \neq [B][A]$$

3. Associative law:

$$[A] \times ([B] \times [C]) = ([A] \times [B]) \times [C]$$

4. If a matrix is multiplied or divided by a scalar quantity, this is equivalent to multiplying or dividing each element of the matrix by the scalar quantity:

$$b \begin{bmatrix} a_{11} & a_{12} \\ a_{21} & a_{22} \end{bmatrix} = \begin{bmatrix} ba_{11} & ba_{12} \\ ba_{21} & ba_{22} \end{bmatrix}$$

or

$$\frac{1}{b} \begin{bmatrix} a_{11} & a_{12} \\ a_{21} & a_{22} \end{bmatrix} = \begin{bmatrix} \dfrac{a_{11}}{b} & \dfrac{a_{12}}{b} \\ \dfrac{a_{21}}{b} & \dfrac{a_{22}}{b} \end{bmatrix}$$

Matrix Inversion

The inverse of a square nonsingular matrix $[K]$, denoted by $[K]^{-1}$, can be defined as an operation in matrix algebra analogous to division in ordinary algebra. Among the many techniques used for inverting a matrix, only two are presented here.

To invert a matrix $[K]$ which must be square and nonsingular, do the following:

1. Form a matrix of cofactors $[C]$ whose elements are found by utilizing Eq. (B.13):

$$c_{ij} = (-1)^{i+j} M_{ij} \qquad (i = 1, 2, \ldots, n; j = 1, 2, \ldots, n) \qquad \text{(B.13)}$$

The new matrix of cofactors $[C]$ is referred to as the *adjoint* of $[K]$:

$$\text{Adj}\,[K] = [C]$$

2. Transpose the adjoint matrix:

$$\text{Transpose}\,[C] = [C]^T$$

3. Find the determinant of $[K]$ by using Eq. (B.11) or (B.12):

$$|K| = \sum_{i=1}^{m} k_{ij} c_{ij} \qquad (j = 1, 2, \ldots, \text{or } m) \qquad \text{(B.11)}$$

or

$$|K| = \sum_{j=1}^{m} k_{ij} c_{ij} \qquad (i = 1, 2, \ldots, \text{or } m) \qquad \text{(B.12)}$$

4. Divide Adj $[K]^T$, or $[C]^T$, by $|K|$ to obtain the inverse:

$$[K]^{-1} = \frac{\text{Adj}\,[K]^T}{|K|} = \frac{[C]^T}{|K|} \qquad \text{(B.21)}$$

To illustrate, consider the following 2×2 matrix:

$$[K] = \begin{bmatrix} 5 & 2 \\ 2 & 1 \end{bmatrix}$$

First, find $[C]$, the adjoint of $[K]$:

$$c_{ij} = (-1)^{i+j} M_{ij}$$

or $\qquad c_{11} = M_{11} \qquad c_{12} = -M_{12} \qquad c_{21} = -M_{21} \qquad c_{22} = M_{22}$

where $\qquad M_{11} = 1 \qquad M_{12} = 2 \qquad M_{21} = 2 \qquad M_{22} = 5$

Therefore, $[C] = \begin{bmatrix} 1 & -2 \\ -2 & 5 \end{bmatrix}$

Second, transpose $[C]$:

$$[C]^T = \begin{bmatrix} 1 & -2 \\ -2 & 5 \end{bmatrix}$$

Third, find the determinant of $[K]$:

$$|K| = 5 \times 1 - 2 \times 2 = 1$$

Fourth, divide $[C]^T$ by $|K|$:

or $\qquad \dfrac{[C]^T}{|K|} = \dfrac{\begin{bmatrix} 1 & -2 \\ -2 & 5 \end{bmatrix}}{1}$

$$= \begin{bmatrix} 1 & -2 \\ -2 & 5 \end{bmatrix}$$

Check the results:

$$[K][K]^{-1} = [I]$$

or $\qquad \begin{bmatrix} 5 & 2 \\ 2 & 1 \end{bmatrix}\begin{bmatrix} 1 & -2 \\ -2 & 5 \end{bmatrix} = \begin{bmatrix} 1 & 0 \\ 0 & 1 \end{bmatrix}$

In matrix inversion by the Choleski method, any square nonsingular matrix can be expressed as the product of an upper triangular matrix $[U]$ and a lower triangular matrix $[L]$. Then the inversion of the original matrix is reduced to inverting two triangular matrices $[U]$ and $[L]$, which is quite simple. The detailed development of this method can be found in Refs. 10 and 24. Restricting ourselves to symmetrical matrices only, we have

$$[K] = [L][U] \tag{B.22}$$

but $\qquad\qquad\qquad [U] = [L]^T \tag{B.23}$

Therefore Eq. (B.22) becomes

$$[K] = [L][L]^T \tag{B.24}$$

The inverse can be written as

$$[K]^{-1} = [L^T]^{-1}[L]^{-1} \tag{B.25}$$

or

$$[K]^{-1} = [S]^T[S] \tag{B.26}$$

where

$$[S] = [L]^{-1} \tag{B.27}$$

The elements of $[L]$ and $[S]$ are from Ref. 10:

$$l_{ij} = \begin{cases} \left(k_{ii} - \sum_{m=1}^{i-1} l_{im}^2\right)^{1/2} & i = j \\[2ex] \dfrac{k_{ij} - \sum_{m=1}^{j-1} l_{im} l_{mj}}{l_{jj}} & i > j \\[2ex] 0 & i < j \end{cases} \tag{B.28}$$

$$s_{ij} = \begin{cases} \dfrac{1}{l_{ii}} & i = j \\[2ex] \dfrac{\sum_{m=j}^{i} s_{im} l_{mj}}{l_{jj}} & i > j \\[2ex] 0 & i < j \end{cases} \tag{B.29}$$

If the inverse of $[K]$ is denoted by $[H]$:

$$[H] = [K]^{-1} = [S]^T[S] = [\bar{S}][S]$$

then the elements of $[H]$ can be determined as follows:

$$h_{ij} = \sum_{m=1}^{n} \bar{s}_{im} s_{mj}$$
$$\bar{s}_{im} = 0 \qquad (i > m) \tag{B.30}$$
$$s_{mi} = 0 \qquad (i < m)$$

Let us consider the following 2×2 symmetric matrix:

$$[K] = \begin{bmatrix} 2 & 3 \\ 3 & 5 \end{bmatrix}$$

From Eq. (B.28),

$$l_{11} = (k_{11})^{1/2} = \sqrt{2}$$
$$l_{21} = \frac{k_{21}}{l_{11}} = \frac{3}{\sqrt{2}}$$

$$l_{22} = \left(k_{22} - \sum_{m=1}^{2-1} l_{2m}^2 \right)^{1/2} = (k_{22} - l_{21}^2)^{1/2}$$

$$= (5 - \tfrac{9}{2})^{1/2} = \frac{1}{\sqrt{2}}$$

From Eq. (B.29),

$$s_{11} = \frac{1}{l_{11}} = \frac{1}{\sqrt{2}}$$

$$s_{22} = \frac{1}{l_{22}} = \sqrt{2}$$

$$s_{21} = -\sum_{m=1}^{2} \frac{s_{2m} l_{m1}}{l_{11}}$$

$$= -\frac{s_{22} l_{21}}{l_{11}} = \frac{-\sqrt{2}(3/\sqrt{2})}{\sqrt{2}}$$

$$= \frac{-3}{\sqrt{2}}$$

$$s_{12} = 0$$

Or

$$[S] = \begin{bmatrix} \dfrac{1}{\sqrt{2}} & 0 \\[3mm] \dfrac{-3}{\sqrt{2}} & \sqrt{2} \end{bmatrix}$$

$$[S]^T = [\bar{S}] = \begin{bmatrix} \dfrac{1}{\sqrt{2}} & \dfrac{-3}{\sqrt{2}} \\[3mm] 0 & \sqrt{2} \end{bmatrix}$$

From Eq. (B.30),

$$h_{11} = \sum_{m=1}^{2} \bar{s}_{1m} s_{m1} = \bar{s}_{11} s_{11} + \bar{s}_{12} s_{21}$$

$$= \frac{1}{\sqrt{2}} \frac{1}{\sqrt{2}} + \left(-\frac{3}{\sqrt{2}} \right)\left(-\frac{3}{\sqrt{2}} \right)$$

$$= 5$$

$$h_{12} = \sum_{m=1}^{2} \bar{s}_{1m} s_{m2} \qquad = \bar{s}_{11} s_{12} + \bar{s}_{12} s_{22}$$

$$= \frac{1}{\sqrt{2}} 0 + \frac{-3}{\sqrt{2}} \sqrt{2} \qquad = -3 = h_{21}$$

$$h_{22} = \sum_{m=1}^{2} \bar{s}_{2m} s_{m2} \qquad = \bar{s}_{21} s_{12} + \bar{s}_{22} s_{22}$$

$$= 0(0) + \sqrt{2}\,(\sqrt{2}) = 2$$

Hence

$$[K]^{-1} = [H] = \begin{bmatrix} 5 & -3 \\ -3 & 2 \end{bmatrix}$$

Properties of the Inverse of a Matrix

1. The inverse of the product of two square matrices $[R]$ and $[K]$ is equal to the inverse of $[K]$ multiplied by the inverse of $[R]$:

$$[[R][K]]^{-1} = [K]^{-1}[R]^{-1} \qquad (B.31)$$

2. The inverse of a symmetric matrix is also symmetric.

Matrix Partitioning

A matrix K can be partitioned (divided) into smaller submatrices by drawing vertical and corresponding horizontal dotted lines as indicated:

$$[K] = \begin{bmatrix} k_{11} & k_{12} & k_{13} & k_{14} & k_{15} \\ k_{21} & k_{22} & k_{23} & k_{24} & k_{25} \\ k_{31} & k_{32} & k_{33} & k_{34} & k_{35} \\ k_{41} & k_{42} & k_{43} & k_{44} & k_{45} \\ k_{51} & k_{52} & k_{53} & k_{54} & k_{55} \end{bmatrix}$$

$$= \begin{bmatrix} [\bar{K}_{11}] & [\bar{K}_{12}] & [\bar{K}_{13}] \\ [\bar{K}_{21}] & [\bar{K}_{22}] & [\bar{K}_{23}] \\ [\bar{K}_{31}] & [\bar{K}_{32}] & [\bar{K}_{33}] \end{bmatrix}$$

where $[\bar{K}_{11}] = \begin{bmatrix} k_{11} & k_{12} \\ k_{21} & k_{22} \end{bmatrix}$

$$[\bar{K}_{12}] = \begin{bmatrix} k_{13} & k_{14} \\ k_{23} & k_{24} \end{bmatrix}$$

$$[\bar{K}_{13}] = \begin{bmatrix} k_{15} \\ k_{25} \end{bmatrix}$$

Properties of Partitioned Matrices

1. Two matrices can be added or subtracted in terms of their partitioned submatrices if the partitioning is done on both matrices in the same manner.
2. Two matrices can be multiplied in terms of their partitioned submatrices if the rule of matrix multiplication applies.
3. The inverse of a matrix can be found in terms of its partitioned submatrices.

B.3 SIMULTANEOUS LINEAR ALGEBRAIC EQUATIONS

The most important application of matrix algebra is in the solution of a large set of linear algebraic equations. For instance, the following is a set of n simultaneous linear algebraic equations with unknowns q_i $(i = 1, 2, \ldots, n)$ expressed in terms of known quantities Q_i $(i = 1, 2, \ldots, n)$ and the coefficients K_{ij} $(i, j = 1, 2, \ldots, n)$:

$$k_{11}q_1 + k_{12}q_2 + \cdots + k_{1n}q_n = Q_1$$

$$k_{21}q_1 + k_{22}q_2 + \cdots + k_{2n}q_n = Q_2$$

$$\cdots\cdots\cdots\cdots\cdots\cdots\cdots\cdots\cdots\cdots\cdots\cdots$$

$$k_{n1}q_1 + k_{n2}q_2 + \cdots + k_{nn}q_n = Q_n$$

or, in matrix algebra notation,

$$\begin{bmatrix} k_{11} & k_{12} & \cdots & k_{1n} \\ k_{21} & k_{22} & \cdots & k_{2n} \\ \cdots\cdots\cdots\cdots \\ k_{n1} & k_{n2} & \cdots & k_{nn} \end{bmatrix} \begin{bmatrix} q_1 \\ q_2 \\ \vdots \\ q_n \end{bmatrix} = \begin{bmatrix} Q_1 \\ Q_2 \\ \vdots \\ Q_n \end{bmatrix} \tag{B.32}$$

In shorthand matrix notation, Eq. (B.32) can be written as

$$[K]\{q\} = \{Q\} \tag{B.33}$$

Upon inverting $[K]$, the unknown column vector $\{q\}$ can be found:

$$\{q\} = [K]^{-1}\{Q\}$$

To illustrate, consider the following simultaneous equations:

$$2q_1 + 3q_2 = 14$$
$$3q_1 + 5q_2 = 22 \tag{a}$$

In matrix form, (a) becomes

$$\begin{bmatrix} 2 & 3 \\ 3 & 5 \end{bmatrix} \begin{bmatrix} q_1 \\ q_2 \end{bmatrix} = \begin{bmatrix} 14 \\ 22 \end{bmatrix} \tag{b}$$

or

$$\begin{bmatrix} q_1 \\ q_2 \end{bmatrix} = \begin{bmatrix} 2 & 3 \\ 3 & 5 \end{bmatrix}^{-1} \begin{bmatrix} 14 \\ 22 \end{bmatrix} \tag{c}$$

The inverse of the matrix is

$$\begin{bmatrix} 2 & 3 \\ 3 & 5 \end{bmatrix}^{-1} = \begin{bmatrix} 5 & -3 \\ -3 & 2 \end{bmatrix}$$

Thus

$$\begin{bmatrix} q_1 \\ q_2 \end{bmatrix} = \begin{bmatrix} 5 & -3 \\ -3 & 2 \end{bmatrix} \begin{bmatrix} 14 \\ 22 \end{bmatrix} = \begin{bmatrix} 4 \\ 2 \end{bmatrix}$$

or

$$q_1 = 4 \quad \text{and} \quad q_2 = 2$$

REFERENCES

1. Bruhn, E. F.: "Anaylsis and Design of Flight Vehicle Structures," Tri-state Offset Co., Cincinnati, Ohio, 1965.
2. Hoff, N. J.: *High Temperature Effects in Aircraft Structures*, Pergamon, New York, 1958.
3. Metallic Materials and Elements for Flight Vehicle Structures, Military Handbook MIL-HDBK-5A, 1968.
4. Wang, C. T.: *Applied Elasticity*, McGraw-Hill, New York, 1953.
5. Boley, B. A., and J. H. Weiner: *Theory of Thermal Stresses*, Wiley, New York, 1960.
6. Von Karman, T. L., L. G. Dunn, and H. S. Tsien: "The Influence of Curvature on the Buckling Characteristics of Structures," *J. Aeronaut. Sci.*, May 1940, p. 276.
7. Cozzone, F. P., and M. A. Melcon: "Non-dimensional Buckling Curves—Their Development and Application," *J. Aeronaut. Sci.*, Oct. 1946, p. 511.
8. Shanley, F. R.: "Inelastic Column Theory," *J. Aeronaut. Sci.*, May 1947, p. 261.
9. Timoshenko, S. P., and J. M. Gere: *Mechanics of Materials*, Van Nostrand, Princeton, N.J., 1972.
10. Azar, J. J.: *Matrix Structural Analysis*, Pergamon, New York, 1972.
11. Sechler, E. E.: *Elasticity in Engineering*, Wiley, New York, 1952.
12. Rivello, Robert M.: *Theory and Analysis of Flight Structures*, McGraw-Hill, New York, 1969.
13. Timoshenko, S. P., and J. N. Goodier: *Theory of Elasticity*, McGraw-Hill, New York, 1951.
14. Chou, P. C., and N. J. Pagano: *Elasticity*, Van Nostrand, Princeton, N.J., 1967.
15. Timoshenko, W. P., and J. M. Gere: *Theory of Elastic Stability*, McGraw-Hill, New York, 1961.
16. Shanley, F. R.: *Weight-Strength Analysis of Structures*, McGraw-Hill, New York, 1962.
17. Bleich, F.: *Buckling Strength of Metal Structures*, McGraw-Hill, New York, 1952.
18. Gerard, G.: *Introduction to Structural Stability Theory*, McGraw-Hill, New York, 1962.
19. Romberg, W., and W. R. Osgood: *Description of Stress-Strain Curves by Three Parameters*, NACA Tech. Note 902, July 1943.
20. Gassner, E.: "Effect of Variable Load and Cumulative Damages on Fatigue in Vehicle and Airplane Structures," *Proceedings of International Conference on Fatigue of Metals*, British Institute of Mechanical Engineers and ASME, London, 1956.
21. Kowalewski, J.: "On the Relations Between Fatigue Lives under Random Loading and under Corresponding Program Loading," in J. F. Plantema et al. (eds.), *ICAF-AGARD Symposium at Amsterdam, 1958*, Pergamon, New York, 1961.
22. Corten, H. J., and J. J. Dolan: "Cumulative Damage Fatigue," *Proceedings of International Conference on Fatigue of Metals*, British Institute of Mechanical Engineers and ASME, London, 1956.
23. Argyris, J. H., and S. Kelsey: *Energy Therms and Structural Analysis*, Butterworth, London, 1960.
24. Przemieniecki, J. S.: *Theory of Matrix Structural Analysis*, McGraw-Hill, New York, 1968.

25. Timoshenko, S. P., and Krieger Woinowsky: *Theory of Plates and Shells*, McGraw-Hill, New York, 1959.

26. Miner, M. A.: "Cumulative Damage in Fatigue," *Trans. ASME*, vol. 67, 1945, p. A-159.

27. Gere, J. M., and W. Weaver, Jr.: *Analysis of Framed Structures*, Van Nostrand, Princeton, N.J., 1965.

28. Beaufait, Fred W., et al.: *Computer Methods of Structural Analysis*, Prentice-Hall, Englewood Cliffs, N.J., 1970.

29. Rubinstein, Mashe F.: *Matrix Computer Analysis of Structures*, Prentice-Hall, Englewood Cliffs, N. J., 1968.

30. Willems, N., and W. M. Lucas, Jr.: *Matrix Analysis for Structural Engineers*, Prentice-Hall, Englewood Cliffs, N.J., 1968.

31. Kardestuner, H.: *Elementary Matrix Analysis of Structures*, McGraw-Hill, New York, 1971.

32. Meek, J. L.: *Matrix Structural Analysis*, McGraw-Hill, New York, 1971.

33. Harrison, H. B.: *Computer Methods in Structural Analysis*, Prentice-Hall, Englewood Cliffs, N.J., 1973.

34. Gallagher, R. H.: *Finite Element Analysis*, Prentice-Hall, Englewood Cliffs, N.J., 1975.

35. Shanley, F. R., and E. I. Ryder: "Stress Ratios," *Aviat. Mag.*, June 1937.

36. Faupel, J. H.: *Engineering Design*, Wiley, New York, 1964.

37. Shanley, F. R.: *Mechanics of Materials*, McGraw-Hill, New York, 1967.

38. Osgood, Carl C.: *Fatigue Design*, Interscience-Wiley, New York, 1970.

39. Marin, J.: *Mechanical Behavior of Engineering Materials*, Prentice-Hall, Englewood Cliffs, N.J., 1962.

40. ASTM Standards, *Methods of Testing Metals*, American Society for Testing and Materials, 1962.

41. Gerard, G.: *Minimum Weight Analysis of Compression Structures*, New York University Press, New York, 1956.

42. Gerard, G., and H. Becker: *Handbook of Structural Stability*, pt. 1: NACA Tech. Note 3783, August 1957.

43. Crandall, S. H.: *Engineering Analysis*, McGraw-Hill, New York, 1956.

44. Williams, D.: *Introduction to the Theory of Aircraft Structures*, E. Arnold, London, 1960.

45. Needham, R. A.: "The Ultimate Strength of Aluminum Alloy Formed Structural Shapes in Compression," *J. Aeronaut. Sci.*, vol. 21, Apr. 1954.

46. Gerard, G., and H. Becker: *Handbook of Structural Stability*, pt. 3: *Buckling of Curved Plates and Shells*, NACA Tech. Note 3784, 1957.

47. Gerard, G.: *Handbook of Structural Stability*, pt. V, Compressive Strength of Flat Stiffened Plates, NACA Tech. Note 3785, 1957.

48. Gerard, G., and H. Becker: *Handbook of Structural Stability*, pt. 6: *Strength of Thin Wing Construction*, NACA Tech. Note D-162, 1959.

49. Gerard, G.: "The Crippling Strength of Compression Elements," *J. Aeronaut. Sci.*, vol. 25, Jan. 1958.

50. Batdorf, S., M. Stein, and M. Schildcrout: *Critical Shear Stress of Curved Rectangular Panels*, NACA Tech. Note 1348, 1947.

51. Kuhn, P., et al.: *A Summary of Diagonal Tension*, pt. 1, vol. 1, NACA Tech. Note 2661, 1952.

52. Kuhn, P., and J. P. Peterson: *Strength Analysis of Stiffened Beam Webs*, NACA Tech. Note 1364, 1947.

53. Levy, S., et al.: *Analysis of Square Shear Web above Buckling Load*, NACA Tech. Note 962, 1945.

54. Levy, S., et al.: *Analysis of Deep Rectangular Shear Web above Buckling Load*, NACA Tech. Note 1009, 1946.

55. Lahde, R., and H. Wagner, *Test for the Determination of the Stress Concentration in Tension Fields*, NACA Tech. Mem. 809, 1936.

56. Stowell, E. Z.: *Compressive Strength of Flanges*, NACA Tech. Note 2020, 1950.

57. Gatewood, B. E.: *Thermal Stresses*, McGraw-Hill, New York, 1957.

58. Nowacki, W.: *Thermoelasticity*, Addison-Wesley, Reading, Mass., 1962.

59. Burgreen, David: *Elements of Thermal Stress Analysis*, C. P. Press, Jamaica, N.Y., 1971.
60. Calcote, Lee R.: *The Analysis of Laminated Composite Structures*, Van Nostrand, New York, 1969.
61. Plantema, F. J.: *Sandwich Construction*, Wiley, New York, 1966.
62. Zienkiewicz, O. C., and G. S. Holister: *Stress Analysis*, Wiley, New York, 1965.
63. Argyris, J. H., and S. Delsey: *Modern Fuselage Analysis and the Elastic Aircraft*, Butterworth, London, 1963.
64. Gallagher, R. H.: *A Correlation Study of Methods of Matrix Structural Analysis*, Pergamon, New York, 1964.
65. Hoff, N. J.: *The Analysis of Structures*, Wiley, New York, 1960.
66. Kuhn, P.: *Stresses in Aircraft and Structures*, McGraw-Hill, New York, 1956.
67. Langhaar, H. L.: *Energy Methods in Applied Mechanics*, Wiley, New York, 1960.
68. Lee, C. W.: "Thermoelastic Stresses in Thick-Walled Cylinders under Axial Temperature Gradient," *J. Appl. Mech.*, vol. 33, 1966, p. 467.
69. Jaeger, J. C.: "On Thermal Stresses in Circular Cylinders," *Phil. Mag.*, ser. 7, vol. 36, 1945, p. 418.
70. Janak, J. F.: "Thermal Expansion in a Constrained Elastic Cylinder," *IBM J. Res. Develop.*, May 1969.
71. Miller, R. E.: *Theory of the Nonhomogeneous Anisotropic Elastic Shells Subjected to Arbitrary Temperature Distribution*, University of Illinois Bulletin 458.
72. Sokolowski, M.: "The Axially Symmetric Thermoelastic Problem of the Infinite Cylinder," *Arch. Mech. Stos.*, vol. 10, 1958, p. 811.
73. Strub, R. A.: "Distribution of Mechanical and Thermal Stresses in Multilayer Cylinders," *ASME Trans.*, Jan. 1953, p. 73.
74. Sun, C. T., and K. C. Valanis: "On the Axially Symmetric Deformation of a Cylinder of Finite Length," Engineering Research Institute, Iowa State University Rep. 48, 1966.
75. Gallagher, R. H.: "Computerized Structural Analysis and Design—The Next Twenty Years," *Computers and Structures*, vol. 7, 1977, pp. 495–501.
76. Norris, C. H., J. B. Wilbur, and S. Utku: *Elementary Structural Analysis*, 3d ed., McGraw-Hill, 1976.
77. Livesley, R. K.: *Matrix Methods of Structural Analysis*, 2d ed., Pergamon, New York, 1976.
78. Schmit, L. A. and B. Farshi: "Minimum Weight Design of Stress Limited Trusses," *Proc. ASCE, J. Struct. Div.*, vol. 100, no. ST1, Jan. 1974, pp. 97–107.
79. Glockner, P. G.: "Symmetry in Structural Mechanics," *Proc. ASCE, J. Struct. Div.*, vol. 99, no. ST1, Jan. 1973, pp. 71–89.
80. Fox, L.: *An Introduction to Numerical Linear Algebra*, Oxford University Press, New York, 1965.
81. Meyer, C.: "Solution of Linear Equations—State of the Art," *Proc ASCE, J. Struct. Div.*, vol. 99, no. ST7, July 1973, pp. 1507–1526.
82. Meyer, C.: "Special Problems Related to Linear Equation Solvers," *Proc. ASCE, J. Struct. Div.*, vol. 101, no. ST4, Apr. 1975, pp. 869–890.
83. Hicks, J. G.: *Welded Joint Design*, Halsted Press, N.J., 1979.
84. Kirby, P. A., and D. A. Nethercot: *Design for Structural Stability*, Halsted, N.J., 1979.
85. Donnell, L. H.: *Beams, Plates, and Shells*, McGraw-Hill, New York, 1976.
86. Jones, R. M.: *Mechanics of Composite Materials*, McGraw-Hill, New York, 1975.
87. Tauchert, T. R.: *Energy Principles in Structural Mechanics*, McGraw-Hill, New York, 1974.
88. Zienkiewicz, O. C.: *The Finite Element Method in Engineering Science*, 3d ed., McGraw-Hill, New York, 1977.
89. Brebbia, C. A., and J. J. Connor: *Fundamentals of Finite Element Techniques for Structural Engineers*, Wiley, New York, 1974.
90. Brick, R. M.: *Structure and Properties of Engineering Materials*, 4th ed., McGraw-Hill, New York, 1977.
91. Brush, D. O., and B. O. Almroth: *Buckling of Bars, Plates and Shells*, McGraw-Hill, New York, 1975.
92. Rockey, K. C.: *The Finite Element Method : A Basic Introduction*, Wiley, New York, 1975.

93. Brooks, C. R.: *Heat Treatment of Ferrous Alloys*, McGraw-Hill, New York, 1979.
94. Cherepanov, G. P.: *Mechanics of Brittle Fracture*, McGraw-Hill, New York, 1979.
95. Cordon, W. A.: *Properties, Evaluation, Control of Engineering Materials*, McGraw-Hill, New York, 1979.
96. Dieter, G. E.: *Mechanical Metallurgy*, 2d ed., McGraw-Hill, New York, 1976.
97. Palmer, A. C.: *Structural Mechanics*, Clarendon Press, Oxford, 1976.
98. Burgreen, A. C.: *Design Methods for Power Plant Structures*, 1st ed., C. P. Press, Jamaica, N.Y., 1975.
99. Fertis, D. G.: *Dynamics and Vibration of Structures*, Wiley, New York, 1973.
100. Lyon, R. H.: *Statistical Energy Analysis of Dynamical Systems; Theory and Applications*, M.I.T. Press, Cambridge, Mass., 1975.
101. Clough, R. W.: *Dynamics of Structures*, McGraw-Hill, New York, 1975.
102. Budynas, R. G.: *Advanced Strength and Applied Stress Analysis*, Rochester Inst. of Technology, Rochester, N.Y., 1977.
103. Shames, I. H.: *Introduction to Solid Mechanics*, Prentice-Hall, Englewood Cliffs, N.J., 1975.
104. Harrison, H. B.: *Structural Analysis and Design: Some Minicomputer Applications*, Pergamon, New York, 1979.

INDEX